Springer Proceedings in Mathematics & Statistics

Volume 480

This book series features volumes composed of selected contributions from workshops and conferences in all areas of current research in mathematics and statistics, including data science, operations research and optimization. In addition to an overall evaluation of the interest, scientific quality, and timeliness of each proposal at the hands of the publisher, individual contributions are all refereed to the high quality standards of leading journals in the field. Thus, this series provides the research community with well-edited, authoritative reports on developments in the most exciting areas of mathematical and statistical research today.

Indra Rajasingh • Sharmila Mary Arul •
Diana Grace Thomas
Editors

Applied Computational Mathematics

ICCE 2023, Chennai, India, December 8-9

Springer

Editors
Indra Rajasingh
SIMATS Engineering
Saveetha Institute of Medical and Technical Sciences
Chennai, Tamil Nadu, India

Sharmila Mary Arul
SIMATS Engineering
Saveetha Institute of Medical and Technical Sciences
Chennai, Tamil Nadu, India

Diana Grace Thomas
SIMATS Engineering
Saveetha Institute of Medical and Technical Sciences
Chennai, Tamil Nadu, India

ISSN 2194-1009 ISSN 2194-1017 (electronic)
Springer Proceedings in Mathematics & Statistics
ISBN 978-3-031-77763-9 ISBN 978-3-031-77764-6 (eBook)
https://doi.org/10.1007/978-3-031-77764-6

Mathematics Subject Classification: 05-XX, 68-XX, 03B52, 90BXX, 76DXX

This Springer imprint is published by the registered company Springer Nature Switzerland AG
The registered company address is: Gewerbestrasse 11, 6330 Cham, Switzerland

Preface

Greetings to you from SIMATS Engineering!

It is with immense pleasure and a profound sense of accomplishment that I present the proceedings of the two-day International Conference on Computational Engineering (ICCE 2023) through this Springer Nature Proceedings book titled Application of Computational Mathematics. This volume stands as a testament to our journey of knowledge and collaboration, meticulously crafted to reflect the diverse insights and discoveries shared during the conference. The conference witnessed an outstanding response. A significant number of participants from both national and international levels converged to explore the frontiers of computational engineering. Additionally, representatives from five journals indexed in prominent databases added further depth to our discourse, ensuring broader dissemination of the groundbreaking research presented at ICCE 2023. We sincerely thank our dynamic directors, Dr. Deepak Nallaswamy and Dr. Ramya Deepak, whose leadership and support have been crucial to our success. Their commitment to fostering academic innovation and nurturing a conducive environment for scholarly pursuits has contributed significantly to the success of ICCE 2023. We are deeply grateful to our supportive principal, Dr. B. Ramesh, and vice-principal, Dr. Srinivasan, for their invaluable guidance and encouragement throughout the journey. Serving as the Implementing Convener of ICCE 2023 alongside Dr. Indra Rajasingh, the Igniting Convener, has been a privilege. Since its inception, our aim has been to establish a platform that fosters interdisciplinary dialogue, showcases cutting-edge research, and inspires the next generation of computational engineers and mathematicians. I am proud to affirm that this goal has been admirably realized. Special thanks go to our dedicated team members, whose efforts were pivotal in bringing this proceedings book to fruition. Dr. Diana Grace Thomas, our Co-Convener and Co-Editor, provided invaluable guidance and expertise, shaping the content with academic rigor and coherence. We extend our deep gratitude to Dr. Swati Kharub for her exceptional dedication and organizational acumen, which were instrumental in orchestrating every aspect of conference planning and proceedings compilation. We also appreciate the unwavering support and expertise of Dr. Simon Raj, which ensured the seamless execution of the conference. A special mention must be made of our Junior Research Fellows-Indhumathi, Libin Bright, and Abigail-whose enthusiasm, dedication, and meticulous attention to detail enriched the depth and quality of this proceedings book. Throughout the conference, it became evident that computational mathematics transcends mere academic study, emerging as a dynamic force driving innovation across various domains. From the inaugural address by Dr. Andrea Semaničová-Feňovčíková to the special address by Dr. Arumugam, each session provided unique perspectives and insights, enriching our understanding of this interdisciplinary field. We owe much of the success of ICCE 2023 to the diligent efforts of our esteemed panel of reviewers, who played a crucial role in ensuring the quality and integrity of the conference proceedings. Each paper underwent two rounds of review, a process that required meticulous attention and scholarly expertise. We extend our heartfelt gratitude to all reviewers, whose dedication and expertise upheld the scholarly standards of ICCE 2023. As we embark on this intellectual journey through the pages of this proceedings book, I sincerely thank all who contributed to the success of ICCE 2023. From the authors whose research papers form the cornerstone of this volume to the reviewers, editors, administrative staff, participants, and representatives from journals, your dedication and expertise have been invaluable. May this proceedings book serve as a lasting testament to our collective pursuit of knowledge and innovation in the realm of computational engineering.

Chennai, India
May 2024

Dr. Sharmila Mary Arul

Contents

Organization

ICCE 2023 is organized by the Department of Mathematics for Excellence, SIMATS Engineering, Saveetha Institute of Medical and Technical Sciences, Chennai.

Organizing Committee

General Chair :	Dr. N M Veeraiyan Chancellor, SIMATS, Chennai
Conference Chair:	Dr. Deepak Nallaswamy, Director Academics, SIMATS, Chennai
	Dr. Ramya Deepak, Director, SIMATS Engineering, SIMATS, Chennai
Chief Editors:	Dr. Indra Rajasingh, Adjunct Professor, Department of Mathematics for Excellence, SIMATS Engineering
	Dr. Sharmila Mary Arul (Corresponding Editor), Professor, Department of Mathematics for Excellence, SIMATS Engineering
	Dr. Diana Grace Thomas, Associate Professor, Department of Mathematics for Excellence, SIMATS Engineering

Referees

Dr M.K. Tamil Selvi	Associate Professor and Head, Department of Mathematics, Alpha College of Engineering, Thirumazhi, Chennai
Dr. M. Joice Punitha	Assistant Professor, Department of Mathematics, Bharathi Women's College, Chennai
Dr. P. Usha	Assistant Professor, Department of Mathematics, Dwaraka Doss Goverdhan Doss Vaishnav College, Chennai
Dr. R. Sundara Rajan	Associate Professor, Department of Mathematics, Hindustan Institute of Technology and Science, Chennai
Dr. K. Subashini	Assistant Professor, Department of Mathematics, Jeppiaar Engineering College, Chennai
Dr. P. Sivagami	Assistant Professor, Department of Mathematics, Jeppiaar Engineering College, Chennai

Dr. S. Little Joice	Assistant Professor, Department of Mathematics, Loyola College, Chennai
Dr. Annamma V.	Assistant Professor, Department of Mathematics, L.N. Government College, Chennai
Dr. M. Helda Mercy	Professor and Head, Department of Information Technology, Panimalar Engineering College, Chennai
Dr. A. S. Shanthakumari	Assistant Professor, Department of Statistics, Presidency College, Chennai
Dr. K. Meenakshi	Assistant professor, Presidency University, Bangalore
Dr. S. Prabhu	Associate Professor, Department of Mathematics, Rajalakshmi Engineering College, Chennai
Dr. P. Venugopal	Associate Professor, Department of Mathematics, Shiv Nadar University, Chennai
Dr. Anithaashri T.P.	Professor, Institute of Computer Science Engineering, SIMATS Engineering, SIMATS, Chennai
Dr. D. Beula David	Professor, Institute of Computer Science Engineering, SIMATS Engineering, SIMATS, Chennai
Dr. M. J. Carmel Belinda	Professor, Institute of Computer Science Engineering, SIMATS Engineering, SIMATS, Chennai
Dr. S. Jesu Jayarin	Professor, Institute of Computer Science Engineering, SIMATS Engineering, SIMATS, Chennai
Dr. J. Mohana	Professor Institute of Electronic and Communication Engineering, SIMATS Engineering, SIMATS, Chennai
Dr. Paul D.	Associate Professor, Department of Mathematics, Sri Sairam Institute of Technology, Chennai
Dr. A. S. Shanthi	Assistant Professor, Department of Mathematics, Stella Maris College, Chennai
Dr. S. Sarah Surya	Assistant Professor, Department of Mathematics, Stella Maris College, Chennai
Dr. Sharon Philomena V	Assistant Professor, Department of Mathematics, Women's Christian College, Chennai

Part I
Graph Theory

Optimizing Wirelength in Embedding Turán Graphs into Complete Binary Trees

Remi Mariam Reji and R. Sundara Rajan

1 Introduction

One essential technique for mapping a guest graph into a host graph is *graph embedding*. Numerous uses are demonstrated in [1]. For example, actions such as allocating subprocesses in a parallel computing system or simulating one architecture using another can be represented by embedding the guest graph into the host network.

Let A and B be two finite, connected graphs. The embedding of A into B is denoted by the pair (ξ, P_ξ) as described in [2], which consists of the following:

1. An injective mapping $\xi : V(A) \to V(B)$
2. An injective mapping $P_\xi : E(A) \to \{P_\xi((uu')) : P_\xi((uu'))$ represents a path joining $\xi(u)$ with $\xi(u')$ in B for $(uu') \in E(A)\}$

To be concise, we will denote (ξ, P_ξ) as ξ.

The edge congestion of $e \in E(B)$, denoted as $EC_\xi(e)$, is defined as the count of edges (uu') in A for which e lies in the path $P_\xi(uu')$ in B [2]. In the process of embedding graph A into graph B through the mapping ξ, the edge congestion $EC_\xi(A, B)$ is determined as

$$EC_\xi(A, B) = \max_{e \in E(B)} EC_\xi(e).$$

The overall edge congestion when A is embedded into B, represented as $EC(A, B)$, is then defined as

R. M. Reji (✉) · R. Sundara Rajan
Department of Mathematics, Hindustan Institute of Technology and Science, Chennai, Tamil Nadu, India

I. Rajasingh et al. (eds.), *Applied Computational Mathematics*,
Springer Proceedings in Mathematics & Statistics 480,
https://doi.org/10.1007/978-3-031-77764-6_1

$$EC(A, B) = \min_{\xi:A\to B} EC_\xi(A, B).$$

When Q is a subset of $E(B)$, then the edge congestion is

$$EC_\xi(Q) = \sum_{e\in Q} EC_\xi(e).$$

The dilation of an edge $(uu') \in E(A)$, as defined in [2], is the length of the path $P_\xi(uu')$ in B, represented as $|P_\xi(uu')|$. The dilation when A is embedded into B by ξ is expressed as

$$dil_\xi(A, B) = \max\{|P_\xi(uu')| : (uu') \in E(A)\}.$$

The overall dilation when A is embedded into B is then described as

$$dil(A, B) = \min_{\xi:A\to B} dil_\xi(A, B).$$

The wirelength for the embedding ξ of A into B as defined in [2] is expressed as

$$WL_\xi(A, B) = \sum_{(uu')\in E(A)} |P_\xi(uu')| = \sum_{e\in E(B)} EC_\xi(e).$$

Subsequently, the wirelength of embedding A into B is expressed as

$$WL(A, B) = \min_{\xi:A\to B} WL_\xi(A, B).$$

For a subset K of $V(A)$,

$$I_A(K) = \{(uu') \in E(A) \mid u, u' \in K\},\ I_A(k) = \max_{K\subseteq V(A),\ |K|=k} |I_A(K)|.$$

The problem of determining a subset $K \subseteq V(A)$, for a given k where $k = 1, 2, \ldots, |V(A)|$, so that $|K| = k$ and $|I_A(K)| = |I_A(k)|$ is called the *maximum subgraph problem* (MSP) [3]. Then the subset K is said to be an *optimal subset*. This problem is NP-complete [4].

Lemma 1 (Modified Congestion Lemma [2]) *Let $\xi : A \to B$ be an embedding. Suppose that S is an edge cut of B, dividing B into two components B_1 and B_2 upon removal of the edges in S. Define $A_1 = A[\xi^{-1}(V(B_1))]$ and $A_2 = A[\xi^{-1}(V(B_2))]$. Additionally, suppose that S satisfies the following conditions:*

1. *$E(P_\xi(uu')) \cap S = \emptyset$ for any $(uu') \in E(A_i)$, where $i = 1, 2$.*
2. *$E(P_\xi(uu')) \cap S = 1$ for every $(uu') \in E(A)$, where $u \in V(A_1)$ and $u' \in V(A_2)$.*
3. *$V(A_1)$ and $V(A_2)$ are optimal sets.*

Then, for all embeddings ξ from A into B, $EC_\xi(S)$ is minimized, and

$$EC_\xi(S) = \sum_{u\in V(A_1)} deg_A(u) - 2|E(A_1)| = \sum_{u\in V(A_2)} deg_A(u) - 2|E(A_2)|.$$

Lemma 2 (Partition Lemma [2]) *Let $\xi : A \to B$ be an embedding. Let $E(B)$ is partitioned into $\{S_1, S_2, \ldots, S_p\}$ with every S_i representing an edge cut of B. If every requirements of Lemma 1 are satisfied by S_i, then*

$$WL_\xi(A, B) = \sum_{i=1}^{p} EC_\xi(S_i).$$

Remark 1 ([3]) Let $\xi : A \to B$ be an embedding that meets the requirements of Lemma 2. Then

$$WL(A, B) = WL_\xi(A, B).$$

2 Results

One of the most notable, easily adaptable, and cost-effective topological forms of interconnection networks is the multipartite graph. Owing to the communication relationships between various subprocesses, the multipartite graph proves valuable in preventing conflicts within the network and facilitating the easy identification of error-prone areas. A complete p-partite graph, denoted as $K_{n_1,\ldots,n_p}$, consists of p disjoint sets of vertices, each containing n_i vertices (where $1 \le i \le p$), and includes all edges possible by connecting vertices from distinct sets. Among these graphs, Turán graphs are a notable category. For positive integers n and p, the Turán graph $T(n, p)$ represents a chromatically distinct p-multipartite graph consisting of n vertices, where the number of vertices in each pair of sets is either identical or differ by precisely one [5].

Property 1 If $T(n, p)$ is a Turán graph, and $T(n', p)$ is its subgraph with $n' < n$, then the set $V(T(n', p))$ is optimal.

Property 2 If $T(n, p)$ is a Turán graph, and a complete p-partite graph $K(n', n', \ldots, n')$ is its subgraph with $n' \le \lfloor \frac{n}{p} \rfloor$, then $V(K(n', n', \ldots, n'))$ is an optimal set.

In various domains, trees serve as fundamental graph-theoretic models. Among them, the binary tree stands out as the most common graph. A complete binary tree is defined as follows.

Definition 1 ([3]) For a nonnegative number n, a complete binary tree T_n of dimension n is defined as a binary tree with n levels. In this tree, every leaf is at the same level, and every internal vertex has precisely two children. Every level s, $1 \le s \le n$, consists of 2^{s-1} vertices, and hence T_n has precisely $2^n - 1$ vertices.A 1-rooted complete binary tree T_n^1 is formed when a pendant vertex is attached with the root of a complete binary tree T_n.

Wirelength Algorithm A

Input Turán graph $T(2^n - 1, 2^p)$ and complete binary tree T_n on $2^n - 1$ vertices.

Algorithm Let the vertices of $T(2^n - 1, 2^p)$ are partitioned into $V_1, V_2, \dots, V_{2^p}$ sets. For $1 \le i \le 2^p - 1$, the vertices in V_i are labeled as $i + m2^p$, where $0 \le m \le 2^{n-p} - 1$. For $i = 2^p$, the vertices in V_{2^p} are labeled as $(m+1)2^p$ where $0 \le m \le 2^{n-p} - 2$. Label the vertices of T_n using inorder labeling [6]. Let $\xi(u) = u$ for all $u \in V(T(2^n - 1, p))$, and for each $(uu') \in E(T_n)$, let $P_\xi(uu')$ be a shortest path from $\xi(u)$ to $\xi(u')$ in T_n.

Output The wirelength of embedding $T(2^n - 1, 2^p)$ into T_n is minimum.

Proof of Correctness Consider the cut edge X_r^s of T_n, where $s = 1, 2, \dots, n-1$ and $r = 1, 2, \dots, 2^{n-s}$. This edge has one end in level $n-s$ and the other end in level $n-s+1$. Consequently, the removal of X_r^s results in the disconnection of T_n into two components, namely, B_r^s and $\overline{B_r^s}$, as illustrated in Fig. 1. Let $A_r^s = A[\xi^{-1}(V(B_r^s))]$ and $\overline{A_r^s} = A[\xi^{-1}(V(\overline{B_r^s}))]$ denote the inverse images of $B(X_r^s)$ and $\overline{B(X_r^s)}$ with respect to ξ, respectively. A_r^s and $\overline{A_r^s}$ are either Turan graph $T(t, 2^p)$, $t < 2^n - 1$ or complete p-partite graphs $K_{t',t',\dots,t'}$, $t' \le 2^{n-p} - 1$. By Property 1 and Property 2, both A_r^s and $\overline{A_r^s}$ are optimal sets in $T(2^n - 1, p)$. Consequently, the cut edge X_r^s meets all the conditions outlined in Lemma 1. As a result, $EC_\xi(X_r^s)$ is minimized for $s = 1, 2, \dots, n-1$ and $r = 1, 2, \dots, 2^{n-s}$. Subsequently, applying Lemma 2, the obtained wirelength is minimized.

Corollary 1 *When $p = 2$, the wirelength of embedding $T(2^n - 1, 2)$ into T_n is*

$$WL(T(2^n - 1, 2), T_n) = 2^{n-1}(n2^n - 3 \cdot 2^n + n + 3).$$

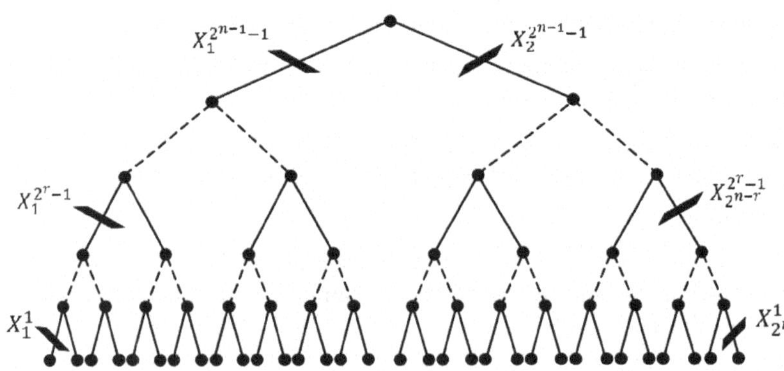

Fig. 1 The cut edges in T_n

Proof Let the vertices of $T(2^n-1,2)$ and T_n are labeled as mentioned in Wirelength Algorithm A. Let $\xi(u)=u$ for all $u\in V(T(2^n-1,2^p))$, and for each $(uu')\in E(T_n)$, let $P_\xi(uu')$ be a shortest path joining $\xi(u)$ with $\xi(u')$ in T_n.

Consider the cut edge X_r^s of T_n, where $s=1,2,\ldots,n-1$ and $r=1,2,\ldots,2^{n-s}$. This edge spans from one end in level $n-s$ to the other end in level $n-s+1$. Consequently, upon removing X_r^s, T_n is split into two components, namely, B_r^s and $\overline{B_r^s}$, as illustrated in Fig. 1. Let $A_r^s=A[\xi^{-1}(V(B_r^s))]$ and $\overline{A_r^s}=A[\xi^{-1}(V(\overline{B_r^s}))]$ denote the inverse images of $B(X_r^s)$ and $\overline{B(X_r^s)}$ with respect to ξ, respectively. A_r^s and $\overline{A_r^s}$ are either a Turan graph $T(t,2)$, $t<2^n-1$, or a complete bipartite graph $K_{t',t'}$, $t'\le 2^{n-1}-1$. By Property 1 and Property 2, both A_r^j and $\overline{A_r^s}$ are optimal sets in $T(2^n-1,2)$. Hence, the cut edge X_r^s meets all the requirements specified in Lemma 1. Consequently, $EC_\xi(X_r^s)$ is minimized for $s=1,2,\ldots,n-1$ and $r=1,2,\ldots,2^{n-s}$ and is expressed as follows:

$$EC_\xi(X_r^s)=\left(2^{s-1}\times(2^{n-1}-1)+(2^{s-1}-1)\times 2^{n-1}\right)-2\times 2^{s-1}\times(2^{s-1}-1).$$

Hence, by Lemma 2,

$$\begin{aligned}WL(T(2^n-1,2),T_n)&=\sum_{s=1}^{n-1}\sum_{r=1}^{2^{n-s}}EC_\xi(X_i^s)\\&=\sum_{s=1}^{n-1}\sum_{r=1}^{2^{n-s}}\Big(\left(2^{s-1}\times(2^{n-1}-1)+(2^{s-1}-1)\times 2^{n-1}\right)\\&\quad-2\times 2^{s-1}\times(2^{s-1}-1)\Big)\\&=2^{n-1}(n2^n-3\cdot 2^n+n+3).\end{aligned}$$

□

3 Comparative Study

Consider the embedding of the complete bipartite graph $K_{2^{n-1},2^{n-1}}$ into a 1-rooted complete binary tree T_n^1. Both $K_{2^{n-1},2^{n-1}}$ and T_n^1 are labeled according to the respective labeling given in Wirelength Algorithm A, where the remaining vertex in both $K_{2^{n-1},2^{n-1}}$ and T_n^1 is labeled as 2^n. The Turán graph $T(2^n-1,2)$ is obtained by removing one vertex from $K_{2^{n-1},2^{n-1}}$. Assume that the vertex having the label $u=2^n$ is removed from (or faulty in) $K_{2^{n-1},2^{n-1}}$. The vertex u is adjacent to all the 2^{n-1} vertices in the other partite. Hence, when removing u, all these 2^{n-1} edges incident to u are also removed. In T_n^1, these 2^{n-1} are mapped to the respective paths joining all the 2^{n-1} leaf vertices to the root vertex; see Fig. 2, and the length of each of these paths is n. Hence, the dilation of each of the edges in $K_{2^{n-1},2^{n-1}}$ incident

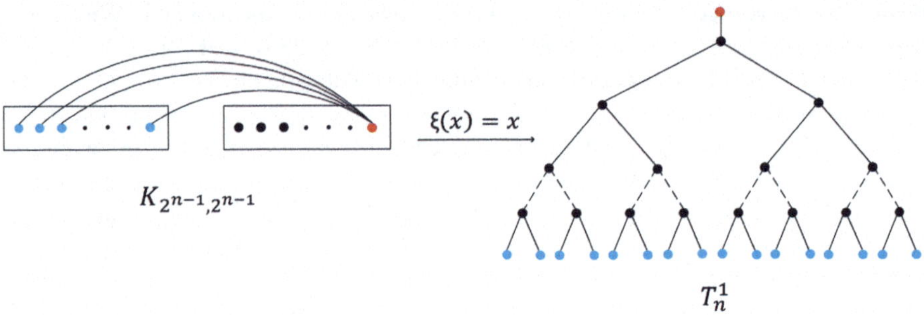

Fig. 2 Embedding of $K_{2^{n-1},2^{n-1}}$ into T_n^1

to the vertex u is equal to n. Consequently, the total dilation of these 2^{n-1} edges is $n2^{n-1}$. Potential locations for introducing the faulty vertex u include the leaf vertices and the root vertex, as removing a vertex from other locations would disconnect the tree T_n^1. Placing u at a leaf vertex leads to a higher total dilation resulting from the paths connecting any 2^{n-1} vertices positioned anywhere T_n^1 compared to the dilation from the paths joining the root vertex to the leaf vertices. Consequently, the total dilation achieved in this scenario is minimized.

In [7], the wirelength obtained when a complete 2^p-partite graph $K_{2^{n-p},2^{n-p},\ldots,2^{n-p}}$ is embedded into a 1-rooted complete binary tree T_n^1 is derived. Consequently, the wirelength for embedding a complete bipartite graph $K_{2^{n-1},2^{n-1}}$ into T_n^1 can be deduced from [7] as follows:

$$WL(K_{2^{n-1},2^{n-1}}, T_n^1) = 2^{n-1}(n2^n - 3 \cdot 2^n + 2n + 3).$$

The minimum wirelength of an embedding is the sum of the minimum edge congestion on each edge in the host graph, and it is also equivalent to the sum of the minimum dilation of each edge in the guest graph. Moreover, we have determined the total dilation contributed by the edges removed when removing the vertex u from $K_{2^{n-1},2^{n-1}}$. Therefore, the wirelength for embedding the Turán graph $T(2^n - 1, 2)$ into T_n can be obtained by subtracting this dilation from the wirelength for embedding $K_{2^{n-1},2^{n-1}}$ into T_n^1. Therefore,

$$\begin{aligned} WL(T(2^n - 1, 2), T_n) &= 2^{n-1}(n2^n - 3 \cdot 2^n + 2n + 3) - n2^{n-1} \\ &= 2^{n-1}(n2^n - 3 \cdot 2^n + n + 3), \end{aligned}$$

which is equivalent to the result obtained in Corollary 1.

When making one vertex in both $K_{2^{n-1},2^{n-1}}$ and T_n^1 faulty, and subsequently calculating the wirelength of embedding $K_{2^{n-1},2^{n-1}}$ into T_n^1, the process involves assessing the impact of the fault on the interconnections between vertices. Before introducing the fault, the wirelength of embedding $K_{2^{n-1},2^{n-1}}$ into T_n^1 reflects the optimal configuration of connections. However, when a vertex is rendered defective, the wirelength computation takes into account the modified structure, accounting for the impact of the fault on the connection pattern. The extent to which the fault influences the interconnections between vertices is indicated by the difference in wirelength before and after making the vertex faulty. This difference captures the additional or reduced wirelength resulting from the faulty vertex, indicating the impact of the fault on the efficiency and overall quality of the embedding. In practical terms, an increase in wirelength after introducing the fault implies that the faulty vertex disrupts the optimal connectivity, potentially leading to longer paths and increased communication distances. Conversely, a decrease in wirelength suggests that the fault has led to a more efficient or direct configuration of connections. Comprehending this variation in wirelength is essential for assessing the resilience and robustness of the embedded graph in the presence of faults, providing insights into the network's performance under various conditions.

4 Applications

The wirelength of embedding a guest graph into a host graph is a crucial parameter in VLSI (Very Large Scale Integration) design that has significant effects on integrated circuit performance and overall efficiency. In essence, the wirelength is the total amount of wiring or connectors needed to join different parts of a chip. In VLSI design, wirelength minimization is a key goal, since it directly affects things like power consumption, signal propagation delays, and the circuit's overall speed and functionality. Shorter interconnects speed up circuits by cutting down on propagation delays and signal travel times. Additionally, since shorter wires have less resistance and capacitance and need less energy to transmit signals, shorter wirelengths also contribute to lower power dissipation. Optimizing the wirelength during the embedding process is therefore essential to producing high-performance, compact VLSI layouts that satisfy the demanding specifications of contemporary electronic products.

The wirelength of embedding a guest graph into a host graph is crucial in figuring out the actual arrangement of components on an integrated circuit (IC) when it comes to hardware description language (HDL) Synthesis. HDL descriptions are converted into a netlist, which shows the connections between different hardware components, as part of the synthesis process. Optimizing the physical location of these components on the chip requires minimizing the wirelength. Reduced signal propagation delays at shorter wirelengths minimize total communication time and improve circuit performance. Effective wirelength optimization in HDL synthesis ensures that the synthesized hardware design satisfies predetermined performance criteria while making the best use of the available chip area. It also improves power consumption, lowers interference, and improves overall reliability of the integrated circuit.

5 Conclusion

This paper establishes the minimality of wirelength when embedding $T(2^n - 1, 2^p)$ into T_n and provides the precise calculation of the wirelength for embedding $T(2^n - 1, 2)$ into T_n. However, determining the wirelength for cases where $p > 2$ presents a challenging task.

Acknowledgments The work of R. Sundara Rajan is supported by the Science and Engineering Research Board (SERB), Government of India for funding under the Teachers Associateship for Research Excellence (TARE) scheme with Project No. TAR/2023/000041.

References

1. Dvořák, T.: Dense sets and embedding binary trees into hypercubes. Discrete Appl. Math. **155**(4), 506–514 (2007). https://doi.org/10.1016/j.dam.2006.09.003
2. Miller, M., Rajan, R.S., Parthiban, N., Rajasingh, I.: Minimum linear arrangement of incomplete hypercubes. Comput. J. **58**(2), 331–337 (2013). https://doi.org/10.1093/comjnl/bxu031
3. Rajan, R.S., Reji, R.M., Rajalaxmi, T.M.: Maximum subgraph problem for 3-regular Knödel graphs and its wirelength. In: Bagchi, A., Muthu, R. (eds.) Algorithms and Discrete Applied Mathematics, vol. 13947, pp. 403–414. Springer, Cham (2023)
4. Garey, M.R., Johnson, D.S.: Computers and Intractability: A Guide to the Theory of NP-Completeness. W. H. Freeman & Co., USA (1979)
5. Shantrinal, A., Klavžar, S., Rajan, R.S., Rajalaxmi, T.M.: An algorithm for embedding Turán graphs into incomplete hypercubes with minimum wirelength. J. Graph Algorithms Appl. **25**, 367–381 (2021). https://doi.org/10.7155/jgaa.00562
6. Rajasingh, I., William, M., Quadras, J., Manuel, P.: Embedding of cycles and wheels into arbitrary trees. Networks **44**, 173–178 (2004). https://doi.org/10.1002/net.20027
7. Shantrinal, A.A., Rajan, R.S., Babu, A.R., Anil, S., Ahmed, M.A.: Embedding complete multipartite graphs into certain trees. J. Comb. Math. Comb. Comput. **112**, 273–286 (2020)

Square Difference Geometric Mean 3-Equitable Labeling of Certain Cycle-Related Graphs

M. I. Jawahar Nisha and V. Annamma

1 Introduction

Let us consider $G(V, E)$ to be a simple, finite, connected, and undirected graph. Graph labeling is an assignment of integers to the vertices or edges, or both subject to certain conditions [1]. A graph G with p vertices and q edges is said to have a geometric mean labeling if there exists an injective function $f : V(G) \rightarrow \{1, 2, \ldots, q + 1\}$ such that the induced edge labeling $f^* : E(G) \rightarrow \{1, 2, \ldots, q\}$ defined by $f^*(uv) = \lceil\sqrt{f(u)f(v)}\rceil$ or $\lfloor\sqrt{f(u)f(v)}\rfloor$ is bijective [2]. A cordial labeling is a binary vertex labeling $f : V(G) \rightarrow \{0, 1\}$ of a graph G that induces an edge labeling f^* of G as follows: $f^*(e) = |f(u) - f(v)|$ for every edge $e = \{uv\} \in E(G)$ such that $|v_f(o) - v_f(1)| \leq 1$ and $|e_{f^*}(0) - e_{f^*}(1)| \leq 1$. A graph which admits cordial labeling is a cordial graph [3]. Many problems in network communication use this cordial labeling for data organization, computational devices, and flow of computation [4]. K. Chitra Lakshmi and K. Nagarajan introduced the concept of geometric mean cordial labeling. Let $G = (V, E)$ be a graph and $f : V(G) \rightarrow \{0, 1, 2\}$ be a mapping. For each edge, uv assign the label $\lceil\sqrt{f(u)f(v)}\rceil$, and f is called a geometric mean cordial labeling if $|v_f(i) - v_f(j)| \leq 1$ and $|e_f(i) - e_f(j)| \leq 1$, where $v_f(x)$ and $e_f(x)$ denote the number of vertices and edges labeled with x where $x \in \{0, 1, 2\}$. A geometric mean cordial graph is a graph with a geometric mean cordial labeling [4].

M. I. Jawahar Nisha (✉)
Department of Mathematics, J.B.A.S College for Women, University of Madras, Chennai, Tamil Nadu, India

V. Annamma
Department of Mathematics, L.N. Government College, University of Madras, Chennai, Tamil Nadu, India

I. Rajasingh et al. (eds.), *Applied Computational Mathematics*, Springer Proceedings in Mathematics & Statistics 480,
https://doi.org/10.1007/978-3-031-77764-6_2

Motivated by these definitions, we defined the new notion called square difference geometric mean (SDGM) 3-equitable labeling. In this paper, we have shown that certain cycle-related graphs such as crown graph, gear graph, triangular cycle graph, and wheel graph are SDGM 3-equitable graph or perfect SDGM 3-equitable graph.

2 Preliminaries

Definition 1 ([5]) The crown graph $Cr_n = [C_n \odot K_1]$ is constructed by attaching absolutely one pendent edge at every vertex of the cycle C_n whenever $n > 2$.

Definition 2 ([6]) The gear graph is constructed from the wheel graph by inserting a vertex between each pair of adjacent vertices of the n-cycle.

Definition 3 ([7]) The triangular cycle graph is a cycle graph C_n in which each of the edges of C_n is replaced by C_3.

Definition 4 ([6]) The wheel graph W_n is join of C_n and K_1, i.e., $W_n = C_n + K_1$.

Definition 5 ([3]) 3-equitable labeling is a function $f : V(G) \to \{0, 1, 2\}$ such that the induced mapping $g : E(G) \to \{0, 1, 2\}$ is defined by $|f(u) - f(v)|, \forall uv \in E(G)$ with the condition $|v_f(i) - v_f(j)| \leq 1$ and $|e_g(i) - e_g(j)| \leq 1$ for all $0 \leq i$, $j \leq 2$.

3 Main Results

Theorem 1 *The crown graph Cr_n is a perfect SDGM 3-equitable graph if $n \equiv 0(mod\ 3)$ and SDGM 3-equitable graph if $n \equiv 1, 2(mod\ 3)$.*

Proof Let $G = (V, E)$ be a Crown graph Cr_n.
Let $V(G) = \{u_i, v_i/1 \leq i \leq n\}$.
Let $E(G) = \{u_i v_i/1 \leq i \leq n\} \cup \{u_i u_{i+1}/1 \leq i \leq n-1\} \cup \{u_n u_1\}$.
Let $|V(G)| = 2n$ and $|E(G)| = 2n$.
Define $f : V(G) \to \{0, 1, 2\}$ as follows:

Case (i): $n \equiv 0(mod\ 3)$

Let $n = 3t, t \geq 1$. $f(u_i) = \begin{cases} 0, & i \equiv 1, 2(mod\ 3) \\ 1, & i \equiv 0(mod\ 3) \end{cases} \forall\, 1 \leq i \leq n$

$$f(v_i) = \begin{cases} 1, & i \equiv 0(mod\ 3) \\ 2, & i \equiv 1, 2(mod\ 3) \end{cases} \quad \forall\, 1 \leq i \leq n$$

Here $|V(G)| = 2n = 6t$ and $|E(G)| = 2n = 6t$.
Then $v_f(0) = v_f(1) = v_f(2) = 2t$ and $e_g(0) = e_g(1) = e_g(2) = 2t$.

Also $(v_f + e_g)(0) = (v_f + e_g)(1) = (v_f + e_g)(2) = 4t$.

Case (ii): $\boldsymbol{n \equiv 1(mod\ 3)}$

Let $n = 3t + 1,\ t \geq 1$.

$$f(u_i) = \begin{cases} 0, & i \equiv 1, 2(mod\ 3) \\ 1, & i \equiv 0(mod\ 3) \end{cases} \quad \forall\, 1 \leq i \leq n$$

$$f(v_i) = \begin{cases} 1, & i \equiv 0(mod\ 3) \\ 2, & i \equiv 1, 2(mod\ 3) \end{cases} \quad \forall\, 1 \leq i \leq n$$

Here $|V(G)| = 2n = 6t + 2$ and $|E(G)| = 2n = 6t + 2$.
Then $v_f(0) = v_f(2) = 2t+1$, $v_f(1) = 2t$ and $e_g(0) = e_g(2) = 2t+1$, $e_g(1) = 2t$.
Also $(v_f + e_g)(0) = (v_f + e_g)(2) = 4t + 2$, $(v_f + e_g)(1) = 4t$.

Case (iii): $\boldsymbol{n \equiv 2(mod\ 3)}$

Let $n = 3t + 2, t \geq 1$.

$$f(u_i) = \begin{cases} 0, & i \equiv 1, 2(mod\ 3) \\ 1, & i \equiv 0(mod\ 3) \end{cases} \quad \forall\, 1 \leq i \leq n - 1 \text{ and } f(u_n) = 1$$

$$f(v_i) = \begin{cases} 1, & i \equiv 0(mod\ 3) \\ 2, & i \equiv 1, 2(mod\ 3) \end{cases} \quad \forall\, 1 \leq i \leq n$$

Here $|V(G)| = 2n = 6t + 4$ and $|E(G)| = 2n = 6t + 4$.
Then $v_f(0) = 2t+2$, $v_f(1) = v_f(2) = 2t+1$ and $e_g(0) = 2t+2$, $e_g(1) = e_g(2) = 2t + 1$.
Also $(v_f + e_g)(0) = 4t + 4$, $(v_f + e_g)(1) = (v_f + e_g)(2) = 4t + 2$.

In case (i), we see that $|v_f(i) - v_f(j)| \leq 1$ and $|e_g(i) - e_g(j)| \leq 1$ for all $0 \leq i, j \leq 2$. Also $|(v_f + e_g)(i) - (v_f + e_g)(j)| \leq 1$ for all $0 \leq i, j \leq 2$ and in cases (ii) and (iii), we see that $|v_f(i) - v_f(j)| \leq 1$ and $|e_g(i) - e_g(j)| \leq 1$ for all $0 \leq i, j \leq 2$.

Hence, crown graph Cr_n is a perfect SDGM 3-equitable graph if $n \equiv 0(mod\ 3)$ and SDGM 3-equitable graph if $n \equiv 1, 2(mod\ 3)$.

Illustration 1 SDGM 3-equitable labeling of crown graph Cr_{26} is shown in Fig. 1.

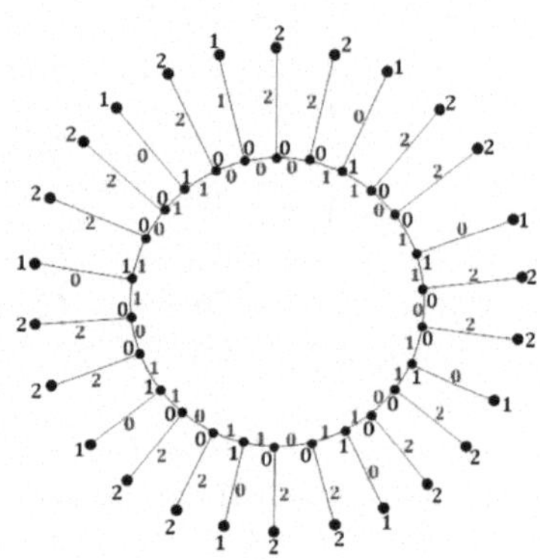

Fig. 1 SDGM 3-equitable labeling of crown graph Cr_{26}

Here $v_f(0) = 18$, $v_f(1) = v_f(2) = 17$ and $e_g(0) = 18$, $e_g(1) = e_g(2) = 17$.
Also $(v_f + e_g)(0) = 36$, $(v_f + e_g)(1) = (v_f + e_g)(2) = 34$.

Therefore, $|v_f(i) - v_f(j)| \leq 1$, $|e_g(i) - e_g(j)| \leq 1$ and $|(v_f + e_g)(i) - (v_f + e_g)(j)| \nleq 1$ for all $0 \leq i, j \leq 2$.

Theorem 2 *The gear graph G_n is a perfect SDGM 3-equitable graph.*

Proof Let $G = (V, E)$ be a gear graph G_n.
Let $V(G) = \{u, u_i/1 \leq i \leq n\}$.
Let $E(G) = \{uu_{2i-1}/1 \leq i \leq n\}$.
Let $|V(G)| = 2n + 1$ and $|E(G)| = 3n$.
Define $f : V(G) \to \{0, 1, 2\}$ as follows: $f(u) = 0$

Case (i): $n \equiv 0(mod\ 3)$
Let $n = 3t, t \geq 1$.

$$f(u_i) = \begin{cases} 0, & i \equiv 1, 3(mod\ 6) \\ 1, & i \equiv 2, 4(mod\ 6)\ \forall\ 1 \leq i \leq n \\ 2, & i \equiv 0, 5(mod\ 6) \end{cases}$$

Here $|V(G)| = 2n + 1 = 6t + 1$ and $|E(G)| = 3n = 9t$.
Then $v_f(0) = 2t + 1$, $v_f(1) = v_f(2) = 2t$ and $e_g(0) = e_g(1) = e_g(2) = 3t$.
Also $(v_f + e_g)(0) = 5t + 1$, $(v_f + e_g)(1) = (v_f + e_g)(2) = 5t$.

Case (ii): $n \equiv 1(mod\ 3)$
Let $n = 3t + 1, t \geq 1$.

$$f(u_i) = \begin{cases} 0, & i \equiv 1, 3(mod\ 6) \\ 1, & i \equiv 2, 4(mod\ 6)\ \forall\ 1 \leq i \leq n-2 \\ 2, & i \equiv 0, 5(mod\ 6) \end{cases}$$

$f(u_{n-1}) = 2$ and $f(u_n) = 1$.
Here $|V(G)| = 2n + 1 = 6t + 3$ and $|E(G)| = 3n = 9t + 3$.
Then $v_f(0) = v_f(1) = v_f(2) = 2t + 1$ and $e_g(0) = e_g(1) = e_g(2) = 3t + 1$.
Also $(v_f + e_g)(0) = (v_f + e_g)(1) = (v_f + e_g)(2) = 5t + 2$.

Case (iii): $n \equiv 2(mod\ 3)$
Let $n = 3t + 2, t \geq 1$.

$$f(u_i) = \begin{cases} 0, & i \equiv 1, 3(mod\ 6) \\ 1, & i \equiv 2, 4(mod\ 6)\ \forall\ 1 \leq i \leq n-2 \\ 2, & i \equiv 0, 5(mod\ 6) \end{cases}$$

$f(u_{n-1}) = 1$ and $f(u_n) = 2$.
Here $|V(G)| = 2n + 1 = 6t + 5$ and $|E(G)| = 3n = 9t + 6$.
Then $v_f(0) = v_f(1) = 2t + 2$, $v_f(2) = 2t + 1$ and $e_g(0) = e_g(1) = e_g(2) = 3t + 2$.
Also $(v_f + e_g)(0) = (v_f + e_g)(1) = 5t + 4$, $(v_f + e_g)(2) = 5t + 3$.
In all the above three cases, we see that $|v_f(i) - v_f(j)| \leq 1$ and $|e_g(i) - e_g(j)| \leq 1$ for all $0 \leq i, j \leq 2$.

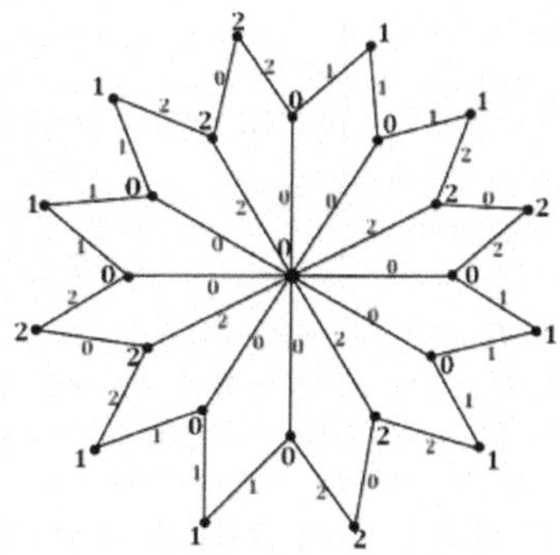

Fig. 2 Perfect SDGM 3-equitable labeling of gear graph G_{12}

Also $|(v_f + e_g)(i) - (v_f + e_g)(j)| \leq 1$ for all $0 \leq i, j \leq 2$.

Hence, gear graph G_n is a perfect SDGM 3-equitable graph.

Illustration 2 Perfect SDGM 3-equitable labeling of gear graph G_{12} is shown in Fig. 2.

Here $v_f(0) = 9$, $v_f(1) = v_f(2) = 8$ and $e_g(0) = e_g(1) = e_g(2) = 12$.

Also $(v_f + e_g)(0) = 21$, $(v_f + e_g)(1) = (v_f + e_g)(2) = 20$.

Therefore, $|v_f(i) - v_f(j)| \leq 1$, $|e_g(i) - e_g(j)| \leq 1$ and $|(v_f + e_g)(i) - (v_f + e_g)(j)| \leq 1$ for all $0 \leq i, j \leq 2$.

Theorem 3 *The wheel graph W_n is a perfect SDGM 3-equitable graph if $n \equiv 0, 2, 3, 4, 5 (mod\ 6)$ and SDGM 3-equitable graph if $n \equiv 1(mod\ 6)$.*

Proof Let $G = (V, E)$ be a wheel graph W_n.

Let $V(G) = \{u, u_i/1 \leq i \leq n\}$.

Let $E(G) = \{uu_i, u_iu_{i+1}/1 \leq i \leq n-1\} \cup \{u_nu_1\}$.

Let $|V(G)| = n + 1$ and $|E(G)| = 2n$.

Define $f : V(G) \rightarrow \{0, 1, 2\}$ as follows:

$f(u) = 0$

Case (i): $n \equiv 0(mod\ 6)$

Let $n = 6t, t \geq 1$.

$$f(u_i) = \begin{cases} 1, & i \equiv 0, 2(mod\ 3) \\ 0, & i \equiv 1(mod\ 3) \end{cases} \quad \forall\, 1 \leq i \leq 3t$$

$$f(u_{3t+1}) = 1$$

$$f(u_i) = \begin{cases} 2, & i \equiv 0, 2(mod\ 3) \\ 0, & i \equiv 1(mod\ 3) \end{cases} \quad \forall\, 3t + 2 \leq i \leq 6t$$

Here $|V(G)| = n + 1 = 6t + 1$ and $|E(G)| = 2n = 12t$.

Then $v_f(0) = v_f(2) = 2t$, $v_f(1) = 2t + 1$ and $e_g(0) = e_g(1) = e_g(2) = 4t$.

Also $(v_f + e_g)(0) = (v_f + e_g)(2) = 6t$, $(v_f + e_g)(1) = 6t + 1$.

Case (ii): $n \equiv 1(mod\ 6)$

Let $n = 6t + 1$, $t \geq 1$.

$$f(u_i) = \begin{cases} 1, & i \equiv 0, 2 (mod\ 3) \\ 0, & i \equiv 1 (mod\ 3) \end{cases} \forall\ 1 \le i \le 3t+1$$

$$f(u_i) = \begin{cases} 2, & i \equiv 0, 2 (mod\ 3) \\ 0, & i \equiv 1 (mod\ 3) \end{cases} \forall\ 3t+2 \le i \le 6t$$

$f(u_{6t+1}) = 2$

Here $|V(G)| = n + 1 = 6t + 2$ and $|E(G)| = 2n = 12t + 2$.

Then $v_f(0) = v_f(2) = 2t + 1$, $v_f(1) = 2t$ and $e_g(0) = e_g(2) = 4t + 1,\ e_g(1) = 4t$.

Also $(v_f + e_g)(0) = (v_f + e_g)(2) = 6t + 2,\ (v_f + e_g)(1) = 6t$.

Case (iii): $\boldsymbol{n \equiv 2(mod\ 6)}$

Let $n = 6t + 2,\ t \ge 1$.

$$f(u_i) = \begin{cases} 1, & i \equiv 0, 2 (mod\ 3) \\ 0, & i \equiv 1 (mod\ 3) \end{cases} \forall\ 1 \le i \le 3t+2$$

$$f(u_i) = \begin{cases} 2, & i \equiv 0, 1 (mod\ 3) \\ 0, & i \equiv 2 (mod\ 3) \end{cases} \forall\ 3t+3 \le i \le 6t+1$$

$f(u_{6t+2}) = 2$

Here $|V(G)| = n + 1 = 6t + 3$ and $|E(G)| = 2n = 12t + 4$.

Then $v_f(0) = v_f(1) = v_f(2) = 2t + 1$ and $e_g(0) = e_g(2) = 4t + 1,\ e_g(1) = 4t + 2$.

Also $(v_f + e_g)(0) = (v_f + e_g)(2) = 6t + 2,\ (v_f + e_g)(1) = 6t + 3$.

Case (iv): $\boldsymbol{n \equiv 3(mod\ 6)}$

Let $n = 6t + 3,\ t \ge 1$.

$$f(u_i) = \begin{cases} 1, & i \equiv 0, 2 (mod\ 3) \\ 0, & i \equiv 1 (mod\ 3) \end{cases} \forall\ 1 \le i \le 3t+2$$

$$f(u_i) = \begin{cases} 2, & i \equiv 0, 1 (mod\ 3) \\ 0, & i \equiv 2 (mod\ 3) \end{cases} \forall\ 3t+3 \le i \le 6t+1$$

$f(u_{6t+2}) = 2$ and $f(u_{6t+3}) = 2$

Here $|V(G)| = n + 1 = 6t + 4$ and $|E(G)| = 2n = 12t + 6$.

Then $v_f(0) = v_f(1) = 2t + 1,\ v_f(2) = 2t + 2$ and $e_g(0) = e_g(1) = e_g(2) = 4t + 2$.

Also $(v_f + e_g)(0) = (v_f + e_g)(1) = 6t + 3,\ (v_f + e_g)(2) = 6t + 4$.

Case (v): $\boldsymbol{n \equiv 4(mod\ 6)}$

Let $n = 6t + 4,\ t \ge 0$.

$$f(u_i) = \begin{cases} 1, & i \equiv 0, 2 (mod\ 3) \\ 0, & i \equiv 1 (mod\ 3) \end{cases} \forall\ 1 \le i \le 3t+3$$

$$f(u_i) = \begin{cases} 2, & i \equiv 1, 2 (mod\ 3) \\ 0, & i \equiv 0 (mod\ 3) \end{cases} \forall\ 3t+4 \le i \le 6t+4$$

Here $|V(G)| = n + 1 = 6t + 5$ and $|E(G)| = 2n = 12t + 8$.

Then $v_f(0) = v_f(1) = 2t + 2,\ v_f(2) = 2t + 1$ and $e_g(0) = 4t + 2,\ e_g(1) = e_g(2) = 4t + 3$.

Also $(v_f + e_g)(0) = (v_f + e_g)(2) = 6t + 4,\ \ (v_f + e_g)(1) = 6t + 5$.

Case (vi): $n \equiv 5(mod\ 6)$

Let $n = 6t + 5,\ t \geq 0$.

$$f(u_i) = \begin{cases} 1,\ i \equiv 0, 2(mod\ 3) \\ 0,\ i \equiv 1(mod\ 3) \end{cases} \forall\ 1 \leq i \leq 3t + 3$$

$$f(u_i) = \begin{cases} 2,\ i \equiv 1, 2(mod\ 3) \\ 0,\ i \equiv 0(mod\ 3) \end{cases} \forall\ 3t + 4 \leq i \leq 6t + 5$$

Here $|V(G)| = n + 1 = 6t + 6$ and $|E(G)| = 2n = 12t + 10$.

Then $v_f(0) = v_f(1) = v_f(2) = 2t + 2$ and $e_g(0) = e_g(1) = 4t + 3,\ e_g(2) = 4t + 4$.

Also $(v_f + e_g)(0) = (v_f + e_g)(1) = 6t + 5,\ \ (v_f + e_g)(1) = 6t + 6$.

In cases (i),(iii), (iv), (v), and (vi), we see that $|v_f(i) - v_f(j)| \leq 1$ and $|e_g(i) - e_g(j)| \leq 1$ for all $0 \leq i, j \leq 2$ and also $|(v_f + e_g)(i) - (v_f + e_g)(j)| \leq 1$ for all $0 \leq i,\ j \leq 2$.

In case (ii), we see that $|v_f(i) - v_f(j)| \leq 1$ and $|e_g(i) - e_g(j)| \leq 1$ for all $0 \leq i,\ j \leq 2$.

Hence, wheel graph W_n is a perfect SDGM 3-equitable graph if $n \equiv 0, 2, 3, 4, 5\ (mod\ 6)$ and SDGM 3-equitable graph if $n \equiv 1(mod\ 6)$.

Illustration 3 Perfect SDGM 3-equitable labeling of wheel graph W_{16} is shown in Fig. 3.

Here $v_f(0) = v_f(1) = 6,\ v_f(2) = 5$ and $e_g(0) = 10,\ e_g(1) = e_g(2) = 11$.

Also $(v_f + e_g)(0) = (v_f + e_g)(2) = 16,\ (v_f + e_g)(1) = 17$.

Therefore, $|v_f(i) - v_f(j)| \leq 1, |e_g(i) - e_g(j)| \leq 1$ and $|(v_f + e_g)(i) - (v_f + e_g)(j)| \leq 1$ for all $0 \leq i,\ j \leq 2$.

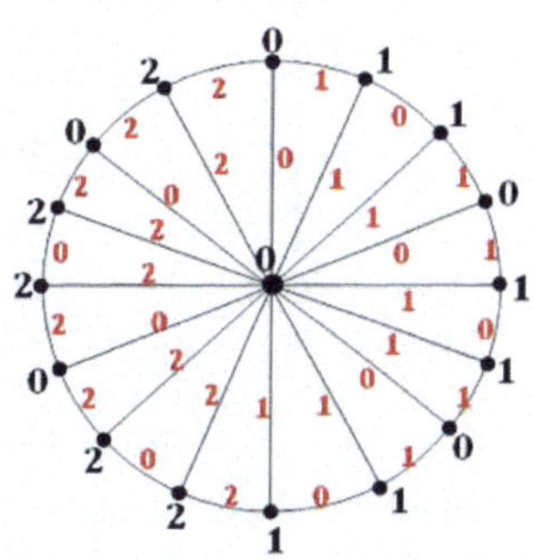

Fig. 3 Perfect SDGM 3-equitable labeling of wheel graph W_{16}

Theorem 4 *The triangular cycle graph TC_n is a perfect SDGM 3-equitable graph when $n \equiv 0(mod\ 2)$.*

Proof Let $G = (V,\ E)$ be a triangular cycle graph TC_n.

Let $V(G) = \{u_i,\ v_i\ /1 \le i \le n\}$.

Let $E(G) = \{u_i u_{i+1},\ u_i v_i,\ u_{i+1} v_i / 1 \le i \le n-1\} \cup \{u_n u_1,\ u_n v_n,\ u_1 v_n\}$.

Let $|V(G)| = 2n$ and $|E(G)| = 3n$.

Define $f : V(G) \to \{0,\ 1,\ 2\}$ as follows:

Case (i): $n \equiv 0, 2(mod\ 6)$

$$f(u_i) = \begin{cases} 0, & i \equiv 1, 2, 4(mod\ 6) \\ 1, & i \equiv 3, 5(mod\ 6) \quad \forall\, 1 \le i \le n \\ 2, & i \equiv 0(mod\ 6) \end{cases}$$

$$f(v_i) = \begin{cases} 0, & i \equiv 4(mod\ 6) \\ 1, & i \equiv 2, 3(mod\ 6) \quad \forall\, 1 \le i \le n \\ 2, & i \equiv 0, 1, 5(mod\ 6) \end{cases}$$

$\boldsymbol{n \equiv 0\,(mod\ 6) \Longrightarrow n = 6t,\ t \ge 1.}$

Here $|V(G)| = 2n = 12t$ and $|E(G)| = 3n = 18t$.

Then $v_f(0) = v_f(1) = v_f(2) = 4t$ and $e_g(0) = e_g(1) = e_g(2) = 6t$.

Also $(v_f + e_g)(0) = (v_f + e_g)(1) = (v_f + e_g)(2) = 10t$.

$\boldsymbol{n \equiv 2\,(mod\ 6) \Longrightarrow n = 6t + 2,\ t \ge 1.}$

Here $|V(G)| = 2n = 12t + 4$ and $|E(G)| = 3n = 18t + 6$.

Then $v_f(0) = 4t + 2,\ v_f(1) = v_f(2) = 4t + 1$ and $e_g(0) = e_g(1) = e_g(2) = 6t + 2$.

Also $(v_f + e_g)(0) = 10t + 4,\ (v_f + e_g)(1) = (v_f + e_g)(2) = 10t + 3$.

Case (ii): $n \equiv 4(mod\ 6)$

Let $n = 6t + 4,\ t \ge 0$.

$$f(u_i) = \begin{cases} 0, & i \equiv 1, 2, 4(mod\ 6) \\ 1, & i \equiv 3, 5(mod\ 6) \quad \forall\, 1 \le i \le n \\ 2, & i \equiv 0(mod\ 6) \end{cases}$$

$$f(v_i) = \begin{cases} 0, & i \equiv 4(mod\ 6) \\ 1, & i \equiv 2, 3(mod\ 6) \quad \forall\, 1 \le i \le n-1 \text{ and } f(v_n) = 2 \\ 2, & i \equiv 0, 1, 5(mod\ 6) \end{cases}$$

Here $|V(G)| = 2n = 12t + 8$ and $|E(G)| = 3n = 18t + 12$.

Then $v_f(0) = v_f(1) = 4t + 3,\ v_f(2) = 4t + 2$ and $e_g(0) = e_g(1) = e_g(2) = 6t + 4$.

Also $(v_f + e_g)(0) = (v_f + e_g)(1) = 10t + 7,\ (v_f + e_g)(2) = 10t + 6$.

In all the above cases, we see that $|v_f(i) - v_f(j)| \le 1$ and $|e_g(i) - e_g(j)| \le 1$ for all $0 \le i,\ j\ \le\ 2$ and also $|(v_f + e_g)(i) - (v_f + e_g)(j) \le 1$ for all $0 \le i,\ j\ \le\ 2$.

Hence, triangular cycle graph TC_n is a perfect SDGM 3-equitable graph if $n \equiv 0(mod\ 2)$.

Fig. 4 Perfect SDGM 3-equitable labeling of triangular cycle graph TC_{12}

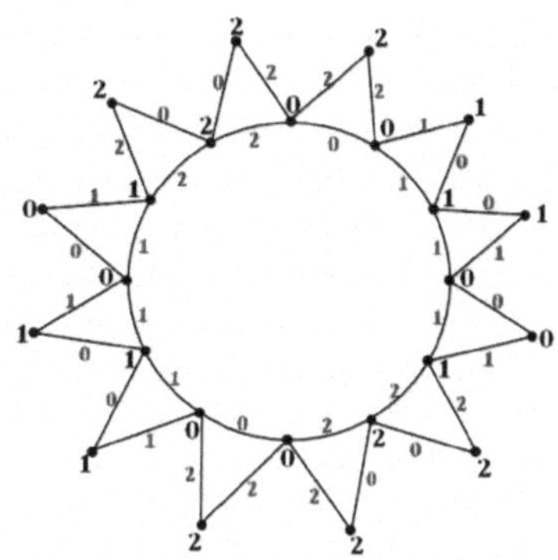

Illustration 4 Perfect SDGM 3-equitable labeling of triangular cycle graph TC_{12} is shown in Fig. 4.

Here $v_f\,(0) = v_f\,(1) = v_f\,(2) = 8$ and $e_g\,(0) = e_g\,(1) = e_g\,(2) = 12$.

Also $\left(v_f + e_g\right)(0) = \left(v_f + e_g\right)(1) = \left(v_f + e_g\right)(2) = 20$.

Therefore, $\left|v_f(i) - v_f(j)\right| \le 1$, $\left|e_g(i) - e_g(j)\right| \le 1$ and $\left|\left(v_f + e_g\right)(i) - \left(v_f + e_g\right)(j)\right| \le 1$ for all $0 \le i,\ j \le 2$.

4 Conclusion

In this paper, we proved the following cycle-related graphs: crown graph, gear graph, triangular cycle graph and wheel graph are the SDGM 3-equitable graph or perfect SDGM 3-equitable graph. The future work includes perfect SDGM 3-equitable labeling of tree-related and ladder-related graphs.

References

1. Gallian, J.A.: A dynamic survey of graph labeling. Electron J Comb. https://doi.org/10.37236/11668 (2023)
2. Deshmukh, U., Shaikh, V.Y.: Geometric mean labeling of some cycle related graphs. Int. J. Math. Sci. Eng. Appl. **11**, 29–38 (2017)
3. Cahit, I.: On cordial and 3-equitable labeling of graphs. Utilitas Math. **37**, 189–198 (1990)
4. Chitra Lakshmi, K., Nagarajan, K.: Geometric mean cordial labeling of graphs. Int. J. Math. Soft Comput. **7**, 75–87 (2017)
5. Maulidia, A.R., Purwanto: Elegant labeling of sun graphs and Helm graphs. J. Phys. Conf. Series **1872**, 1–7 (2021)
6. Le Bras, R., Gomes, C.P., Selman, B.: Double-wheel graphs are graceful. In: Proceedings of the Twenty-Third International Joint Conference on Artificial Intelligence, pp. 587–593 (2013)
7. Zeen El Deen, M.R.: Edge δ-graceful labeling for some cyclic-related graphs. Adv. Math. Phys. **2020**, 1–18 (2020)

Domination Number in Sierpiński Networks

J. Anitha, Indra Rajasingh, and Jane Olive Sharon

AMS Subject Classification 68R10, 81Q30, 82B20, 82C20

1 Proposed Results

In this section, we determine the DN for divide-and-swap cube networks and Sierpiński networks.

Let G be a graph with two non-empty sets. Open neighbourhood v is defined as $N(p) = \{q \in V : (p, q) \in E\}$ and the closed neighbourhood of $N[p] = \{q \in V : (p, q) \in E\} \cup \{p\}$ [1].

A set $K \subseteq V(G)$ is a DS of G, such that every node in the graph is either in the DS K or neighbouring to at least one node in K. The DN of G is $\gamma(G)$ [1].

1.1 Sierpiński Networks

Sierpiński network S_t^r , $t, r \geq 1$ defined on the node set $\{1, \dots, t\}^t$, two different nodes $p = (p_1 p_2 \dots p_r)$ and $q = (q_1 q_2 \dots q_r)$ being neighbours if there exists an $h \in \{1 \dots h\}$ such that

J. Anitha (✉)
Department of Mathematics, Rajalakshmi Institute of Technology, Chennai, India

I. Rajasingh
Department of Mathematics for Excellence, Saveetha School of Engineering, Saveetha Institute of Medical and Technical Sciences, Chennai, India

J. O. Sharon
Department of Computing Technologies, SRM Institute of Science and Technology, Chennai, India
e-mail: janeolip@srmist.edu.in

I. Rajasingh et al. (eds.), *Applied Computational Mathematics*,
Springer Proceedings in Mathematics & Statistics 480,
https://doi.org/10.1007/978-3-031-77764-6_3

1. $p_s = q_s$, for $s = 1 \ldots h-1$;
2. $p_h \neq q_h$ and ;
3. $p_t = q_h$ and $p_t = q_h$ for $s = h+1 \ldots r$.. See Fig. 1.

An extreme node (EN) is a node of S_t^r with form $(ii \ldots i)$. Observe that $|V(S_t^r)| = t^r$ and that t extreme nodes. Let iS_t^{r-1} be the subgraph of S_t^r induced by the nodes of form $(iq_2 \ldots q_r)$ for $r \geq 2$ and $i \in \{1 \ldots t\}$. More generally, we designate the subgraph of S_t^r induced by the nodes of the form $(i_1 \ldots i_s v_{s+1} \ldots v_r)$ by $i_1 \ldots i_k S_t^{r-1}$ for given $i_1 \ldots i_k \in \{1, \ldots, t\}$. You should take note that iS_t^{r-1} is isomorphic to S_t^{r-1}. In a broader sense, $i_1 \ldots i_s S_t^{r-s}$ is isomorphic to S_t^{r-s}. A clique edge is defined as an edge of S_t^r of the type $(p_1p_2 \ldots p_{r-1}i)(p_1p_2 \ldots p_{r-1}j)$, $i \neq j$. A unique subgraph K_t contains a clique edge.

An extreme node(EN) node of S_t^r is the form $(ii \ldots i)$. Observe that $|V(S_t^r)| = t^r$ has t EN. Let iS_t^{r-1} be the subgraph of S_t^r induced by the nodes of form $(iq_2 \ldots q_r)$ for $r \geq 2$ and $i \in \{1 \ldots t\}$. Generally, for given the subgraph of S_t^r induced by the nodes of the form $(i_1 \ldots i_s v_{s+1} \ldots v_r)$ by $i_1 \ldots i_k S_t^{r-1}$ for given $i_1 \ldots i_k \in \{1, \ldots, t\}$. Note that iS_t^{r-1} is isomorphic to S_t^{r-1}. In a broader sense, $i_1 \ldots i_s S_t^{r-s} \cong S_t^{r-s}$. A clique edge is defined as an edge of S_t^r of the type $(p_1p_2 \ldots p_{r-1}i)(p_1p_2 \ldots p_{r-1}j)$, $i \neq j$. A unique subgraph K_t contains a clique edge.

The remaining edges as non-clique edges. Assume that $i \neq j$. Then the unique edge between iS_t^{r-1} and jS_t^{s-1} is $(ijj \ldots j)(jii \ldots i)$. The notation for it is $e_{ij}^{(r)} = e_{ji}^{(r)}$. The subgraph $i_1 \ldots i_s S_t^{r-s}$ of S_t^r is to be considered. Then, $(i_1 \ldots i_s e_{jl}^{(r-s)})$ will represent the edge between $(i_1 \ldots i_r jl \ldots l)$ and $(i_1 \ldots i_r lj \ldots j)$ [2].

Remark 1 In this paper, S_t^r, when $k = 4$ is taken up for study.

(i) S_4^r has 4^r nodes, four of which are of degree 2 and the rest $4^r - 4$ nodes are of degree 3.
(ii) For $2 \leq i \leq r$, each dimension i has 4^{i-t} node distinct copies of H_j, $1 \leq j \leq i$. This yields that there are 4^{r-2} node distinct copies of S_4^2 in S_4^r.

Notation Let L be a set of labels. Then $i \| L$ is defined as iL where every label in L is prefixed label i.

In the sequel, we require the ordering of nodes of S_4^r as follows:

Labeling of S_4^r:

The nodes of S_4^r, $r \geq 1$ labelled as follows:

(i) S_4^1 named as 1-digit Z_3 number, say 0, 1, 2, 3.
(ii) S_4^2 named as 2-digit Z_3 number, $0S_4^1, 1S_4^1, 2S_4^1, 3S_4^1$.
(iii) In general, S_4^r named as r-digit Z_3 number, tS_4^{r-1}, $0 \leq t \leq 3$. See Fig. 1.

1.2 Domination in Sierpiński Network

In this section, we obtain the $\gamma(G)$ for Sierpiński networks S_4^r.

Theorem 1 ([1]) *Let G be any graph of order r with maximum degree $\Delta(G)$,* $\gamma(G) \geq \left\lceil \frac{r}{\Delta(G)+1} \right\rceil$.

Theorem 1 yields the following result.

Lemma 1 *For the Sierpiński network S_4^r, we have $\gamma(S_4^r) \geq 4^{r-1}$.*

Proof By Theorem 1, $\gamma(S_4^r) \geq 4^{r-1}$.

We proceed to give an algorithm to get 4^{r-1} as an upper bound for $\gamma(S_4^r)$.

Domination Algorithm for Sierpiński Network S_4^r
Input: Sierpiński network, S_t^r, $r \geq 2$ with Z_3 number.

Algorithm

(i) Choose $S_2 = \{01, 13, 23, 30\}$ in S_4^2.

(ii) Let $S_3 = \bigcup_{t=0}^{3} tS_2$ in S_4^3.

(iii) Inductively select $S_r = \bigcup_{t=0}^{3} tS_{r-1}$ in S_4^r.

Output $\gamma(G) \leq 4^{r-1}$.

Proof of Correctness Let S_r be a DS of S_4^r. S_4^1 is a cycle on 4 vertices labelled as $Z_4 = \{0, 1, 2, 3\}$. Let S_4^2 composed of four copies of S_4^1, say iS_4^1, $0 \leq i \leq 3$. Let $S_2 = \{01, 13, 23, 30\}$. We claim that any two vertices, p and q in S_2, are such that $d(p, q) \geq 3$ if $|N[p] \cap N[q]| = \phi$. Vertices ij selected in S_2, ij dominate $\{ii, ji, i(j-1), i(j+2), i(j+1)\}$, $0 \leq i, j \leq 3$ and hence $S_2 = V(S_4^2)$. See Fig. 1a. For $r = 3$, S_4^3 composed of four copies of S_4^2. Let $S_3 = \bigcup_{i=0}^{3} iS_2$. Vertices

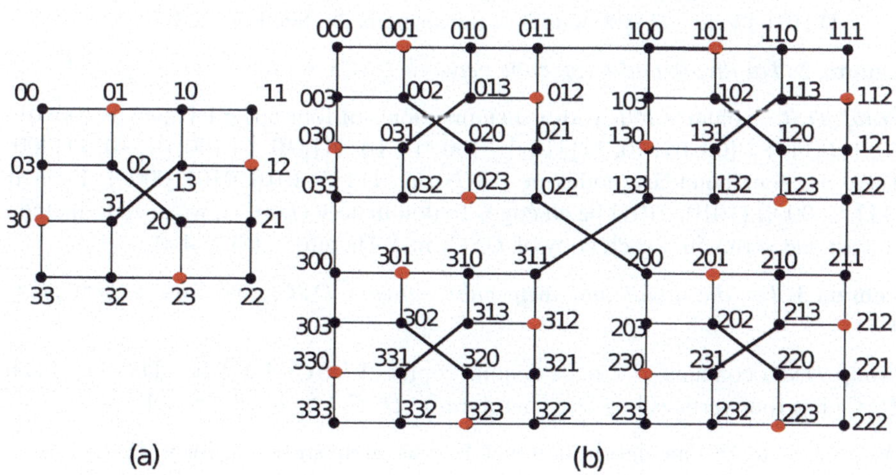

Fig. 1 Red vertices indicate the DS of **(a)** S_4^2 **(b)** S_4^3

in S_3 dominate $i\|\{ii, ji, i(j-1), i(j+2), i(j+1)\}$, $0 \le i, j \le 3$ and hence $S_3 = V(S_4^3)$. See Fig. 1b. Inductively, let $S_r = \bigcup_{i=0}^{3} i S_{r-1}$. S_t^r composed of four copies of S_t^{r-1}. Then vertices in S_r dominate $ii \ldots i(r-2)times\|\{ii, ji, i(j-1), i(j+2), i(j+1)\}$, $0 \le i, j \le 3$. Thus, $S_r = V(S_4^r)$. Thus, $\gamma(S_t^r) \le 4^{r-1}$.

Domination Algorithm and Lemma 1 yield the following theorem.

Theorem 2 *For the Sierpiński Network S_t^r, $r \ge 2$, we have $\gamma(S_4^r) = 4^{r-1}$.*

1.3 Divide-and-Swap Cube Network DSC_n

A DSC_n is a variant of hypercube-variant networks with a recursive structure which is isomorphic to Sierpiński network $S_4^r, r \ge 1$ with additional edges by joining extreme nodes denoted by DSC_n. See Fig. 2a, b [3].

1.4 Domination in Divide-and-Swap Cube Network

In this section, we obtain the lower bound using Theorem 1 for divide-and-swap cube network.

$2r$ **Bit String Labeling**
The nodes of DSC_r, $r \ge 1$ are labelled as follows:

(i) DSC_1 as 00, 01, 10, 11 length of the string is 2.
(ii) DSC_2, as 4 bit string, say, $\{00, 01, 10, 11\}\|11 \bigcup \{00, 01, 10, 11\}\|01 \bigcup \{00, 01, 10, 11\}\|10 \bigcup \{00, 01, 10, 11\}\|00$ length of the string is 4.
(iii) In general, DSC_r, as $2r$ bit string, say, $V_1(r-1)\|11 \bigcup V_1(r-1)\|01 \bigcup V_1(r-1)\|10 \bigcup V_1(r-1)\|00$ length of the string is $2r$. See Fig. 2a, b.

Lemma 2 *For divide-and-swap cube network DSC_2, we have $\gamma(DSC_2) \ge 4$.*

Proof DSC_2 contains four vertex disjoint copies of four-cycle labelled as $\{00, 01, 10, 11\}\|11 \bigcup \{00, 01, 10, 11\}\|10 \bigcup \{00, 01, 10, 11\}\|10 \bigcup \{00, 01, 10, 11\}\|00$. There are four boundary nodes in DSC_2 say, $\{1111, 1010, 0101, 0000\}$. Even if $\{1111, 0000\}$, $\{1010, 0101\}$ be an edge. To dominate $V(DSC_2)$, we need to include at least one vertex in S each copy of DSC_1 in S Therefore, $|S| \ge 4$.

Lemma 3 *For the divide-and-swap cube network DSC_r, we have $\gamma(DSC_r) \ge 4^{r-1}$.*

Proof DSC_r contains 4^{r-2} node disjoint copies of DSC_2. Let S be a DS of G. Each DSC_2 has four vertices in a Ds. Therefore, $\gamma(DSC_r) \ge 4 \times 4^{r-2} = 4^{r-1}$.

We proceed to give an algorithm to get 4^{r-1} as an upper bound for $\gamma(DSC_r)$.

Domination Algorithm Divide-and-Swap Cube Networks

Input Divide-and-Swap Cube Networks, DSC_r, $r \geq 2$, with $2r$-bit string labelling.

Algorithm
(i) Choose $S_2 = \{0010, 1100, 0111, 1001\}$ in DSC_2.
(ii) Let $S_3 = \{0010, 1100, 0111, 1001\}\|11 \bigcup \{0010, 1100, 0111, 1001\}\| 10 \bigcup \{0010, 1100, 0111, 1001\}\|01 \bigcup \{0010, 1100, 0111, 1001\}\|00$ in DSC_3.
(iii) Inductively select $S_r = V(S_{r-1})\|11 \bigcup V(S_{r-1})\|10 \bigcup V(S_{r-1}) \|01 V(S_{r-1})\|00$ in DSC_r.

Output $\gamma(G) \leq 4^{r-1}$.

Proof of Correctness Let S_r be a Ds of DSC_r. DSC_1 is a cycle on 4 vertices labelled as $00, 11, 01, 10$. Let DSC_2 composed of four copies of DSC_1, say $V(DSC_1)\|11 \bigcup V(DSC_1)\|01 \bigcup V(DSC_1)\|10 \bigcup V(DSC_1)\|00$. Let $S_2 = \{0010, 1100, 0111, 1001\}$ be a DS of DSC_2. We claim that any two vertices p and q in S_2 are such that $d(p, q) \geq 3$ iff $|N[p] \cap N[q]| = \phi$. Vertices $\{0010, 1100, 0111, 1001\}$ in S_2 dominate $\{1010, 1000, 1110, 0100, 0000, 0011, 1011, 1111, 1101, 1110, 0101, 0110\}$ and hence $S_2 = V(DSC_2)$. See Fig. 2a.

For $n = 3$, DSC_3 composed of four copies of DSC_2. Let $S_3 = \{0010, 1100, 0111, 1001\}\|11 \bigcup \{0010, 1100, 0111, 1001\}\|10 \bigcup \{0010, 1100, 0111, 1001\}\| 01 \bigcup \{0010, 1100, 0111, 1001\}\|00$ be a DS of DSC_3. We claim that any two vertices p and q in S_3 are such that $d(p, q) \geq 3$ iff $|N[p] \cap N[q]| = \phi$. Vertices $\{0010, 1100, 0111, 1001\}\|11 \bigcup \{0010, 1100, 0111, 1001\}\|10 \bigcup \{0010, 1100, 0111, 1001\}\|01 \bigcup \{0010, 1100, 0111, 1001\}\|00$ selected in S_3 dominate $\{1010, 1000, 1110, 0100, 0000, 0011, 1011, 1111, 1101, 1110, 0101, 0110\}\|11 \bigcup \{1010, 1000, 1110, 0100, 0000, 0011, 1011, 1111, 1101, 1110, 0101, 0110\}\|10 \bigcup \{1010, 1000, 1110, 0100, 0000, 0011, 1011, 1111, 1101, 1110, 0101, 0110\}\| 01 \bigcup \{1010, 1000, 1110, 0100, 0000, 0011, 1011, 1111, 1101, 1110, 0101, 0110\} \| 00$ and hence $S_3 = V(DSC_3)$. See Fig. 2b. Inductively, $S_n = V(S_{r-1})\|11 \bigcup V(S_{r-1})\|10 \bigcup V(S_{r-1})\|01 \bigcup V(S_{r-1})\|00$ be a dominating set of DSC_r. DSC_r composed of four copies of DSC_{r-1}. Inductively, Then the vertices in S_r dominate $V(DSC_r)$. Thus, $S_n = V(DSC_r)$. Thus $\gamma(DSC_r) \leq 4^{r-1}$.

Domination Algorithm and Lemma 3 yield the following theorem.

Theorem 3 *For the divide-and-swap cube network DSC_r, $r \geq 1$, we have $\gamma(DSC_r) = 4^{r-1}$.*

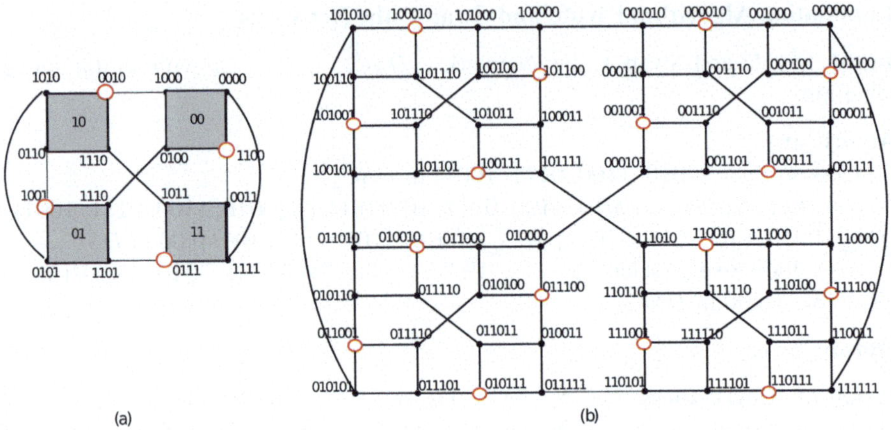

Fig. 2 DS of (**a**) DSC_2 (**b**) DSC_3

2 Conclusion

In this paper, we have solved the DN for S_k^n when $k = 4$ and the divide-and-swap networks. The discussion is open for obtaining domination number for S_k^n, $k \geq 5$.

References

1. Haynes, T.W., Hedetniemi, S.M., Hedetniemi, S.T., Henning, M.A.: Power domination in graphs applied to electrical power networks. SIAM J. Discrete Math. **15**(4), 519–529 (2002)
2. Klavžar, S., Peterin, I., Zemljic, S.S.: Hamming dimension of a graph—the case of Sierpiński graphs. Eur. J. Comb. **34**, 460–473 (2013)
3. Arulperumjothi, M., Klavžar, S., Prabhu, S.: Redefining fractal cubic networks and determining their metric dimension and fault-tolerant metric dimension. Appl. Math. Comput. **452**, 128037 (2023)

Vertex Prime and Edge Ternion Sum Labeling of Certain Centralized Networks

N. Hameeda Begum and V. Annamma

1 Introduction

A network is a collection of nodes or vertices that are linked with one another. The links uniting the nodes are called links or edges. In this research article, we have introduced one-edge varied ternion sum, k-edge varied ternion sum, and k-edge varied ternion sum with l-variations, and we investigate that the centralized path network, centralized cycle network, centralized fan network, centralized triangular path network admit vertex prime and edge ternion sum labeling. $V(G)$ and $E(G)$ stands for vertex set and edge set with cardinality $|V(G)|$ and $|E(G)|$, respectively. The networks have vast application in mass communication, data sharing, alliance among the devices that are connected to a network, etc.

2 Preliminaries

Definition 1 Centralized graph of a graph G is constructed by taking m copies $G_1, G_2, G_3, \ldots, G_m$ of the graph G and merging one of their vertices of G_1 with the corresponding copies of that vertex from each $G_2, G_3, \ldots, G_m$ graph.

Definition 2 ([1]) The path graph $P_n, n \geq 2$ is constructed by forming a route from one vertex to another by virtue of placing edges linking successive vertices of the path.

N. Hameeda Begum (✉) · V. Annamma
Department of Mathematics, University of Madras, Chennai, India

I. Rajasingh et al. (eds.), *Applied Computational Mathematics*,
Springer Proceedings in Mathematics & Statistics 480,
https://doi.org/10.1007/978-3-031-77764-6_4

Definition 3 The centralized path network is constructed by taking m copies of path P_n and merging first vertex of each path.

Definition 4 ([2]) The cycle graph C_n is defined as a route, using at least one edge, that returns to its starting point but does not repeat any vertex except for the initial and the last vertex.

Definition 5 The centralized cycle network $C_{m,n}$ is constructed by taking m copies of the cycle graph C_n and merging one of their vertices from each graph.

Definition 6 ([3]) A fan graph $F_n = P_n + K_1$ is formed by connecting a single vertex to all vertices of a path P_n.

Definition 7 The Centralized Fan Network $F_{m,n} = mP_n + K_1$ is constructed by connecting a single vertex to all vertices of m copies of path P_n.

Definition 8 ([4]) Triangular path graph TP_n is constructed by replacing every edge of a path P_n by C_3.

Definition 9 Centralized triangular path network $TP_{m,n}$ is constructed by taking m copies of triangular path TP_n and merging all the first vertex of each triangular path.

Definition 10 ([5]) A graph $G = (V(G), E(G))$ is said to admit prime labeling. The graph vertices are labeled with unique integers from $[1, |V|]$ so that for every edge $m_i n_j$, the end vertices m_i and n_j are allocated labels such that $gcf(m_i, n_j) = 1$.

3 Main Definitions

Definition 11 Let $G(V, E)$ be a finite, connected, and non-oriented graph. Let $V(G)$ and $E(G)$ be the vertex set and edge set with cardinality $|V(G)|$ and $|E(G)|$, respectively. We define the labeling of the vertices as a bijective mapping, $f : V(G) \rightarrow \{1, 2, 3, \ldots, |V(G)|\}$ such that for each edge $uv \in E(G)$, $gcf(f(u), f(v)) = 1$, and we also define the labeling of the edges as an injective mapping, $g : E(G) \rightarrow \{1, 2, 3, \ldots, |V(G)| + |E(G)|\}$ such that each edge has a distinct label and for an edge $uv \in E(G)$, the ternion sum denoted by $T(uv)$ is defined by $T(uv) = f(u) + f(v) + g(uv)$.

Definition 12 The **Uniform ternion sum** of a graph G is the ternion sum $f(u) + f(v) + g(uv)$, which is a unique constant c for all the edges $uv \in E(G)$. The uniform ternion sum is denoted by $\mu(G) = c$.

Definition 13 The **1-Edge varied ternion sum** of a graph G is the ternion sum $f(u) + f(v) + g(uv)$ which is a unique constant c for all $|E(G)| - 1$ edges and varies for only one edge of the graph G. The 1-Edge varied ternion sum is denoted by, $\mu_1(G) = \begin{cases} c & \text{for } |E(G)| - 1 \text{ edges} \\ c_1 & \text{for 1 edge} \end{cases}$

Definition 14 The ***k*-Edge varied ternion sum** of a graph G is the ternion sum $f(u)+f(v)+g(uv)$ which is a unique constant c for $|E(G)|-k$ edges and another varied constant c_1 which is same for the remaining k edges of the graph G such that $|\mu_e(G)| \geq k$, where $\mu_e(G)$ denotes the set of edges that constitute unique ternion sum c and $|\mu_e(G)|$ denotes the number of edges of the set $\mu_e(G)$. The k-Edge varied ternion sum is denoted by

$$\mu_k^1(G) = \begin{cases} c & \text{for } |E(G)| - 1 \text{ edges} \\ c_1 & \text{for } k \text{ edges} \end{cases} \quad where\ k > 1$$

Definition 15 The ***k*-Edge varied ternion sum with *l*-variations** of a graph G is the ternion sum $f(u) + f(v) + g(uv)$ which is a unique constant c for $|E(G)| - k$ edges where $k = \sum_{i=1}^{l} k_i$ and another varied constant c_i which is same for the remaining k_i edges of the graph G such that $|\mu_e(G)| \geq k_i$ for $1 \leq i \leq l$ where $\mu_e(G)$ denotes the set of edges that constitute unique ternion sum c and $|\mu_e(G)|$ denotes the number of edges of the set $\mu_e(G)$. The k-Edge varied ternion sum with l-variations are denoted by

$$\mu_k^1(G) = \begin{cases} c & \text{for } |E(G)| - 1 \text{ edges} \\ c_1 & \text{for } k \text{ edges} \end{cases} \quad where\ k_i > 1 and\ 1 \leq i \leq l,\ l \geq 1$$

Definition 16 A graph $G(V, E)$ is said to admit **Vertex Prime and Edge Ternion Sum Labeling** if it satisfies the following conditions:

1. The graph vertices are labeled with unique integers from $[1, |V|]$, so that for every edge $m_i n_j$, the end vertices m_i and n_j are allocated labels such that $gcf(m_i, n_j) = 1$.
2. The edges of the graph are assigned labels with distinct integral values from $[1, |V| + |E|]$ such that ternion sum is uniform ternion sum or 1-Edge varied ternion sum or k-Edge varied ternion sum with l-variations

Note 1 In particular, the k-Edge varied ternion sum is also known as k-Edge varied ternion sum with 1-variation.

Note 2 If $|\mu_e(G)| < k_i$ for $1 \leq i \leq l$, then the graph G is only a vertex prime graph and not an edge ternion sum graph.

4 Main Results

4.1 Vertex Prime and Edge Ternion Sum Labeling of Centralized Path Network

Theorem 1 *Centralized (m, n)-path network $P_{m,n}$ admits vertex prime and edge ternion sum labeling with uniform ternion sum, $\mu(P_{m,n}) = 2(mn + 1)$ for all $m.n \geq 2$.*

Proof Let $P_{m,n}$ denote the centralized (m, n)-path network, for all $m.n \geq 2$.

The vertex set $V(P_{m,n}) = \{u, v_{i,j}/1 \leq i \leq m, 1 \leq j \leq n\}$ and $|V(P_{m,n})| = mn + 1$. The edge set $E(P_{m,n}) = \{uv_{i,1}/1 \leq i \leq m, \} \cup \{v_{i,j}v_{i,j+1}/1 \leq i \leq m, 1 \leq j \leq n-1\}$ and $|E(P_{m,n})| = mn$.

We define the labeling of the vertices as a bijective mapping, $f : V(P_{m,n})) \rightarrow \{1, 2, 3, \ldots mn+1\}$ by $f(u) = 1$, $f(v_{i,j}) = n(i-1)+j+1$, $1 \leq i \leq m$, $1 \leq j \leq n$.

We define the labeling of the edges as an injective mapping, $g : E(P_{m,n}) \rightarrow \{1, 2, 3, \ldots 2mn + 1\}$ by $g(uv_{i,1}) = n(2m - i + 1) - 1$, for $1 \leq i \leq m$

$g(v_{i,j}v_{i,j+1}) = 2(n(m - i + 1) - j) - 1$, for $1 \leq i \leq m$, $1 \leq j \leq n - 1$,

We observe that for the edges $uv_{i,1}$ and $v_{i,j}\ v_{i,j+1}$

$gcf(f(u), f(v_{i,1})) = gcf(1, f(v_{i,j})) = 1$, $1 \leq i \leq m$ and

$gcf(f(v_{i,j})), f(v_{i,j+1})) = 1$, $1 \leq i \leq m$, $1 \leq j \leq n - 1$, respectively.

We also observe that for $1 \leq i \leq m$, the ternion sum for the edges $uv_{i,1}$ is $T(uv_{i,1}) = f(u)+f(v_{i,1})+g(uv_{i,1}) = (1)+(n(i-1)+2)+(n(2m-i+1)-1) = 2(mn+1)$

We also observe that for $1 \leq i \leq m$, $1 \leq j \leq n-1$ the ternion sum for the edges $v_{i,j}v_{i,j+1}$ is $T(v_{i,j}v_{i,j+1}) = f(v_{i,j}) + f(v_{i,j+1}) + g(v_{i,j}v_{i,j+1}) = (n(i-1)+j+1) + (n(i-1)+j+2) + (2(n(m-i+1)-j)-1) = 2(mn+1)$

Therefore, we observe that the ternion sum of all the edges of $P_{m,n}$ is a unique constant with uniform ternion sum $\mu(P_{m,n}) = 2(mn + 1)$ for all $m.n \geq 2$. Hence, the centralized (m, n)-path network $P_{m,n}$ admits vertex prime and edge ternion sum labeling. The centralized (m, n)-path network $P_{m,n}$ and its vertex prime and edge ternion sum labeling are as in the following Fig. 1a and b, respectively.

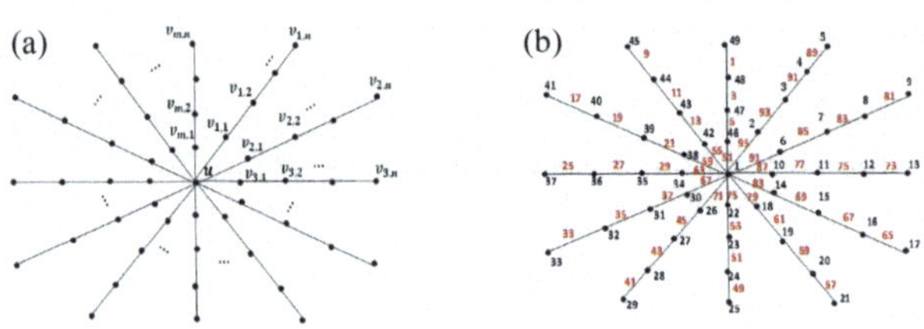

Fig. 1 (**a**) centralized path network and (**b**) Vertex prime and edge ternion sum labeling of centralized path network $P_{12,4}$ with uniform ternion sum $\mu(P_{12,4}) = 98$

4.2 Vertex Prime and Edge Ternion Sum Labeling of Centralized Triangular Path Network

Theorem 2 *Centralized (m, n)-triangular path network $TP_{m,n}$ admits vertex prime and edge ternion sum labeling with 1-Edge varied ternion sum* $\mu_1(G) = \begin{cases} 4mn+2 & \text{for } |E(G)|-1 \text{ edges} \\ 2n+7 & \text{for 1 edge} \end{cases}$ *for all* $m \geq 3$, $n \geq 2$.

Proof Let $TP_{m,n}$ denote the centralized (m, n)-triangular path network, for all $m \geq 3, n \geq 2$.

The vertex set $V(TP_{m,n}) = \{u, v_{i,j}/1 \leq i \leq m, 1 \leq j \leq 2n\}$ and $|V(TP_{m,n})| = 2mn + 1$

The edge set $E(TP_{m,n}) = \{uv_{i,j}/1 \leq i \leq m, j = 1, 2\} \cup \{v_{i,j}v_{i,j+1}/1 \leq i \leq m, 1 \leq j \leq 2n\} \cup \{v_{i,j}v_{i,j+2}/1 \leq i \leq m, 1 \leq j \leq 2n-2$ and $j \equiv 0(mod2)\}$ and $|E(P_{m,n})| = 3mn$

We define the labeling of the vertices as a bijective mapping, $f : V(TP_{m,n}) \rightarrow 1, 2, 3, \ldots, 2mn + 1$ by $f(u) = 1$,

$f(v_{i,j}) = 2n(i-1) + j + 1, 1 \leq i \leq m, 1 \leq j \leq n$

We define the labeling of the edges as an injective mapping, $g : E(P_{m,n}) \rightarrow \{1, 2, 3, \ldots, 5mn + 1\}$ by $g(uv_{i,j}) = 2n(2m - i + 1) - j$, for $1 \leq i \leq m$, $j = 1, 2$,

$g(uv_{2,1}) = 4$

$g(v_{i,j}v_{i,j+1}) = 4n(m - i + 1) - 2j - 1$, for $1 \leq i \leq m, 1 \leq j \leq 2n$,

$g(v_{i,j}v_{i,j+2}) = 4n(m - i + 1) - 2j - 3$, for $1 \leq i \leq m, 1 \leq j \leq 2n - 2$ and $j \equiv 0(mod2)$.

We observe that for the edges $uv_{i,1}$ and $v_{i,j}\ v_{i,j+1}$

$gcf(f(u), f(v_{i,j})) = gcf(1, f(v_{i,j})) = 1, 1 \leq i \leq m$ and $j = 1, 2$

$gcf(f(v_{i,j}), f(v_{i,j+1})) = 1, 1 \leq i \leq m, 1 \leq j \leq n$

$gcf(f(v_{i,j}), f(v_{i,j+2})) = 1, 1 \leq i \leq m, 1 \leq j \leq 2n-2, j \equiv 0\ (mod\ 2)$, respectively.

We also observe that for $1 \leq i \leq m$, $m \neq 2$ and $j = 1, 2$ the ternion sum for the edges $uv_{i,j}$ is $T(uv_{i,j}) = f(u) + f(v_{i,j}) + g(uv_{i,j}) = (1) + (n(i-1) + 2) + (n(2m - i + 1) - 1) = 4mn + 2$. We also observe that for $1 \leq i \leq m, 1 \leq j \leq n$ the ternion sum for the edges $v_{i,j}v_{i,j+1}$ is $T(v_{i,j}v_{i,j+1}) = f(v_{i,j}) + f(v_{i,j+1}) + g(v_{i,j}v_{i,j+1}) = 4mn + 2$. We also observe that for $1 \leq i \leq m, 1 \leq j \leq 2n - 2$ and $j \equiv 0(mod\ 2)$, the ternion sum for the edges $v_{i,j}\ v_{i,j+2}$ is $T(v_{i,j}v_{i,j+2}) = f(v_{i,j}) + f(v_{i,j+2}) + g(v_{i,j}v_{i,j+2}) = 4mn + 2$

For the edge $uv_{2,1}$, the ternion sum is $T(uv_{2,1}) = f(u) + f(v_{2,1}) + g(uv_{2,1}) = 2n + 7$

Therefore, we observe that the ternion sum of all the edges of $TP_{m,n}$ except one edge is a unique constant.

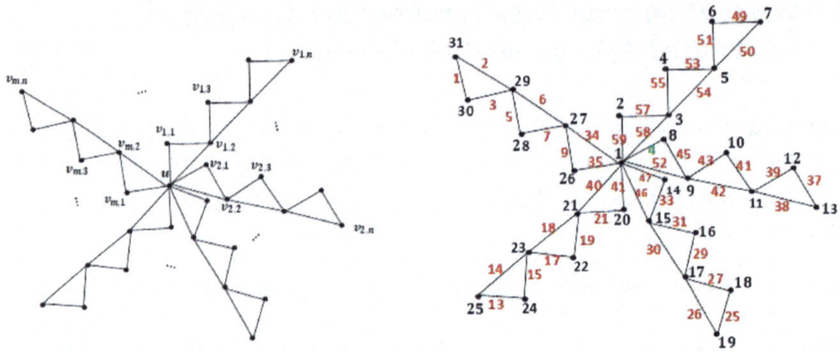

Fig. 2 (**a**) centralized triangular path network and (**b**) vertex prime and edge ternion sum labeling of centralized triangular path network $TP_{5,3}$ with 1-Edge varied ternion sum $\mu_1(G) = \begin{cases} 62 & \text{for 44edges} \\ 13 & \text{for 1 edge} \end{cases}$

Therefore, the centralized (m, n)-triangular path network $TP_{m,n}$, for all $m \geq 3$, $n \geq 2$ admits vertex prime and edge ternion sum labeling with 1-Edge varied ternion sum $\mu_1(G) = \begin{cases} 4mn + 2 & \text{for } |E(G)| - 1 \text{ edges} \\ 2n + 7 & \text{for 1 edge} \end{cases}$

The centralized (m, n)- triangular path network $TP_{m,n}$ and its vertex prime and edge ternion sum labeling are in Fig. 2a and b, respectively.

4.3 Vertex Prime and Edge Ternion Sum Labeling of Centralized Cycle Network

Theorem 3 *Centralized (m, n)-cycle network admits vertex prime and edge ternion sum labeling with k-edge varied ternion sum*

$$\mu_k^1(G) =$$

$$\begin{cases} 2mn + 2 & for \| E(G)| - k \ edges \\ \begin{cases} (m-1) + 5 & if\, m, n \equiv 0 (mod\ 2) \\ n(m-1) + 4 & if\, m \equiv 0 (mod\ 2) and\, n \equiv 1 (mod\ 2) \\ mn + 4 & if\, m, n \equiv 1 (mod\ 2) \\ n(m-2) + 5 & if\, m \equiv 1 (mod\ 2) and\, n \equiv 0 (mod\ 2) \end{cases} & for\ k\ edges \end{cases}$$

where $k > 1$, for all $m, n \geq 2$.

Proof Let $C_{m,n}$ denote the centralized (m, n)-cycle network.

The vertex set $V(C_{m,n}) = \{u, v_{i,j}/1 \leq i \leq m, j = 1, n\}$ and $|V(C_{m,n})| = mn + 1$

The edge set $E(C_{m,n}) = \{uv_{i,j}/1 \leq i \leq m, j = 1, n\} \cup \{v_{i,j}v_{i,j+1}/1 \leq i \leq m, 1 \leq j \leq n - 1\}$ and $|E(C_{m,n})| = m(n + 1)$

We define the labeling of the vertices as a bijective mapping, $f : V(C_{m,n}) \to \{1, 2, 3, \ldots, mn + 1\}$ by $f(u) = 1$, $f(v_{i,j}) = n(i - 1) + j + 1$, $1 \leq i \leq m$, $1 \leq j \leq n$.

We define the labeling of the edges as an injective mapping, $g : E(C_{m,n}) \to \{1, 2, 3, \ldots, m(2n + 1) + 1\}$ by

For $1 \leq i \leq m$, $j = 1$, $g(uv_{i,j}) = 2m - n(i - 1) - 1$

For $1 \leq i \leq m$, $i \equiv 0(mod\ 2)$, $j = n$, $g(uv_{i,j}) = n(2m - i)$

For $1 \leq i \leq m$, $i \equiv 1(mod\ 2)$, $j = n$,

$$g(uv_{i,j}) = \begin{cases} n(m-1)+5 & \text{if } m, n \equiv 0(mod\ 2) \\ n(m-1)+4 & \text{if } m \equiv 0(mod\ 2) \text{and } n \equiv 1(mod\ 2) \\ mn+4 & \text{if } m, n \equiv 1(mod\ 2) \\ n(m-2)+5 & \text{if } m \equiv 1(mod\ 2) \text{and } n \equiv 0(mod\ 2) \end{cases}$$

For $1 \leq i \leq m$, $1 \leq j \leq n - 1$, $g(v_{i,j}\ v_{i,j+1}) = 2n(m - i + 1) - 2j - 1$

We observe that for $1 \leq i \leq m$, $j = 1, n$, the edges $uv_{i,j}$

$gcf(f(u), f(v_{i,j})) = gcf(1, f(v_{i,j})) = 1$ And for $1 \leq i \leq m$, $1 \leq j \leq n - 1$, $gcf(f(v_{i,j}), f(v_{i,j+1})) = 1$. We also observe that the ternion sum for the edges $v_{i,j}v_{i,j+1}$ for $1 \leq i \leq m$, $1 \leq j \leq n - 1$ is $T(v_{i,j}v_{i,j+1}) = f(v_{i,j}) + f(v_{i,j+1}) + g(v_{i,j}v_{i,j+1}) = [n(i-1)+j+1]+[n(i-1)+j+2]+[2n(m-i+1)-2j-1] = 2mn + 2$

We also observe that the ternion sums for the edges $uv_{i,j}$, $1 \leq i \leq m$, $j = 1, n$ are $T(uv_{i,j}) = f(v_{i,j}) + f(v_{i,j+1}) + g(v_{i,j}v_{i,j+1}) = 2mn + 2$ for all $i \equiv 0(mod\ 2)$ and $j = 1, n$

Also $T(uv_{i,j}) = f(v_{i,j}) + f(v_{i,j+1}) + g(v_{i,j}v_{i,j+1}) = 2mn + 2$ for all $i \equiv 1(mod\ 2)$ and $j = 1$

And $T(uv_{i,j}) = f(v_{i,j}) + f(v_{i,j+1}) + g(v_{i,j}v_{i,j+1})$

$$= \begin{cases} n(m-1)+5 & \text{if } m, n \equiv 0(mod\ 2) \\ n(m-1)+4 & \text{if } m \equiv 0(mod\ 2) \text{and } n \equiv 1(mod\ 2) \\ mn+4 & \text{if } m, n \equiv 1(mod\ 2) \\ n(m-2)+5 & \text{if } m \equiv 1(mod\ 2) \text{and } n \equiv 0(mod\ 2) \end{cases} \quad \begin{matrix} \text{for all } i \equiv 1(mod\ 2) \\ \text{and} j = n \end{matrix}$$

Hence, centralized (m, n)-cycle network admits vertex prime and edge ternion sum labeling. The centralized (m, n)-cycle network $C_{m,n}$ and its vertex prime and edge ternion sum labeling are in Fig. 3.

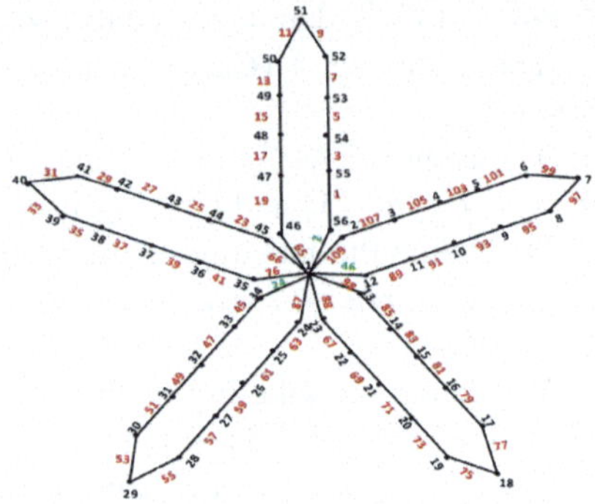

Fig. 3 Vertex prime and edge ternion sum labeling of centralized cycle network $C_{5,9}$

4.4 *Vertex Prime and Edge Ternion Sum Labeling of Centralized Fan Network*

Theorem 4 *The centralized (m, n)- fan network admits vertex prime and edge ternion sum labeling with k-edge varied ternion sum*

$$\mu_k^1(F_{m,n}) = \begin{cases} 2mn+2 & for\ (|E(F_{m,n})|-k) edges \\ \begin{cases} nm+3 & if n \equiv 0 (mod\ 2) \\ nm+4 & if n \equiv 1 (mod\ 2) \end{cases} & for\ k\ edges \end{cases} \quad where$$

$k > 1$, *for all* $m \geq 1$ *and* $n \geq 3$.

Proof Let $F_{m,n} = mP_n + K_1$ denote the centralized (m, n)-Fan network for all $m \geq 1$ and $n \geq 3$, The vertex set $V(F_{m,n}) = u, v_{i,j}/1 \leq i \leq m, 1 \leq j \leq n$ and $|V(F_{m,n})| = mn + 1$

The edge set $E(F_{m,n}) = \{uv_{i,j}/1 \leq i \leq m, 1 \leq j \leq n\} \cup \{v_{i,j}v_{i,j+1}/1 \leq i \leq m, 1 \leq j \leq n-1\}$ and $|E(F_{m,n})| = m(2n - 1)$ We define the labeling of the vertices as a bijective mapping, $f : V(F_{m,n}) \to \{1, 2, 3, \ldots, mn+1\}$ by $f(u) = 1$, $f(v_{i,j}) = n(i-1) + j + 1, 1 \leq i \leq m, 1 \leq j \leq n$

We define the labeling of the edges as an injective mapping, $g : E(F_{m,n}) \to \{1, 2, 3, \ldots, m(3n - 1) + 1\}$ by $g(v_{i,j}v_{i,j+1}) = 2n(m - i + 1) - 2j - 1$, for $1 \leq i \leq m, 1 \leq j \leq n - 1$

For the edges $uv_{i,j}$, we have two cases $g(uv_{i,j}) = n(2m - i + 1) - j$, for $1 \leq i \leq m, 1 \leq j \leq n$ where $j = 1$, and $j \equiv 0(mod\ 2)$

$$g(uv_{i,j}) = \begin{cases} n(m-i+1) - j + 1, n \equiv 0(mod\ 2) \\ n(m-i+1) - j + 2, n \equiv 1(mod\ 2) \end{cases} \text{for } 1 \leq i \leq m, 3 \leq j \leq$$

n and $j \equiv 1(mod\ 2)$. We observe that for $1 \leq i \leq m$, $1 \leq j \leq n$, the edges $uv_{i,j}$ $gcf(f(u), f(v_{i,j})) = gcf(1, n(i-1) + j + 1) = 1$

$gcf(f(v_{i,j}), f(v_{i,j+1})) = gcf(n(i-1) + j + 1, n(i-1) + j + 2) = 1$

We also observe that the ternion sum for the edges $v_{i,j}v_{i,j+1}$ for $1 \leq i \leq m$, $1 \leq j \leq n - 1$ is $T(v_{i,j}v_{i,j+1}) = f(v_{i,j}) + f(v_{i,j+1}) + g(v_{i,j}v_{i,j+1}) = [n(i -$

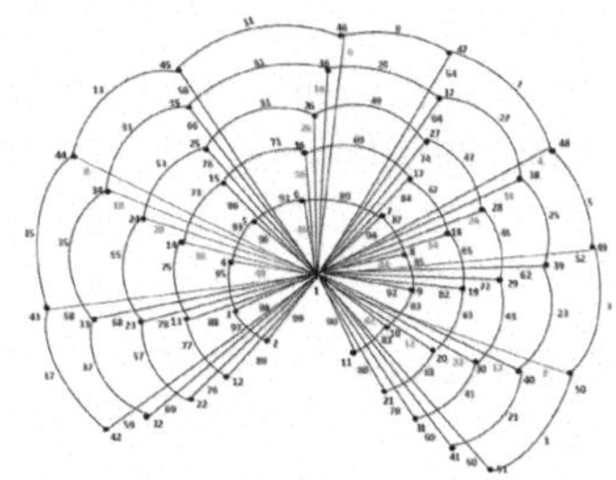

Fig. 4 Vertex prime and edge ternion sum labeling of centralized fan network $F_{5,10}$

$1) + j + 1] + [n(i-1) + j + 2] + [2n(m-i+1) - 2j - 1] = 2mn + 2$, which is a unique constant.

We also observe that for $1 \leq i \leq m$, $1 \leq j \leq n$ and $j \equiv 0(mod\ 2)$, $j = 1$, the ternion sum for the edges $uv_{i,j}$ is $T(uv_{i,j}) = f(u) + f(v_{i,j}) + g(uv_{i,j}) = [1] + [n(i-1) + j + 1] + [n(2m-i+1) - j] = 2mn + 2$, which is a unique constant.

We also observe that for $1 \leq i \leq m, 3 \leq j \leq n$ and $j \equiv 1(mod\ 2)$, there are two cases for the ternion sum for the edges $uv_{i,j}$.

Case (1): $n \equiv 0(mod\ 2)$

$T(uv_{i,j}) = f(u) + f(v_{i,j}) + g(uv_{i,j}) = [1] + [n(i-1) + j + 1] + [n(m-i+1) - j + 1] = nm + 3$, which is a varied constant

Case (2): $n \equiv 1(mod\ 2)$

$T(uv_{i,j}) = f(u) + f(v_{i,j}) + g(uv_{i,j}) = [1] + [n(i-1) + j + 1] + [n(m-i+1) - j + 2] = nm + 4$, which is a varied constant We also infer that, $\mu_e(F_{m,n}) = \{v_{i,j}v_{i,j+1}/1 \leq i \leq m, 1 \leq j \leq n-1\} \cup \{uv_{i,j}/1 \leq i \leq m, 1 \leq j \leq n$ and $j \equiv 0(mod\ 2), j = 1\}$ and $|\mu_e(F_{m,n})| = \begin{cases} \frac{3mn}{2}\ if\ n \equiv 0(mod\ 2) \\ \frac{m(3n-1)}{2}\ if\ n \equiv 1(mod\ 2) \end{cases}$

And $\mu'_e(F_{m,n}) = \begin{cases} \{uv_{i,j}/n \equiv 0(mod\ 2)\} \\ \{uv_{i,j}/n \equiv 1(mod\ 2)\} \end{cases}$ $for\ 1 \leq i \leq m, 3 \leq j \leq n\ and\ j \equiv 1(mod\ 2)\ and\ k = |\mu'_e(F_{m,n})| = \begin{cases} \frac{m(n-2)}{2}\ if\ n \equiv 0(mod\ 2) \\ \frac{m(n-1)}{2}\ if\ n \equiv 0(mod\ 2) \end{cases}$

We observe that in both the cases, $|\mu_e(F_{m,n})| > k$. Therefore, centralized (m, n)-fan network admits vertex prime and edge ternion sum labeling. The centralized (m, n)- fan network $F_{m,n}$ and its vertex prime and edge ternion sum labeling are in Fig. 4.

5 Conclusion

In this research article, we have introduced 1-Edge varied ternion sum, k-Edge varied ternion sum, and k-Edge varied ternion sum with l-variations, and we investigate that the centralized path network, centralized cycle network, centralized fan network, and centralized triangular path network admit vertex prime and edge ternion sum labeling.

References

1. Carballosa, W., Rodríguez, J.M., Sigarreta, J.M.: Hyperbolicity in the corona and join of graphs. Aequationes Mathematicae **89**. 1311–1327 (2015)
2. Daoud, S.N.: Edge odd graceful labeling of some path and cycle related graphs. AKCE Int. J. Graphs Comb. **14**, 178–203 (2017)
3. Maulana, R.J.N.R., Wijaya, K., Santoso, K.A.: On chromatic polynomial of a Fan graph. Majalah Ilmiah Matematika dan Statistika **18**, 55–60 (2018)
4. El Deen, M.R.Z.: Edge δ- Graceful Labeling for Some Cyclic-Related Graphs. In: Hindawi Advances in Mathematical Physics, vol. 2020. Wiley.
5. Gallian, J.A.: A dynamic survey of graph labeling. Electron J Comb. https://doi.org/10.37236/11668 (2022)

Vertex Odd Mean(VOM) Labeling of Super Subdivision of Some Graphs

D. Devakirubanithi, J. Jeba Jesintha, and E. Mercy Gunaseeli

1 Introduction

In graph theory, the branch of graph labeling plays a crucial role in organizing and interpreting the complex relationship within diverse domains. This paper specifically examines undirected graphs that are both simple and finite. In 1967, Rosa [4] initiated the idea of graph labeling, which involves assigning integers to vertices, edges, or both, based on certain specified conditions. Gallian [1] offers the most recent overview of graph labeling, covering over 350 techniques discussed in a wealth of literature, comprising 3600 or more papers.

In 2003, Somasundaram and Ponraj [7] are credited with introducing the notion of "Mean labeling." Revathi [5] established the vertex odd and even mean labeling. Another approach to graph construction, termed super subdivision of graphs, was presented by Sethuraman and Selvaraju. [6] demonstrated that arbitrary super subdivisions of paths and cycle exhibit gracefulness. The odd mean labeling [8] is applied on a graph (network) in order to enhance fastness, efficient communication, and various issues. Implementing odd mean labeling on a sufficiently connected graph enhances secure communication protocols. Enhancing the safety of transmissions in areas like cellular telephony, Wi-Fi, and security systems is achieved through efficient odd mean labeling methods. Utilizing channel labeling aids in determining the communication timing for sensors. This paper establishes that super subdivision of star and ladder graph admits for vertex odd mean labeling (VOM).

D. Devakirubanithi (✉)
Department of Mathematics, St. Thomas College of Arts and Science, Chennai, India

J. Jeba Jesintha · E. Mercy Gunaseeli
Department of Mathematics, Women's Christian College, University of Madras, Chennai, India

I. Rajasingh et al. (eds.), *Applied Computational Mathematics*,
Springer Proceedings in Mathematics & Statistics 480,
https://doi.org/10.1007/978-3-031-77764-6_5

2 Preliminaries

In this section, we present a few definitions essential for the main results.

2.1 *Definition*

The *star graph* [2] denoted as $k_{1,n}$ is a tree with one vertex adjacent to every other vertex as n number of pendant vertices are linked to one vertex.

2.2 *Definition*

The *ladder graph* [3] represented as $L_n = P_n \times P_2$ is formed by taking the Cartesian product of paths P_n and P_2. In this graph, there are $2n$ vertices and $3n - 2$ edges.

2.3 *Definition*

The *super subdivision* [3] of G is derived by replacing each edge by a complete bipartite graph $k_{2,m}$, and it is denoted by $SS(G)$

3 Main Results

In this section, we demonstrate that super subdivision $SS(G)$ of star and ladder graph admits for vertex odd mean labeling.

3.1 *Theorem*

Super subdivision of star graph admits vertex odd mean labeling (VOM).

Proof Let $G = SS_m(k_{1,n})$ be the super subdivision of star graph formed by replacing each edge of $k_{1,n}$ by a complete bipartite graph, where m denotes the number of times of super subdivision of star. The graph G is described as follows: Let u_0 be the central vertex and u_i $(1 \leq i \leq n)$ denotes the pendant vertices of star. Let v_j^i $(1 \leq i \leq n$ and $1 \leq j \leq m)$ represent the vertices of m times of super subdivision of star graph.

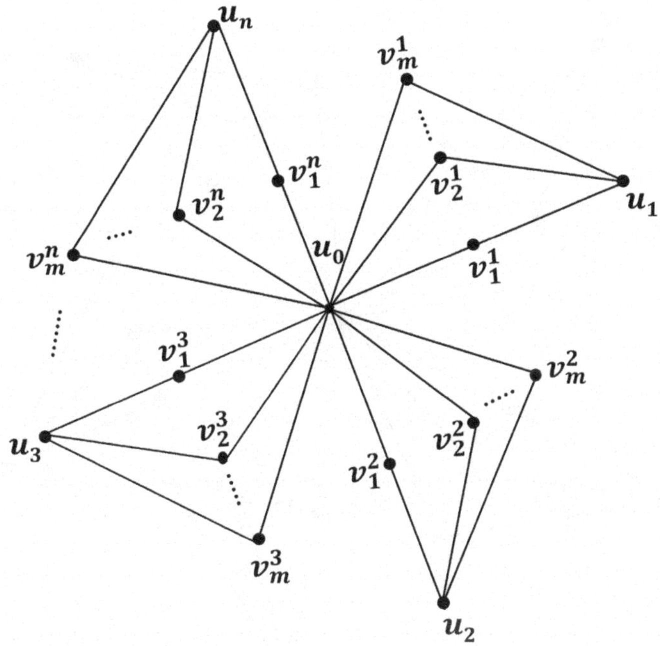

Fig. 1 Super subdivision of star graph $SS_m(k_{1,n})$

Let $V(G) = \{u_0, u_i, v_j^i : 1 \leq i \leq n \text{ and } 1 \leq j \leq m\}$
Let $E(G) = \{u_0 v_j^i, v_j^i u_i : 1 \leq i \leq n \text{ and } 1 \leq j \leq m\}$
The graph has $n + 1 + mn$ vertices and $2mn$ edges.
The generalized super subdivision of star graph $SS_m(k_{1,n})$ is depicted in Fig. 1.
We define an injective mapping ϕ, and the vertex labels for the graph $SS_m(k_{1,n})$ are defined as follows:

$$\phi(u_0) = 1$$

$$\phi(u_i) = 2mn + 2i + 1, 1 \leq i \leq n$$

$$\phi(v_j^i) = 2mi + 2j - 2m + 1, 1 \leq i \leq n, 1 \leq j \leq m$$

Clearly, all the vertex labels from the set $\phi : V(G) \to \{1, 3, 5, ..., (2q - 1)\}$ are distinct odd numbers.

ϕ^* is defined as follows:

$$\phi^*(u_0 v_j^i) = \frac{2m(i-1) + 2j + 2}{2}, 1 \leq i \leq n, 1 \leq j \leq m$$

$$\phi^*(v_j^i u_i) = \frac{2m(i-1) + 2mn + 2i + 2j + 2}{2}, 1 \leq i \leq n, 1 \leq j \leq m$$

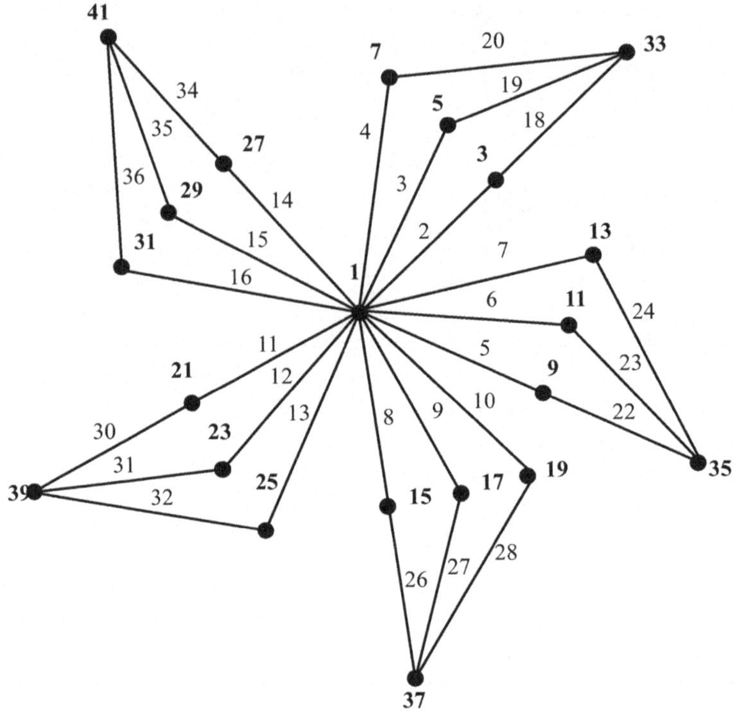

Fig. 2 Super subdivision of star graph $SS_3(k_{1,5})$

From the above computed edge labels, we observe that the labels for the edges are positive and distinct. Hence, $SS_m(k_{1,n})$ admits VOM.

The illustration for the above theorem is given in Fig. 2.

3.2 *Theorem*

Super subdivision of ladder graph admits vertex odd mean labeling(VOM).

Proof Let $G = SS_m(L_n)$ denote the super subdivision of ladder graph is formed by replacing each edge of L_n by a complete bipartite graph, where m denotes the number of times of super subdivision of ladder. The graph G is described as follows: Let u_i $(1 \le i \le n)$ and υ_i $(1 \le i \le n)$ represent the vertices of first and second path. Let u_j^{κ} $(1 \le \kappa \le n-1$ and $1 \le j \le m)$ represent the vertices of $SS(G)$ of first path associated with u_i and υ_j^{κ} $(1 \le \kappa \le n-1$ and $1 \le j \le m)$ represent the vertices of $SS(G)$ of second path associated with υ_i. Let w_j^i $(1 \le i \le n$ and $1 \le j \le m)$ represent the vertices of $SS(G)$ of pendant edges connecting each vertex of the path u_i and υ_i respectively.

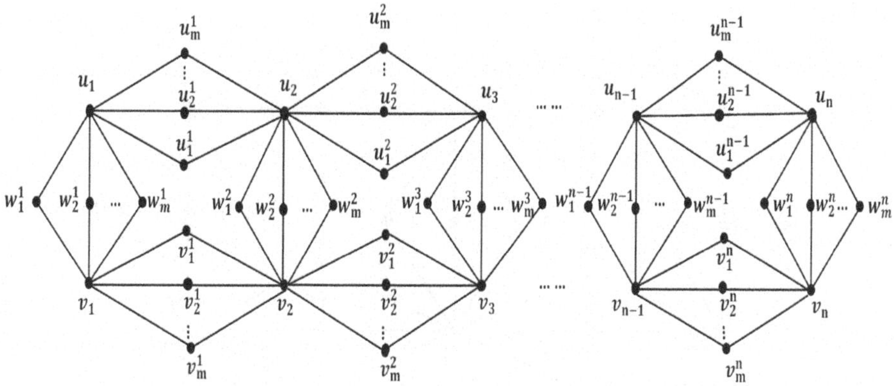

Fig. 3 Super subdivision of ladder graph $SS_m(L_n)$

Let the vertex set be $V(G) = \{u_i, u_j^\kappa, \upsilon_i, \upsilon_j^\kappa, w_j^i : 1 \leq i \leq n, 1 \leq j \leq m$ and $1 \leq \kappa \leq n-1\}$

Let the edge set be $E(G) = \{u_i u_j^\kappa, u_i w_j^i, w_j^i \upsilon_i, \upsilon_i \upsilon_j^\kappa : 1 \leq i \leq n, 1 \leq j \leq m$ and $1 \leq \kappa \leq n-1\}$

The graph G has $3mn + 2m - 2n$ vertices and $6mn - 4m$ edges.

The generalized super subdivision of ladder graph is depicted in Fig. 3.

We define an injective mapping ϕ, and the vertex labels for the graph $SS_m(L_n)$ are defined as follows:

$$\phi(u_i) = 2m(i-1) + 2i - 1, 1 \leq i \leq n$$

$$\phi(u_j^\kappa) = 2m(\kappa - 1) + 2\kappa + 2j - 1, 1 \leq \kappa \leq n-1, 1 \leq j \leq m$$

$$\phi(\upsilon_i) = 2(2m+2)(n-i) + 2mn + 2n + 1, 1 \leq i \leq n$$

$$\phi(\upsilon_j^\kappa) = 2(2m+2)(n-\kappa) + 2m(n-1) + 2n + 2j - 1, 1 \leq \kappa \leq n-1, 1 \leq j \leq m$$

$$\phi(w_j^i) = 3(2m+2)(n-i) + 2m(n-1) + 2n + 2j - 1, 1 < i < n, 1 \leq j \leq m$$

Clearly, all the vertex labels from the set $\phi : V(G) \rightarrow \{1, 3, 5, ..., (2q-1)\}$ are distinct odd numbers.

ϕ^* is defined as follows:

$$\phi^*(u_i u_j^\kappa) = \frac{2m(i+\kappa-1) + 2(i+j+\kappa-2)}{2}, 1 \leq i \leq n, 1 \leq \kappa \leq n-1, 1 \leq j \leq m$$

$$\phi^*(u_i w_j^i) = \frac{8n(m+1) - 4m(i+1) - 2(2i - j + 2)}{2}, 1 \leq i \leq n, 1 \leq j \leq m$$

$$\phi^*(w_j^i \upsilon_i) = \frac{14n(m+1) - 2i(m+5) + 2(j - m - 1)}{2}, 1 \leq i \leq n, 1 \leq j \leq m$$

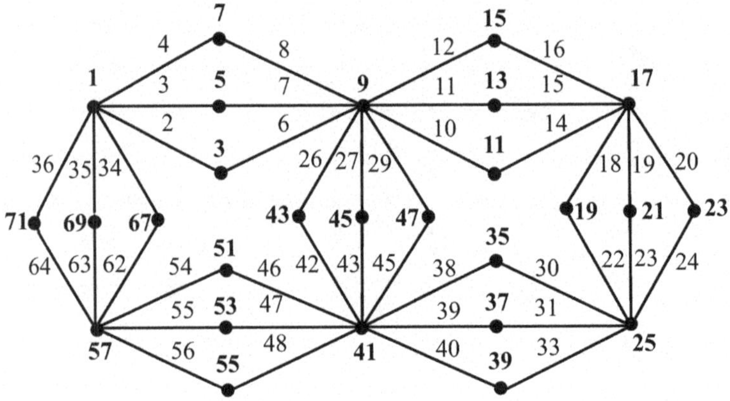

Fig. 4 Super subdivision of ladder graph $SS_3(L_3)$

$$\phi^*(\upsilon_i \upsilon_{\rm j}^{\kappa}) = \frac{12n(m+1) + 4m(i+\kappa) + 2(n+{\rm j}-m-2\kappa)}{2},$$

$$1 \leq i \leq n,\ 1 \leq \kappa \leq n-1,\ 1 \leq {\rm j} \leq m$$

From the above computed edge labels, we observe that the labels for the edges are positive and distinct. Hence, $SS_m(L_n)$ admits VOM.

The illustration for the above theorem is given in Fig. 4.

4 Conclusion

This paper demonstrates that super subdivision of star graph $SS_m(k_{1,n})$ and super subdivision of ladder graph $SS_m(L_n)$ permit vertex odd mean labeling.

References

1. Gallian, J.A.: A dynamic survey of graph labeling. Electron. J. Comb. 1(DynamicSurveys), DS6, 1–644 (2023)
2. Jesintha, J.J., Devakirubanithi, D., Sherlin, J.M.: Odd graceful labeling of super subdivision of few graphs. Bull. Pure Appl. Sci. Math. Stat. **41**(2), 167–171 (2022)
3. Jeyanthi, P., Philo, S., Siddiqui, M.K.: Odd harmonious labeling of super subdivision graphs. Proyecciones J. Math. **38**(1), 1–11 (2019)
4. Rosa A.: On certain valuations of the vertices of a graph. In: Theory of Graphs (International Symposium, Rome, July 1966), Gordon and Breach, New York and Dunod Paris, pp.349–355 (1967)

5. Revathi, N.: Vertex odd mean and even mean labeling of some graphs. IOSR J. Math. **11**(2), 70–74, (2015)
6. Sethuraman, G., Selvaraju, P.: Gracefulness of arbitrary super subdivisions of graphs. Indian J. Pure Appl. Math. **32**(7), 1059–1064 (2001)
7. Somasundaram, S., Ponraj, R.: Mean labelings of graphs. Natl. Acad. Sci. Lett. **26**(7–8), 210–213 (2003)
8. Sudhakar, S., Francis, S., Balaji, V.: Odd mean labeling for two star graph. Appl. Math. Nonlinear Sci. **2**(1), 195–200 (2017)

Burning Neural Networks with Harary Graphs for Optimal Connectivity Control

R. Sundara Rajan, V. Kamala Kannan, and M. Navaneetan

Mathematics Subject Classification 05C30, 05C35, 05C90

1 Introduction

Graph burning serves as a discrete-time progress for modeling contagion spread within social networks. During pandemics, governments typically release statistical reports detailing individuals affected by the disease, often organized districtwise or zonewise. Representing this real-time problem as a network model involves treating people as vertices, connected by edges if they have a relationship. The key question revolves around the speed at which the disease spreads among individuals. In [1], Bonato et al. introduced a streamlined concept of this challenge termed graph burning. A graph is used to represent relationships between pairs of objects. The closed neighborhood of a node v up to the k-th level is symbolized as $N_k[v]$, denoting the set $u \in V(G) : d(u, v) \leq k$, where $N_1[v]$ is abbreviated as $N[v]$. For any positive integer α, α_{odd} denotes the smallest odd number that is greater than or equal to α, whereas α_{even} represents the smallest even number greater than or equal to α. Graph burning is applied to a finite connected graph G commences with all vertices in an unburned state. At time step $t = 1$, a single vertex is selected and set on fire. Subsequently, two events transpire: (1) the fire extends to encompass all neighbors of a previously burned vertex, causing them to become burned, and (2) another vertex is chosen for incineration. Once a vertex is ignited, it remains permanently burned. The process concludes when all vertices within the graph are burned. The objective is to minimize the number of time steps, and the minimum

R. Sundara Rajan · V. Kamala Kannan (✉)
Department of Mathematics, Hindustan Institute of Technology and Science, Chennai, Tamil Nadu, India

M. Navaneetan
Department of Computer Science and Engineering, Hindustan Institute of Technology and Science, Chennai, Tamil Nadu, India

I. Rajasingh et al. (eds.), *Applied Computational Mathematics*,
Springer Proceedings in Mathematics & Statistics 480,
https://doi.org/10.1007/978-3-031-77764-6_6

number of steps required is denoted as the burning number of G, represented as $b(G)$. About [1], Bonato et al. established a key property of the burning number, stating that $(y_1, y_2, \ldots, y_r)$ forms a burning sequence for graph G if and only if,

$$N_{r-1}[y_1] \cup N_{r-2}[y_2] \cup \cdots \cup N_0[y_r] = V(G). \tag{1}$$

The configuration of Harary graphs, denoted as $H_{k,m}$, relies on the parity of both k and m, where k signifies the vertex connectivity, m represents the number of vertices, and $m > k$. Consideration 1: When k is an even number and m is a positive integer. Suppose $k = 2n$, then $H_{2n,m}$ is constructed as follows. It comprises vertices labeled $0, 1, 2, 3, \ldots, m-1$, and two vertices are linked if $i - n \leq j \leq i + n$ (with addition computed modulo m). Consideration 2: When k is odd and m is even. Let $k = 2n+1$, then $H_{2n+1,m}$ is formed by initially creating $H_{2n,m}$ and subsequently adding edges connecting vertex j to vertex $j + \frac{m}{2}$ for $1 \leq j \leq \frac{m}{2}$. Consideration 3: When both k and m are odd. Assume $k = 2n+1$; then $H_{2n+1,m}$ is constructed by first drawing $H_{2n,m}$ and then introducing edges connecting vertex 0 to vertices $\frac{m-1}{2}$ and $\frac{m+1}{2}$, along with edges from vertex j to $j + \frac{m+1}{2}$ for $1 \leq j \leq \frac{m-1}{2}$. For clarity, see Fig. 1. In the context of reference [2], for any positive integer n where $n \geq 2$, a pencil graph, denoted PC_n, consists of $2n+2$ vertices. It is defined by the following vertex set and edge set:

$$\begin{aligned} V(PC_n) =& \{u_i, v_i | i \in [0, n]\} \\ E(PC_n) =& \{u_i u_{i+1}, v_i v_{i+1} | i \in [1, n-1]\} \cup \{u_i v_i | i \in [0, n]\} \\ & \cup \{u_1 u_0, v_1 u_0, u_n v_0, v_n v_0\} \end{aligned}$$

For clarity, see Fig. 2.

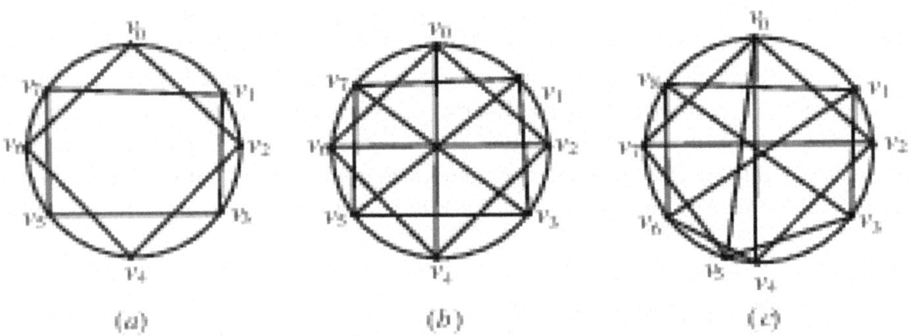

Fig. 1 Harary graphs: **(a)** $H_{4,8}$, **(b)** $H_{5,8}$ and **(c)** $H_{5,9}$

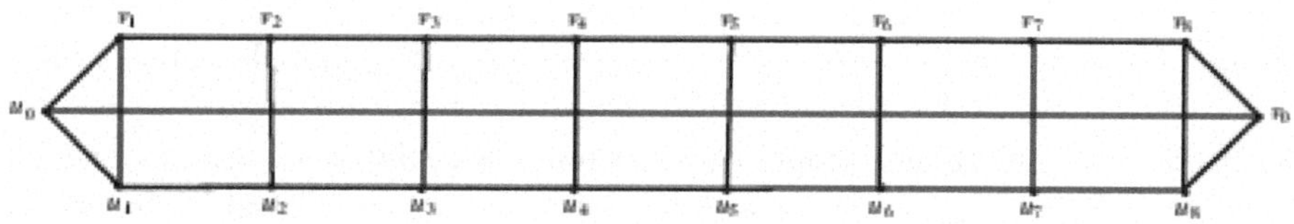

Fig. 2 The pencil graph PC_8

2 Main Results

Under this portion, we determine $b(H_{k,m})$ and $b(PC_n)$.

Theorem 1 *Let $H_{k,m}$ be the Harary graph. If k is an even number, $k \geq 2$ and m are any positive integer, $m > k$, then*

$$b(H_{k,m}) = \left\lceil \frac{(k-2) + \sqrt{(k-2)^2 + 8mk}}{2k} \right\rceil.$$

Proof Let $v_o, v_1, v_2, v_3, ..., v_{m-1}$ be the vertices in $H_{k,m}$ and $(x_1, x_2, \ldots, x_r)$ be its optimal burning sequence. Since the graph is vertex-transitive, for any m, k and k is even, without loss of generality, we choose $x_1 = v_0$, and x_i is chosen arbitrarily, $2 \leq i \leq r$. From the construction of the Harary graph, we have $|N_i[x_{r-i}]| \leq ik + 1$ for $0 \leq i \leq r - 1$. Let us assume that the total number of vertices burned at the end of rth-time step will be at least m, and from Eq. 1, we get

$$\sum_{i=0}^{r-1}(ik+1) = \frac{kr^2 - (k-2)r}{2} \geq m.$$

Since r is the minimum burning sequence for $H_{k,m}$, we get

$$r = b(H_{k,m}) \geq \left\lceil \frac{(k-2) + \sqrt{(k-2)^2 + 8mk}}{2k} \right\rceil.$$

Now let us assume that $r = \left\lceil \frac{(k-2) + \sqrt{(k-2)^2 + 8mk}}{2k} \right\rceil$ and choose the optimal burning sequence as follows: $x_i = v_\alpha$, where $\alpha = (i-1) + \frac{k(i-1)(2r-i-1)}{2}$ and $1 \leq i \leq r$. Then

$$\bigcup_{i=0}^{r-1} N_i[x_{r-i}] = V(H_{k,m}).$$

Thus, $b(H_{k,m}) = r = \left\lceil \dfrac{(k-2)+\sqrt{(k-2)^2+8mk}}{2k} \right\rceil$. This completes the proof.

When $k = 2$, the Harary graph $H_{k,m}$ becomes a cycle on m vertices and by Theorem 1, $b(H_{2,m}) = \lceil \sqrt{m} \rceil$, which has been already established by Bonato et al. in [1].

Theorem 2 *Let $H_{k,m}$ be the Harary graph. If k is an odd number, $k \geq 3$ and m are an even number, $m > k$, then $b(H_{k,m}) = r = min\{r_{odd}, r_{even}\}$, where*

$$r_{odd} = \left\lceil \frac{(k-2)+\sqrt{(k-2)^2+(k-1)(m+2)}}{k-1} \right\rceil \text{ and}$$

$$r_{even} = \left\lceil \frac{(k-2)+\sqrt{1+(k-1)m}}{k-1} \right\rceil.$$

Proof As proceeded in Theorem 1, there arises two cases as follows:
Case(i). If the burning number is odd, then we have

$$|N_{r-i}[x_i]| \leq 2(r-i-1)(k-1)+k+1 \text{ for } i = 1, 3, 5, ..., r-2;$$
$$|N_{r-i}[x_{i-1}]| \leq 2(r-i)(k-1)+k+1 \text{ for } i = 3, 5, ..., r-2; \text{ and}$$
$$|N_0[x_{r-1}]| = |N_0[x_r]| \leq 1.$$

Let us assume that the total number of vertices burned at the end of rth-time step will be at least m, and from Eq. 1, we get

$$\sum_{i=1,3,5,...}^{r-2} [2(r-i-1)(k-1)+k+1] + \sum_{i=3,5,7,...}^{r-2} [2(r-i)(k-1)+k+1]$$
$$+ \mid N_0[x_{r-1}] \mid + \mid N_0[x_r] \mid = (k-1)r^2 - 2(k-2)r - 2 \geq m.$$

Since r is the minimum burning sequence for $H_{k,m}$, we get

$$r = r_{odd} \geq \left\lceil \frac{(k-2)+\sqrt{(k-2)^2+(k-1)(m+2)}}{k-1} \right\rceil.$$

Case (ii). If the burning number is even, then we have

$$|N_{r-i}[x_i]| \leq 2(r-i-1)(k-1)+k+1 \text{ for } i = 1, 3, 5, \ldots, r-1 \text{ and}$$
$$|N_{r-i}[x_{i-1}]| \leq 2(r-i)(k-1)+k+1 \text{ for } i = 3, 5, 7, \ldots, r-1.$$

Let us assume that the total number of vertices burned at the end of rth-time step will be at least m, and from Eq. 1, we get

$$\sum_{i=1,3,5,\ldots}^{r-1} [2(r-i-1)(k-1)+k+1] + \sum_{i=3,5,7,\ldots}^{r-1} [2(r-i)(k-1)+k+1]$$
$$= (k-1)r^2 - 2(k-2)r + (k-3) \geq m$$

Since r is the minimum burning sequence for $H_{k,m}$, we get

$$r = r_{even} \geq \left\lceil \frac{(k-2)+\sqrt{1+(k-1)m}}{k-1} \right\rceil.$$

To prove the equality of the above two cases, let us choose the optimal burning sequence as follows:

$$x_1 = v_0;$$
$$x_2 = v_\alpha;$$
$$x_3 = v_\beta;$$
$$x_{4i} = v_\gamma, \text{ for all } i \leq \frac{r}{4};$$
$$x_{4i+1} = v_\delta, \text{ for all } i \leq \frac{r-1}{4};$$
$$x_{4i+2} = v_\lambda, \text{ for all } i \leq \frac{r-2}{4};$$
$$x_{4i+3} = v_\xi, \text{ for all } i \leq \frac{r-3}{4}.$$

where

$$\alpha = \frac{m}{2} + (r-2)(k-1) + 1;$$

$$\beta = \frac{m}{2} - (r-2)\left(\frac{k-1}{2}\right) - (r-3)\left(\frac{k-1}{2}\right) - 1;$$

$$\gamma = m - (r-1)\left(\frac{k-1}{2}\right) - 4\left(\frac{k-1}{2}\right)\sum_{j=1}^{i}(r-4j) + (r-4i)\left(\frac{k-1}{2}\right) - 2i$$

for all $i \leq \frac{r}{4}$;

$$\delta = 1 + (r-1)\left(\frac{k-1}{2}\right) + 2\left(\frac{k-1}{2}\right)\sum_{j=2}^{2i+1}[r-(2j+1)]$$
$$-[r-(4i+1)]\left(\frac{k-1}{2}\right) + 2i$$

$$\text{for all } i \leq \frac{r-1}{4};$$

$$\lambda = \alpha + (r-2)\binom{\frac{k-1}{2}} + 4\binom{\frac{k-1}{2}}\sum_{j=2}^{i+1}[r-(4j-2)]$$

$$-[r-(4i+2)]\binom{\frac{k-1}{2}} + 2i$$

$$\text{for all } i \leq \frac{r-2}{4};$$

$$\xi = \beta - (r-3)\binom{\frac{k-1}{2}} - 2\binom{\frac{k-1}{2}}\sum_{j=2}^{2i+1}[r-(2j+1)]$$

$$+[r-(4i+3)]\binom{\frac{k-1}{2}} - 2i$$

$$\text{for all } i \leq \frac{r-3}{4}.$$

Consequently, the burning process for $H_{k,m}$ can be accomplished precisely in $r = \min\{r_{\text{odd}}, r_{\text{even}}\}$-steps, employing the burning sequence $(x_1, x_2, \ldots, x_r)$. Therefore, the burning number of $H_{k,m}$ is determined as $b(H_{k,m}) = r = \min\{r_{\text{odd}}, r_{\text{even}}\}$. This concludes the proof.

Continuing similarly to Theorem 2, we present the following outcome.

Theorem 3 *Let $H_{k,m}$ be the Harary graph. If k is an odd number, $k \geq 3$ and m are an odd number, $m > k$, then $b(H_{k,m}) = r = min\{r_{odd}, r_{even}\}$, where*

$$r_{odd} = \left\lceil \frac{(k-2) + \sqrt{k(k-3) + m(k-1) + 3}}{k-1} \right\rceil$$

$$r_{even} = \left\lceil \frac{(k-2) + \sqrt{m(k-1) + 2 - k}}{k-1} \right\rceil.$$

Theorem 4 *Consider a positive integer n, where $n \geq 2$, then $b(PC_n) = \left\lceil \frac{1+\sqrt{4n+3}}{2} \right\rceil$.*

Proof Let us consider a pencil graph with $n \geq 2$ vertices, and let the vertex set for PC_n be $\{v_0, v_1, \ldots v_n, u_0, u_1, \ldots u_n\}$. Let $(x_1, x_2, \ldots, x_r)$ be the optimal burning sequence for PC_n. From the construction of pencil graph, we have $|N_{r-i}[x_i]| \leq 4(r-i)$ for $1 \leq i \leq r-1$. Let us assume that the total number of vertices burned at the end of rth-time step will be at least $2n+2$, and from Eq. 1, we get

$$\sum_{i=1}^{r-1}(4(r-i))+1=2r^2-2r+1\geq 2n+2.$$

Since r is the minimum burning sequence for PC_n, we get

$$r=b(PC_n)\geq\left\lceil\frac{1+\sqrt{4n+3}}{2}\right\rceil.$$

Now let us assume that $r=\left\lceil\frac{1+\sqrt{4n+3}}{2}\right\rceil$ and choose the optimal burning sequence as follows:

$$\begin{aligned}
x_1 &= u_{\lceil\frac{n}{2}\rceil};\\
x_2 &= v_0;\\
x_{4i-1} &= v_{\lceil\frac{n}{2}\rceil-[(4i-1)(r-2i)-r+2]} \text{ for all } i\leq\frac{r+1}{4};\\
x_{4i} &= v_{\lceil\frac{n}{2}\rceil+[(4i-1)(r-2i)-r+1]} \text{ for all } i\leq\frac{r}{4};\\
x_{4i+1} &= u_{\lceil\frac{n}{2}\rceil+[(4i+1)(r-2i-1)-r+1]} \text{ for all } i\leq\frac{r-1}{4};\\
x_{4i+2} &= u_{\lceil\frac{n}{2}\rceil-[(4i+1)(r-2i-1)-r+2]} \text{ for all } i\leq\frac{r-2}{4}.
\end{aligned}$$

Then,

$$\bigcup_{i=1}^{r}N_{r-i}[x_i]=V(PC_n).$$

Thus, $b(PC_n)=r=\left\lceil\frac{1+\sqrt{4n+3}}{2}\right\rceil$. This completes the proof.

3 Applications in Neural Connectivity

The application of Harary graphs in neural connectivity through the burning number involves using this graph-theoretical concept to model and analyze the spread of neural activity within the brain. The burning number denotes the smallest count of initially activated neurons needed to ultimately activate the entire neural network. The pencil graphs, denoted as PC_n, may not have direct applications in the medical field as a specific graph structure but could potentially be used as part of larger graph-based models.

Disease Spread and Intervention About the context [3], Harary graphs with the burning number can simulate the progression of diseases like Alzheimer's or Parkinson's. The burning number helps identify critical nodes, representing regions of the brain, where interventions could be applied to limit the spread of the disease.

Brain Stimulation Optimization In reference to the context [4], understanding the burning number in Harary graphs helps optimize the placement of stimuli to influence the entire neural network. This is relevant in brain-computer interfaces (BCIs) or therapeutic interventions where precise and efficient activation of neurons is crucial.

Network Resilience Analysis About the context [5], by applying the burning number to Harary graphs, researchers can evaluate the resilience of neural networks. Identifying critical nodes that, when activated, influence the entire network helps assess how resilient the network is to disruptions or damage.

Neural Network Plasticity Studies About the context [6], Harary graphs with the burning number can be used to study how neural networks adapt and reorganize in response to stimuli or learning experiences. This can have implications for understanding neural plasticity and the dynamics of information processing.

4 Concluding Remarks

In this study, we have calculated the burning number for both Harary graphs and pencil graphs under various scenarios. Additionally, the applications of Harary graphs in neural connectivity through the burning number were discussed in general.

References

1. Bonato, A.., Janssen, J., Roshanbin, E.: How to burn a graph. Internet Math. **12**, 85–100 (2016)
2. Simamora, D.N.S., Salman, A.N.M.: The Rainbow (vertex) connection number of pencil graphs. In: International Conference On Graph Theory And Information Security, Procedia Computer Science, vol. 74, pp. 138–142 (2015)
3. Houghton, P.J., Howes, M.J.: Natural products and derivatives affecting neurotransmission relevant to Alzhaimer's and Parkinson's disease. Neurosignals **14**, 6–22 (2005)
4. Gonzalez-Astudillo, J., Cattai, T., Bassignana, G., Corsi, M.C., Fallani, F.D.V.: Network-based brain-computer interfaces: principles and applications. J. Neural Eng. **18**, 011001 (2021)
5. Gao, J., Liu, X., Li, D., Havlin, S.: Recent Progress on the resilience of complex networks. Energies **8**, 12187–12210 (2015)
6. Nakamura, T., Hillary, F.G., Biswal, B.B.: Resting network plasticity following brain injury. PLoS ONE **4**, e8220 (2009)

k-Prime Total Labeling of Union of Graphs

S. Teresa Arockiamary and G. Vijayalakshmi

1 Introduction

Graph theory is a branch of discrete mathematics, used in the analysis of pairwise interactions between finite or discrete elements. A graph is a nonlinear data structure composed of points and lines. One of the wide areas of research in graph theory is graph labeling which originated in midway through the 1960s [1]. Rosa [5] originally proposed the graceful labeling in 1967, and thereafter a wide range of various graph labeling strategies have been developed. A graph labeling is an assignment of integers to the vertices or edges, or both, subject to certain conditions. A comprehensive assessment of different graph labeling findings is provided by Gallian [3], who has provided a survey on an abundance of papers focusing on numerous forms of graph labeling over the last three decades.

Teresa Arockiamary et al. [7] pioneered the idea of k-prime total labeling. They proved some results on tree-related graphs such as path, star, bistar, $(1, 4)$-bi-regular caterpillar graph CP_n, and generalized star graphs for odd k is k-prime total. Further, the study was on some shell-related graphs such as shell, uniform shell-butterfly, and shell-flower that conform to k-prime total labeling [8]. The shell graph was originally constructed by Limaye and Deb [2]. Meena and Amuda provided the study of shell graph to be prime labeling [4].

S. Teresa Arockiamary · G. Vijayalakshmi (✉)
Department of Mathematics, Stella Maris College (Autonomous), Affiliated to University of Madras, Chennai, India

I. Rajasingh et al. (eds.), *Applied Computational Mathematics*,
Springer Proceedings in Mathematics & Statistics 480,
https://doi.org/10.1007/978-3-031-77764-6_7

2 Preliminaries

We begin with the definition of k-prime total labeling.

Definition 1 ([7]) A one-one and onto function $g : V(G) \cup E(G) \rightarrow \{k, k+1, k+2, \ldots, k + |V(G)| \cup |E(G)| - 1\}$ is called a k-prime total labeling if for the given positive integer k and for any edge $uv \in E(G)$; $g(u)$, $g(v)$, $g(uv)$ are comparatively prime in pairs. A graph G which admits a k-prime total labeling is called a k-prime total graph.

Definition 2 ([6]) A graph G in which a vertex is distinguished from other vertices is called a rooted graph, and the vertex is called the root of G. Let G be a rooted graph. The graph $G^{(n)}$ obtained by identifying the roots of n copies of G is called the *one point union of n copies of the graph* G.

Definition 3 ([2]) The shell graph is defined as a cycle C_n with $(n - 3)$ chords sharing a common end point called the apex. Shell graphs are denoted as Sh_n.

3 Main Results

Construction of One Point Union of Shell Graph One point union of shell graph is a graph obtained by joining t number of shells of width n with a central vertex known as apex vertex y_0. The graph is denoted as Sh_n^t. One point union of shell graph is constructed using collection of vertices denoted as $(y_{1,1}, y_{2,1}, \ldots, y_{n-1,1})$ in the first shell, $(y_{1,2}, y_{2,2}, \ldots, y_{n-1,2})$ in the second shell,..., and $(y_{1,t}, y_{2,t}, \ldots, y_{n-1,t})$ in the tth shell.

Theorem 1 *The one-point union of shell graph Sh_n^t is k-prime total for $t > 1$ and $n \geq 3$.*

Proof Let $G = Sh_n^t$ be the one-point union of shell graph of order $t(n-1)+1$ and size $t(2n-3)$. Let y_0 be the apex vertex of G. Let $V(G) = \{y_0\} \cup \{y_{r,s} : 1 \leq r \leq n-1, 1 \leq s \leq t\}$

$$E(G) = \begin{cases} y_0 y_{r,s} : & 1 \leq r \leq n-1,\ 1 \leq s \leq t \\ y_{r,s} y_{r+1,s} : & 1 \leq r \leq n-2,\ 1 \leq s \leq t \end{cases}$$

Refer Fig. 1a. Let l be the largest prime number from k to $k + t(3n-4)$. A one-one and onto function $f : V(G) \cup E(G) \rightarrow \{k, k+1, \ldots, k+t(3n-4)\}$ is defined as follows.
Let us assume that $|g(y_{r,s}) - g(y_0 y_{r,s})| \not\equiv 0$(mod (multiple of any factor of $g(y_{r,s})$)) for $1 \leq r \leq n-1$, $1 \leq s \leq t$.

Case 1. k equals 1

Subcase 1. For odd n
$g(y_0) = 1$

For $1 \le r \le n-1, 1 \le s \le t$,

$$g(y_{r,s}) = \begin{cases} (3s-3)(n-1)+2r-(s-2): & \text{for } s \text{ odd} \\ (3s-3)(n-1)+2r-(s-1): & \text{for } s \text{ even} \end{cases}$$

For $1 \le r \le n-2, 1 \le s \le t$,

$$g(y_{r,s}y_{r+1,s}) = \begin{cases} (3s-3)(n-1)+2r-(s-3): & \text{for } s \text{ odd} \\ (3s-3)(n-1)+2r-(s-2): & \text{for } s \text{ even} \end{cases}$$

$g(y_0y_{1,s}) = (3s-3)n - 2(2s-3)$, for s odd

For $1 \le s \le t$,

$$g(y_0y_{a,b}) = \begin{cases} s(3n-4)-r+3: 2 \le r \le n-1, & \text{for } s \text{ odd} \\ s(3n-4)-r+2: 1 \le r \le n-1, & \text{for } s \text{ even} \end{cases}$$

Next, we justify that every label on edge and on its corresponding end vertices are comparatively prime in pairs.

For each edge

$$y_0y_{r,s} \in E(G),$$

$$gcd(g(y_0), g(y_{r,s})) = \begin{cases} gcd(1, (3s-3)(n-1)+2r-(s-2)) = 1: \\ \quad 2 \le r \le n-1, & \text{for s odd} \\ gcd(1, (3s-3)(n-1)+2r-(s-1)) = 1 \\ \quad 1 \le r \le n-1, & \text{for s even} \end{cases}$$

$$gcd(g(y_0), g(y_0y_{r,s})) = \begin{cases} gcd(1, s(3n-4)-r+3) = 1: \\ \quad 2 \le r \le n-1, & \text{for } s \text{ odd} \\ gcd(1, s(3n-4)-r+2) = 1: \\ \quad 1 \le r \le n-1, & \text{for } s \text{ even} \end{cases}$$

$$gcd(g(y_0y_{r,s}),\ g(y_{r,s})) = \begin{cases} gcd(s(3n-4)-r+3, \\ (3s-3)(n-1)+2r-(s-2)) = 1 \\ : 2 \le r \le n-1, \text{ for } s \text{ odd} \\ gcd(s(3n-4)-r+2, \\ (3s-3)(n-1)+2r-(s-1)) = 1 \\ : 1 \le r \le n-1, \text{ for } s \text{ even} \end{cases}$$

since $|g(y_{r,s}) - g(y_0y_{r,s})| \not\equiv 0(\text{mod (multiple of any factor of } g(y_{r,s})))$ for $1 \le r \le n-1, 1 \le s \le t$.

For each edge $y_{r,s}y_{r+1,s} \in E(G)$,

$$gcd(g(y_{r,s}), g(y_{r+1,s})) = \begin{cases} gcd((3s-3)(n-1)+2r-(s-2), \\ (3s-3)(n-1)+2r+2-(s-2)) = 1 \\ : 2 \le r \le n-1, \text{ for } s \text{ odd} \\ gcd((3s-3)(n-1)+2r-(s-1), \\ (3s-3)(n-1)+2r+2-(s-1)) = 1 \\ : 1 \le r \le n-1, \text{ for } s \text{ even} \end{cases}$$

$$gcd(g(y_{r,s}y_{r+1,s}), g(y_{r+1,s})) = \begin{cases} gcd(((3s-3)(n-1)+2r-(s-3)), \\ (3s-3)(n-1)+2r+2-(s-2)) \\ = 1 : 2 \leq r \leq n-1, \text{ for } s \text{ odd} \\ gcd((3s-3)(n-1)+2r-(s-2)), \\ (3s-3)(n-1)+2r+2-(s-1)) \\ = 1 : 1 \leq r \leq n-1, \text{ for } s \text{ even} \end{cases}$$

$$gcd(g(y_{r,s}), g(y_{r,s}y_{r+1,s})) = \begin{cases} gcd((3s-3)(n-1)+2r-(s-2), \\ (3s-3)(n-1)+2r-(s-3)) = 1 \\ : 2 \leq r \leq n-1, \text{ for } s \text{ odd} \\ gcd((3s-3)(n-1)+2r-(s-1), \\ (3s-3)(n-1)+2r-(s-2)) = 1 \\ : 1 \leq r \leq n-1, \text{ for } s \text{ even} \end{cases}$$

Subcase 2. *For even n*
$g(y_0) = 1$
$g(y_{r,s}) = (3s-3)(n-1)+2r-(s-2), 1 \leq r \leq n-1, 1 \leq s \leq t$
$g(y_{r,s}y_{r+1,s}) = (3s-3)(n-1)+2r-(s-3), 1 \leq r \leq n-2, 1 \leq s \leq t$
$g(y_0y_{1,s}) = (3s-3)n-2(2s-3), 1 \leq s \leq t$
$g(y_0y_{r,s}) = s(3n-4)-r+3, 2 \leq r \leq n-1, 1 \leq s \leq t$
Next we justify that every label on edge and on its corresponding end vertices are comparatively prime in pairs.

For each edge $y_0y_{r,s} \in E(G)$, $gcd(g(y_0),g(y_{r,s})) = gcd(1, (3s-3)(n-1)+2r-(s-2)) = 1$, $gcd(g(y_0), g(y_0y_{r,s})) = gcd(1, s(3n-4)-r+3) = 1$, $gcd(g(y_0y_{r,s}),g(y_{r,s})) = gcd(s(3n-4)-r+3, (3s-3)(n-1)+2r-(s-2)) = 1$. For each edge $y_{r,s}y_{r+1,s} \in E(G)$, $gcd(g(y_{r,s}),g(y_{r+1,s})) = gcd((3s-3)(n-1)+2r-(s-2), (3s-3)(n-1)+2r+2-(s-2)) = 1$ since $(3s-3)(n-1)+2r-(s-2)$ and $(3s-3)(n-1)+2r+2-(s-2)$ are consecutive odd integers, $gcd(g(y_{r,s}),g(y_{r,s}y_{r+1,s})) = gcd((3s-3)(n-1)+2r-(s-2),(3s-3)(n-1)+2r-(s-3)) = 1$, $gcd(g(y_{r,s}y_{r+1,s}),g(y_{r+1,s})) = gcd((3s-3)(n-1)+2r-(s-3),(3s-3)(n-1)+2r+2-(s-2)) = 1$ since $(3s-3)(n-1)+2r-(s-2)$ and $(3s-3)(n-1)+2r-(s-3)$ are consecutive integers.

Case 2. *k* equals *l*

Subcase 1. For odd *n*

$g(y_0) = l$
For $1 \leq r \leq n-1, 1 \leq s \leq t$,

$$g(y_{r,s}) = \begin{cases} l+(3s-3)(n-1)+2r-(s-1) : & \text{for } s \text{ odd} \\ l+(3s-3)(n-1)+2r-s : & \text{for } s \text{ even} \end{cases}$$

For $1 \leq r \leq n-2, 1 \leq s \leq t$,

$$g(y_{r,s}y_{r+1,s}) = \begin{cases} l+(3s-3)(n-1)+2r-(s-2) : \text{for } s \text{ odd} \\ l+(3s-3)(n-1)+2r-(s-1) : \text{for } s \text{ even} \end{cases}$$

$g(y_0y_{1,s}) = l+(3s-3)n-4s+5$, for s odd

For $1 \le s \le t$,
$$g(y_0 y_{r,s}) = \begin{cases} l + s(3n-4) - r + 2 : 2 \le r \le n-1, & \text{for } s \text{ odd} \\ l + s(3n-4) - r + 1 : 1 \le r \le n-1, & \text{for } s \text{ even} \end{cases}$$

Subcase 2. For even n

$g(y_0) = l$
$g(y_{r,s}) = l + (3s-3)(n-1) + 2r - (s-1), 1 \le r \le n-1, 1 \le s \le t$
$g(y_{r,s} y_{r+1,s}) = l + (3s-3)(n-1) + 2r - (s-2), 1 \le r \le n-2, 1 \le s \le t$
$g(y_0 y_{1,s}) = l + (3s-3)n - 4s + 5, 1 \le s \le t$
$g(y_0 y_{r,s}) = l + s(3n-4) - r + 2, 2 \le r \le n-1, 1 \le s \le t$
Analogously as shown in the above case, we can justify that the labels assigned by the function to vertices and edges are comparatively prime in pairs.

Case 3. $k + t(3n-4)$ equals l

Subcase 1. For odd n
(*a*) When t is even, the labeling k starts with odd.
$g(y_0) = l$
For $1 \le r \le n-1, 1 \le s \le t$,
$$g(y_{r,s}) = \begin{cases} k + (3s-3)(n-1) + 2r - (s+1) : & \text{for } s \text{ odd} \\ k + (3s-3)(n-1) + 2r - s : & \text{for } s \text{ even} \end{cases}$$
For $1 \le r \le n-2, 1 \le s \le t$,
$$g(y_{r,s} y_{r+1,s}) = \begin{cases} k + (3s-3)(n-1) + 2r - s : & \text{for } s \text{ odd} \\ k + (3s-3)(n-1) + 2r - (s-1) : & \text{for } s \text{ even} \end{cases}$$
$g(y_0 y_{1,s}) = k + (3s-3)n - 4(s-1)$, for s even
For $1 \le s \le t$,
$$g(y_0 y_{r,s}) = \begin{cases} k + s(3n-4) - r : \quad 1 \le r \le n-1, & \text{for } s \text{ odd} \\ k + s(3n-4) - r + 1 : 2 \le r \le n-1, & \text{for } s \text{ even} \end{cases}$$
(*b*) When t is odd, the labeling k starts with even.
$g(y_0) = l$
For $1 \le r \le n-1, 1 \le s \le t$,
$$g(y_{r,s}) = \begin{cases} k + (3s-3)(n-1) + 2r - s : & \text{for } s \text{ odd} \\ k + (3s-3)(n-1) + 2r - (s+1) : & \text{for } s \text{ even} \end{cases}$$
For $1 \le r \le n-2, 1 \le s \le t$,
$$g(y_{r,s} y_{r+1,s}) = \begin{cases} k + (3s-3)(n-1) + 2r - s + 1 : & \text{for } s \text{ odd} \\ k + (3s-3)(n-1) + 2r - s : & \text{for } s \text{ even} \end{cases}$$
$g(y_0 y_{1,s}) = k + (3s-3)n - 4(s-1)$, for s odd
For $1 \le s \le t$,
$$g(y_0 y_{r,s}) = \begin{cases} k + s(3n-4) - r + 1 : 2 \le r \le n-1, & \text{for } s \text{ odd} \\ k + s(3n-4) - r : \quad 1 \le r \le n-1, & \text{for } s \text{ even} \end{cases}$$

Subcase 2. For even n
$g(y_0) = l$
$g(y_{r,s}) = k + (3s-3)(n-1) + 2r - (s+1), 1 \le r \le n-1, 1 \le s \le t$
$g(y_{r,s} y_{r+1,s}) = k + (3s-3)(n-1) + 2r - s, 1 \le r \le n-2, 1 \le s \le t$
$g(y_0 y_{r,s}) = k + s(3n-4) - r, 1 \le r \le n-1, 1 \le s \le t$

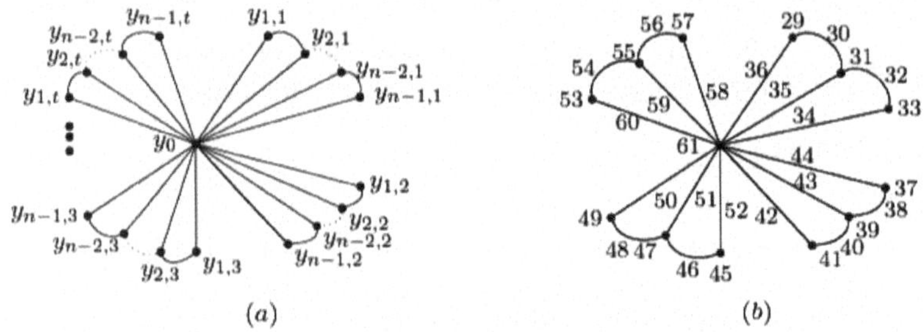

Fig. 1 (**a**) One-point union of shell graph Sh_n^t (**b**) k-prime total labeling of one-point union of shell graph Sh_4^4 for $k = 29$

From the above discussed case, one can justify that the labels assigned by the function to the vertices and edges are comparatively prime in pairs.
Thus, one-point union of shell graph Sh_n^t is k-prime total for $k \geq 1$.
An illustration for subcase 2 of case 3 is depicted in Fig. 1b.

Remark 1 It is clear from the first case of the Theorem 1 that the one-point union of shell graph Sh_n^t admits an edge vertex prime labeling.

Theorem 2 *The cycle union of t copies of shell graph $C(t.Sh_n)$ is k-prime total for $t > 2$ and $n \geq 3$.*

Proof Let $G = C(t.Sh_n)$ be the cycle union of t copies of shell graph of order tn and size $2t(n-1)$. Let l_r $(1 \leq r \leq t)$ be the vertices of the cycle. Each vertex l_r is replaced by Sh_n such that l_r $(1 \leq r \leq t)$ is the apex vertex of rth copy of Sh_n graph. Assume that each Sh_n has at least one prime number, and let l_1 be the largest prime number of first Sh_n, l_2 be the largest prime number of second $Sh_n, \ldots$, and l_r be the largest prime number of rth Sh_n where $k + (3r-1)(n-1) - 1 \leq l_r \leq k + 3r(n-1) - 1$. Let $V(G) = \{l_r : 1 \leq r \leq t\} \cup \{u_{x,y} : 1 \leq x \leq n-1, 1 \leq y \leq t\}$

$$E(G) = \begin{cases} l_r u_{x,y} : & 1 \leq r \leq t, \quad 1 \leq x \leq n-1,\ 1 \leq y \leq t \\ u_{x,y} u_{x+1,y} : & 1 \leq x \leq n-2, \quad 1 \leq y \leq t \\ l_1 l_t \cup l_r l_{r+1} : & 1 \leq r \leq t-1 \end{cases}$$

Refer Fig. 2a. A one-one and onto function $g : V(G) \cup E(G) \rightarrow \{k, k+1, \ldots, k + t(3n-2) - 1\}$ is defined as follows.

Case 1. $n \equiv 0 (\text{mod } 2)$

Subcase 1. When $k + t(3n-2) - 1$ is even, the labeling k starts with odd.
For $1 \leq x \leq n-1, \quad 1 \leq y \leq t$,

$$g(u_{x,y}) = \begin{cases} k + (3y-3)(n-1) + 2x - 2 : \text{ for } y \text{ odd} \\ k + (3y-3)(n-1) + 2x - 1 : \text{ for } y \text{ even} \end{cases}$$

For $1 \leq x \leq n-2, 1 \leq y \leq t$,

$$g(u_{x,y}u_{x+1,y}) = \begin{cases} k + (3y-3)(n-1) + 2x - 1: & \text{for } y \text{ odd} \\ k + (3y-3)(n-1) + 2x: & \text{for } y \text{ even} \end{cases}$$

$g(l_r u_{1,y}) = k + (3y-3)n - 3(y-1)$, for y even

$g(l_r)$ = largest prime number of each Sh_n where $k + (3r-1)(n-1) - 1 \leq l_r \leq k + 3r(n-1) - 1$.

$g(l_1 l_t) = k + t(3n-2) - 1$

$g(l_r l_{r+1}) = k + 3t(n-1) + (r-1), 1 \leq r \leq t-1$

For $1 \leq y \leq t$,

$$g(l_r u_{x,y}) = \begin{cases} k + (3r-1)(n-1) - 1 \text{ to } k + 3r(n-1) - 1 \\ \text{other than the largest prime number} : \text{ for } y \text{ odd} \\ k + (3r-1)(n-1) \text{ to } k + 3r(n-1) - 1 \\ \text{other than the largest prime number} : \text{ for } y \text{ even} \end{cases}$$

Next we justify that every label on edge and on its corresponding end vertices is comparatively prime in pairs.

For each edge $l_r u_{x,y} \in E(G)$,

$$gcd(g(l_r),\ g(u_{x,y})) = \begin{cases} gcd(\text{largest prime number}, \\ k + (3y-3)(n-1) + 2x - 2) = 1: \text{ for } y \text{ odd} \\ gcd(\text{largest prime number}, \\ k + (3y-3)(n-1) + 2x - 1) = 1: \text{ for } y \text{ even} \end{cases}$$

$$gcd(g(l_r u_{x,y}), g(u_{x,y})) = \begin{cases} gcd(g(l_r u_{x,y}), k + (3y-3)(n-1) + 2x - 2) = 1: \\ \text{for } y \text{ odd} \\ gcd(g(l_r u_{x,y}), k + (3y-3)(n-1) + 2x - 1) = 1: \\ \text{for } y \text{ even} \end{cases}$$

$$gcd(g(l_r), g(l_r u_{x,y}))) = \begin{cases} gcd(\text{largest prime number}, \\ f(l_r u_{x,y})) = 1: \quad \text{for } y \text{ odd} \\ gcd(\text{largest prime number}, \\ f(l_r u_{x,y})) = 1: \quad \text{for } y \text{ even} \end{cases}$$

For each edge $u_{x,y}u_{x+1,y} \in E(G)$,

$$gcd(g(u_{x,y}), g(u_{x+1,y})) = \begin{cases} gcd(k + (3y-3)(n-1) + 2x - 2, \\ k + (3y-3)(n-1) + 2x) = 1: \\ \text{for } y \text{ odd} \\ gcd(k + (3y-3)(n-1) + 2x - 1, \\ k + (3y-3)(n-1) + 2x + 1) = 1: \\ \text{for } y \text{ even} \end{cases}$$

$$gcd(g(u_{x,y}), g(u_{x,y}u_{x+1,y})) = \begin{cases} gcd(k + (3y-3)(n-1) + 2x - 2, \\ k + (3y-3)(n-1) + 2x - 1) = 1: \\ \text{for } y \text{ odd} \\ gcd(k + (3y-3)(n-1) + 2x - 1, \\ k + (3y-3)(n-1) + 2x) = 1: \\ \text{for } y \text{ even} \end{cases}$$

$$gcd(g(u_{x,y}u_{x+1,y}), g(u_{x+1,y})) = \begin{cases} gcd(k+(3y-3)(n-1)+2x-1, \\ k+(3y-3)(n-1)+2x) = 1 : \\ \text{for } y \text{ odd} \\ gcd(k+(3y-3)(n-1)+2x, \\ k+(3y-3)(n-1)+2x+1) = 1 : \\ \text{for } y \text{ even} \end{cases}$$

For each edge $l_1 l_t \in E(G), gcd(g(l_1), g(l_t)) = gcd$(largest prime number of $C(1.Sh_n)$, largest prime number of $C(t.Sh_n)) = 1$, $gcd(g(l_1), g(l_1 l_t)) = gcd$(largest prime number of $C(1.Sh_n), k+t(3n-2)-1) = 1$, $gcd(g(l_1 l_t), g(l_t)) = gcd(k+t(3n-2)-1$, largest prime number of $C(t.Sh_n)) = 1$. For each edge $l_r l_{r+1} \in E(G), gcd(g(l_r), g(l_{r+1})) = gcd$(largest prime number of $C(r.Sh_n)$, largest prime number of $C((r+1).Sh_n)) = 1$, $gcd(g(l_r), g(l_r l_{r+1})) = gcd$(largest prime number of $C(r.Sh_n), k+3t(n-1)+(r-1)) = 1$, $gcd(g(l_r l_{r+1}), g(l_{r+1})) = gcd(k+3t(n-1)+(r-1)$, largest prime number of $C((r+1).Sh_n)) = 1$.

Subcase 2. When $k+t(3n-2)-1$ is odd, the labeling k starts with even.
For $1 \leq x \leq n-1, \quad 1 \leq y \leq t$,

$$g(u_{x,y}) = \begin{cases} k+(3y-3)(n-1)+2x-1 : \text{ for } y \text{ odd} \\ k+(3y-3)(n-1)+2x-2 : \text{ for } y \text{ even} \end{cases}$$

For $1 \leq x \leq n-2, 1 \leq y \leq t$,

$$g(u_{x,y}u_{x+1,y}) = \begin{cases} k+(3y-3)(n-1)+2x : \quad \text{for } y\text{odd} \\ k+(3y-3)(n-1)+2x-1 : \text{for } y \quad \text{even} \end{cases}$$

$g(l_1 u_{1,y}) = k$
$g(l_r u_{1,y}) = k+(3y-3)n-3(y-1)$, for y odd
$g(l_r)$ = largest prime number of each Sh_n where $k+(3r-1)(n-1)-1 \leq l_r \leq k+3r(n-1)-1$.
$g(l_1 l_t) = k+t(3n-2)-1$
$g(l_r l_{r+1}) = k+3t(n-1)+(r-1), 1 \leq r \leq t-1$
For $1 \leq j \leq t$,

$$f(l_r u_{x,y}) = \begin{cases} k+(3r-1)(n-1)-1 \text{ to } k+3r(n-1)-1 \\ \text{other than the largest prime number} : \text{ for } y \text{ odd} \\ k+(3r-1)(n-1) \text{ to } k+3r(n-1)-1 \\ \text{other than the largest prime number} : \text{ for } y \text{ even} \end{cases}$$

Case 2. $n \equiv 1 (\text{mod } 2)$
$g(u_{x,y}) = k+(3y-3)(n-1)+2x-2, 1 \leq x \leq n-1, 1 \leq y \leq t$
$g(u_{x,y}u_{x+1,y}) = k+(3y-3)(n-1)+2x-1, 1 \leq x \leq n-2, 1 \leq y \leq t$
$g(l_r)$ =largest prime number of each Sh_n where $k+(3r-1)(n-1)-1 \leq l_r \leq k+3r(n-1)-1$.
$g(l_1 l_t) = k+t(3n-2)-1$
$g(l_r l_{r+1}) = k+3t(n-1)+(r-1), 1 \leq r \leq t-1$
$g(l_r u_{x,y})$ = label the edges from $k+(3r-1)(n-1)-1$ to $k+3r(n-1)-1$ other than the largest prime number.

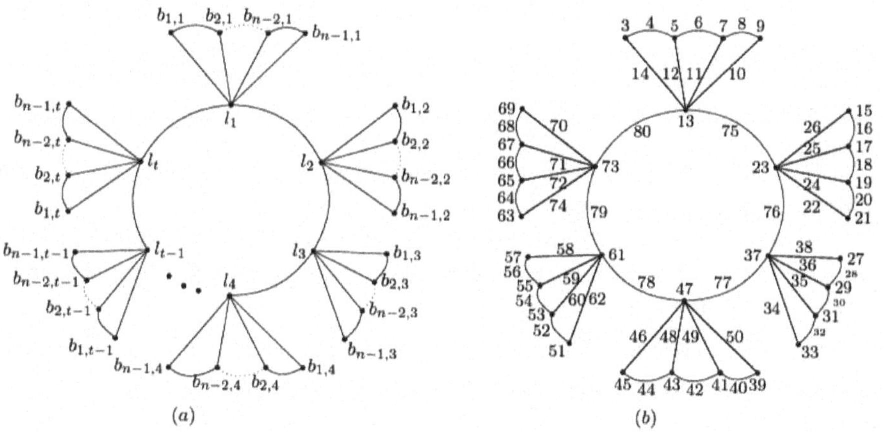

Fig. 2 (**a**) Cycle union of t copies of shell $C(t.Sh_n)$ (**b**) k-prime total labeling of cycle union of 6 copies of shell $C(6.Sh_5)$ for $k = 3$

As discussed above, one can check that the labels assigned to the vertices and edges are comparatively prime in pairs.

Thus, the graph cycle union of t copies of shell $C(t.Sh_n)$ is k-prime total for $k \geq 1$.

An illustration for case 2 is depicted in Fig. 2b.

Theorem 3 *The union of two copies of even cycle* C_n *is k-prime total for* $n > 3$ *and odd k.*

Proof Let $G(V, E) = C_n \cup C_n$ be the graph with $2n$ vertices and $2n$ edges. Let $V(G) = \{v_1, v_2, \ldots, v_n, v_{n+1}, \ldots, v_{2n}\}$ be the vertex set of G. Let $E(G) = \{E_1 \cup E_2 \cup E_3\}$ where $E_1(G) = \{v_x v_{x+1} : 1 \leq x \leq n-1\}$, $E_2(G) = \{v_x v_{x+1} : n+1 \leq x \leq 2n-1\}$, $E_3(G) = \{v_1 v_n, v_{n+1} v_{2n}\}$ be the edge set of G. Define a one-one and onto function $g : V(G) \cup E(G) \rightarrow \{k, k+1, \ldots, k+4n-1\}$ as follows:

Case 1. $(n-1)$ and $(2n-1)$ are prime, $k \not\equiv 0(\text{mod } (n-1))$ and $k+2n \not\equiv 0(\text{mod } (n-1))$, $k \not\equiv 0(\text{mod } (2n-1))$ and $k+2n \not\equiv 0(\text{mod } (2n-1)$
$g(v_x) = k + 2x - 2, 1 \leq x \leq 2n$
$g(v_x v_{x+1}) = k + 2x - 1, 1 \leq x \leq n-1$
$g(v_n v_1) = k + 2n - 1$
$g(v_x v_{x+1}) = k + 2x - 1, n+1 \leq x \leq 2n-1$
$g(v_n + 1 v_2 n) = k + 4n - 1$
For any edge $v_x v_{x+1} \in E_1(G) \cup E_2(G)$, $gcd(g(v_x), g(v_{x+1})) = gcd(k+2x-2, k+2x) = 1$ since $k+2x-2$ and $k+2x$ are consecutive odd integers, $gcd(g(v_x v_{x+1}), g(v_{x+1})) = gcd(k+2x-1, k+2x) = 1$ since $k+2x-1$ and $k+2x$ are consecutive positive integers and $gcd(g(v_x), g(v_x v_{x+1})) = gcd(k+2x-2, k+2x-1) = 1$, since $k+2x-2$ and $k+2x-1$ are successive positive integers. For the edge $v_n v_1 \in E_3(G)$, $gcd(g(v_n), g(v_1)) = gcd(k+2n-2, k) = 1$

since k and $k + 2n \not\equiv 0(\text{mod } (2n - 1)$ and k and $k + 2n \not\equiv 0(\text{mod } (n - 1))$, $gcd(g(v_n), g(v_n v_1)) = gcd(k+2n-2, k+2n-1) = 1$ since $k+2n-2$ and $k+2n-1$ are successive positive integers, $gcd(g(v_n v_1), g(v_1)) = gcd(k + 2n - 1, k) = 1$ since k and $k + 2n \not\equiv 0(\text{mod } (2n - 1))$ and k and $k + 2n \equiv 0(\text{mod } (n - 1))$. For the edge $v_{n+1}v_{2n} \in E_3(G)$, $gcd(g(v_{n+1}), g(v_{2n})) = gcd(k + 2n, k + 4n - 2) = 1$ since k and $k + 2n \not\equiv 0(\text{mod } (2n - 1)$ and k and $k + 2n \not\equiv 0(\text{mod } (n - 1))$, $gcd(g(v_{2n}), g(v_{2n}v_{n+1})) = gcd(k + 4n - 2, k + 4n - 1) = 1$ since $k + 4n - 2$ and $k + 4n - 1$ are successive positive integers, $gcd(g(v_{2n}v_{n+1}), g(v_{n+1})) = gcd(k + 4n - 1, k + 2n) = 1$ since k and $k + 2n \not\equiv 0(\text{mod } (2n - 1))$ and k and $k + 2n \not\equiv 0(\text{mod } (n - 1))$.

Case 2. $(n - 1)$ and $(2n - 1)$ are not prime and for k and $k + 2n$ not a multiple of any factor of $(n - 1)$, k and $k + 2n$ not a multiple of any factor of $(2n - 1)$
$g(v_x) = k + 2x - 2, 1 \leq x \leq 2n$
$g(v_x v_{x+1}) = k + 2x - 1, 1 \leq x \leq n - 1$
$g(v_n v_1) = k + 2n - 1$
$g(v_x v_{x+1}) = k + 2x - 1, n + 1 \leq x \leq 2n - 1$
$g(v_n + 1v_2n) = k + 4n - 1$
Similarly as discussed above in case 1, one can justify that the labels assigned by the function to the vertices and edges are comparatively prime in pairs.

Case 3. $(n - 1)$ prime and $(2n - 1)$ not prime and $k \not\equiv 0(\text{mod } (n - 1))$, $k + 2n$ not a multiple of any factor of $(2n - 1)$
$g(v_x) = k + 2x - 2, 1 \leq x \leq 2n$
$g(v_x v_{x+1}) = k + 2x - 1, 1 \leq x \leq n - 1$
$g(v_n v_1) = k + 2n - 1$
$g(v_x v_{x+1}) = k + 2x - 1, n + 1 \leq x \leq 2n - 1$
$g(v_n + 1v_2n) = k + 4n - 1$
Similarly as discussed above in case 1, one can justify that the labels assigned by the function to the vertices and edges are comparatively prime in pairs.

Case 4. $(n - 1)$ not prime and $(2n - 1)$ prime and for $k \not\equiv 0(\text{mod } (2n - 1))$ and $k + 2n \not\equiv 0(\text{mod } (2n - 1))$, k not a multiple of any factor of $(n - 1)$
$g(v_x) = k + 2x - 2, 1 \leq x \leq 2n$
$g(v_x v_{x+1}) = k + 2x - 1, 1 \leq x \leq n - 1$
$g(v_n v_1) = k + 2n - 1$
$g(v_x v_{x+1}) = k + 2x - 1, n + 1 \leq x \leq 2n - 1$
$g(v_{n+1} v_2 n) = k + 4n - 1$
Thus from the case analyzed earlier, we can conclude that the labels assigned by the function to vertices and edges are comparatively prime in pairs.
Hence, union of two copies of even cycle C_n admits k-prime total for $n > 3$ and odd k.

4 Conclusion

In this paper, k-prime total labeling has been discussed for the class of union of graphs for some positive integer k. To examine the existence of k-prime total labeling in certain classes of graphs obtained through graph operations such as corona of graphs, superimposing of graphs is the analysis of future.

References

1. Bondy, J.A., Murty, U.S.R.: Graph Theory with Application. Macmillon, London (1982)
2. Deb, P., Limaye, N.B: On harmonious labeling of some cycle related graphs. Ars Comb. **65**, 177–197 (2002)
3. Gallian, J.A.: A dynamic survey of graph labeling. Electron. J. Comb. **18**, DS6 (2023)
4. Meena, S., Amuda, G.: Some results on prime labeling of graphs. Int. J. Innovative Sci. Eng. Technol. **3**, 630–638 (2016)
5. Rosa, A.: On certain valuations of the vertices of a graph. In: Theory of Graphs (International Symposium, Rome, July 1966). Gordon and N. Y Breach and Dunod Paris, pp. 349–355 (1967)
6. Shee, S.-C., Ho, Y.-S.: The cordiality of one-point union of n copies of a graph. Discrete Math. **117**, 225–243 (1993)
7. Teresa Arockiamary, S., Baskar Babujee, J., Vijayalakshmi, G.: k-prime total labeling of graphs. Adv. Appl. Discrete Math. **35**, 45–59 (2022)
8. Teresa Arockiamary, S., Vijayalakshmi, G.: k-prime total labeling of shell related graphs. Adv. Appl. Discrete Math. **39**(1), 1–27 (2023)

Optimizing Wirelength in Graph Embedding: Folded Hypercube into Fan and Windmill Networks: A Comparative Study

Rini Dominic, R. Sundara Rajan, and T. M. Rajalaxmi

1 Introduction

Interconnection networks, depicted as graphs in integrated circuit research, are significant. Graph embedding in parallel computing is crucial for assessing networks, with wirelength, dilation, and edge congestion as key metrics [1]. Research has focused on various graph embedding techniques, including mapping hypercubes to necklace graphs [2], arranging hierarchical cubic networks on k-rooted complete binary trees with minimal wirelength [3], and placing complete multipartite graphs onto path, cycle, wheel, and circulant graphs [4].

Consider a graph A with vertex set V and edge set E. Let T be a subset of vertices in V that is not empty. We define the subgraph of A induced by T, denoted $A[T]$, as the subgraph that contains only the vertices in T and the edges that connect pairs of vertices within T.

Let U be an edge cut in graph A. Removing edges in $E(A) \setminus U$ splits A into components $H_1, H_2, \ldots, H_l$. A component H_i $(1 \leq i \leq l)$ is convex if all paths between its vertices stay within H_i.

The pair (ω, P_ω) represents the mapping of graph R into graph S. Here, ω is a one-to-one correspondence from the vertices of R to those of S, and P_ω maps the edges of R to paths $P_\omega(wz)$ in S, where $P_\omega((wz))$ is a path from $\omega(w)$ to $\omega(z)$ in S for each edge $(wz) \in E(R)$. For simplicity, we denote this pair as ω [5].

The edge congestion, denoted by $EC_\omega(e)$ for an edge $e \in E(S)$, is determined by counting the number of edges $(wz) \in E(R)$ for which e is part of the path $P_\omega(wz)$

R. Dominic (✉) · R. Sundara Rajan
Department of Mathematics, Hindustan Institute of Technology and Science, Chennai, Tamil Nadu, India

T. M. Rajalaxmi
Department of Mathematics, Sri Sivasubramaniya Nadar College of Engineering, Chennai, India

I. Rajasingh et al. (eds.), *Applied Computational Mathematics*, Springer Proceedings in Mathematics & Statistics 480,
https://doi.org/10.1007/978-3-031-77764-6_8

between $\omega(w)$ and $\omega(z)$ in S. This measure determines the overall edge congestion for the mapping ω from R to S as:

$$EC_{\omega}(R, S) = \max_{e \in E(S)} EC_{\omega}(e). \tag{1}$$

The edge congestion for the mapping of R into S, denoted as $EC(R, S)$, is given by Rajan et al. [6]:

$$EC(R, S) = \min_{\omega: R \to S} EC_{\omega}(R, S), \tag{2}$$

the minimum is computed across all mappings $\omega : R \to S$. Moreover, for any subset Q of $E(S)$, the edge congestion on Q is defined as:

$$EC_{\omega}(Q) = \sum_{e \in Q} EC_{\omega}(e). \tag{3}$$

The wirelength of the mapping ω from R into S is given by Reji et al. [7]:

$$WL_{\omega}(R, S) = \sum_{e \in E(S)} EC_{\omega}(e). \tag{4}$$

Thus, the wirelength of the mapping R into S is formally defined as:

$$WL(R, S) = \min_{\omega: R \to S} WL_{\omega}(R, S). \tag{5}$$

Lemma 1 (Modified Congestion Lemma [8]) *An embedding ω of graph R into graph S is characterized by an edge cut T in S that partitions S into two components, S_1 and S_2, when the edges in T are removed. The subgraph R_1 of R is induced by the vertices of R that are mapped by ω to $V(S_1)$, while R_2 is induced by the vertices mapped to $V(S_2)$. This edge cut T satisfies these conditions.*

1. *For all $(wz) \in E(R_i)$, where $i = 1, 2$, the path $P_{\omega}(wz)$ contains no edges from T.*
2. *For all $(wz) \in E(R)$, where $w \in V(R_1)$ and $z \in V(R_2)$, the path $P_{\omega}(wz)$ includes exactly one edge from T.*
3. *The vertex sets $V(R_1)$ and $V(R_2)$ are optimal.*

Under these conditions, $EC_{\omega}(T)$ achieves its minimum for all embeddings ω from R into S. The expression for $EC_{\omega}(T)$ is given by:

$$EC_{\omega}(T) = \sum_{a \in V(R_1)} deg_R(a) - 2|E(R_1)| = \sum_{a \in V(R_2)} deg_R(a) - 2|E(R_2)|. \tag{6}$$

Lemma 2 (Partition Lemma [8]) *In the embedding denoted by ω of graph R into graph S, suppose the edges in $E(S)$ are divided into sets $T_1, T_2, \ldots, T_p$, where each T_i represents an edge cut in S. If each T_i fulfills the requirements of Lemma 1, then the wirelength of the embedding ω for R into S is given by:*

$$WL_\omega(R, S) = \sum_{i=1}^{p} EC_\omega(T_i). \tag{7}$$

Moreover, it is true that $WL(R, S) = WL_\omega(R, S)$.

2 Embedding onto Folded Hypercubes

The folded hypercube, widely employed in interconnecting processors within parallel computing systems like supercomputers and clusters, portrays each processor as a node, with edges symbolizing communication channels.

In the s-dimensional hypercube Q^s, $y = x$ means x and $\overline{x}$ have complementary addresses. The s-dimensional folded hypercube FQ^s is an $(s+1)$-regular graph that extends Q^s, including all Q^s vertices and adding complementary edges between vertices with complementary addresses, enriching its connectivity structure [9].

Lemma 3 ([10]) *For $k = 2^{u_1} + 2^{u_2} + \ldots + 2^{u_l}$ with $s \geq u_1 > u_2 > \cdots > u_l \geq 0$, the expression for $|E(FQ(L_k))|$ is given by:*

$$|E(FQ(L_k))| = \begin{cases} |E(Q^s(L_k))| & \text{if } k \leq 2^{n-1} \\ |E(Q^s(L_k))| + k - 2^{n-1} & \text{if } k > 2^{n-1} \end{cases}$$

where

$$|E(Q^s(L_k))| = \left[u_1 \cdot 2^{u_1-1} + u_2 \cdot 2^{u_2-1} + \cdots + u_l \cdot 2^{u_l-1}\right] + \left[2^{u_2} + 2 \cdot 2^{u_3} + \cdots + (l-1)2^{u_l}\right].$$

2.1 Wirelength of Embedding Folded Hypercube into Fan Graph

The graph labeled as F_s is a specific kind of graph containing s vertices. It takes the form of a path with $s - 1$ vertices, and each vertex within this path is connected to a central vertex referred to as the core vertex [1].

Embedding Algorithm A

Input The s-dimensional folded hypercube FQ^s and F_{2^s}, $s \geq 2$.

Algorithm In the context of a host graph being a fan graph on 2^s vertices, the labeling of vertices in F_{2^s} follows a specific pattern similar to a Gray code sequence from 1 to $2^s - 1$:

1. The core vertex is labeled as 2^s.
2. The outer path vertices are labeled in clockwise order as $1, 2, \ldots, 2^s - 1$.

For each vertex $x \in V(R)$, let $g(x) = x$. For any edge $(uv) \in E(R)$, $P_g(uv)$ is the shortest path from $g(u)$ to $g(v)$ in F_{2^s}.

Output The embedding $g(x) = x$ of FQ^{2^s} into F_{2^s} achieves minimum wirelength.

Proof of Correctness Define the edge cut T_j in F_{2^s} for $1 \leq j \leq 2^s - 4$ as $T_j = (j, j+1), (j+2, j+3), (2^s, j+1), (2^s, j+2)$. When $j = 2^s - 3$, $T_j = (2^s - 3, 2^s - 2), (2^s, 2^s - 2), (2^s, 2^s - 1)$. For $j = 2^s - 2$, $T_j = (2, 3), (2^s, 1), (2^s, 2)$. If $j = 2^s - 1$, $T_j = (2^s - 2, 2^s - 1), (2^s, 2^s - 1)$. Finally, for $j = 2^s$, $T_j = \{(1, 2), (2^s, 1)\}$.

The set of edges $\{(j, j+1), (j+2, j+3), (2^s, j+1), (2^s, j+2) : 1 \leq j \leq 2^s - 4\} \cup T_{2^s-3} \cup T_{2^s-2} \cup T_{2^s-1} \cup T_{2^s}$ includes all edges in F_{2^s} twice. Thus, $T_1, T_2, \ldots, T_{2^s}$ forms a partition of $[2E(F_{2^s})]$.

For $1 \leq j \leq 2^s - 3$, removing T_j from $E(F_{2^s})$ splits the graph into S_j and $\overline{S_j}$. Define $V(S_j) = j+1, j+2$. Let $R_j = R[V(g^{-1}(S_j))]$ and $\overline{R_j} = R[V(g^{-1}(\overline{S_j}))]$. Both S_j and $\overline{S_j}$ are convex components, and R_j and $\overline{R_j}$ are optimal sets with cardinalities 2 and $2^s - 2$, respectively. Thus, T_j meets the requirements of Lemma 1, resulting in $EC_g(T_j)$ being minimized for $1 \leq j \leq 2^s - 3$. For $j = 2^s - 2$, removing T_j from $E(F_{2^s})$ creates components S_j and $\overline{S_j}$. Define $V(S_j) = 1, 2$. Let $R_j = R[V(g^{-1}(S_j))]$ and $\overline{R_j} = R[V(g^{-1}(\overline{S_j}))]$. Both S_j and $\overline{S_j}$ are convex components, and R_j and $\overline{R_j}$ are optimal sets with cardinalities 2 and $2^s - 2$, respectively. Thus, T_j meets the requirements of Lemma 1, leading to $EC_g(T_j)$ being minimized for $j = 2^s - 2$.

For $j = 2^s - 1, 2^s$, removing T_j from $E(F_{2^s})$ creates components S_j and $\overline{S_j}$. Define $V(S_j) = j$ modulo 2^s. Let $R_j = R[V(g^{-1}(S_j))]$ and $\overline{R_j} = R[V(g^{-1}(\overline{S_j}))]$. Both S_j and $\overline{S_j}$ are convex components and R_j and $\overline{R_j}$ are optimal sets with cardinalities 1 and $2^s - 1$, respectively. Thus, T_j meets the requirements of Lemma 1, leading to $EC_g(T_j)$ being minimized for $j = 2^s - 1, 2^s$.

Theorem 1 *Consider the graph R, which represents the s-dimensional folded hypercube (FQ^s) with $s \geq 2$. Let S be the fan graph on 2^s vertices, denoted F_{2^s}. Then,*

$$WL(R, S) = s \cdot 2^s - s + 1. \tag{8}$$

Proof By using Lemma 1, for $1 \leq j \leq 2^s - 2$, $EC_g(T_j) = 2(s+1) - 2$ and for $j = 2^s - 1, 2^s$, $EC_g(T_j) = (s+1)$.

Therefore, the wirelength is minimized, as indicated by the Partition Lemma, and we have

$$WL(R, S) = \frac{1}{2}\left[\sum_{j=1}^{2^s-2} EC_g(T_j) + \sum_{j=2^s-1}^{2^s} EC_g(T_j)\right]$$
$$= \frac{1}{2}\left[\sum_{j=1}^{2^s-2}(2(s+1)-2) + \sum_{j=2^s-1}^{2^s}(s+1)\right]$$
$$= s \cdot 2^s - s + 1.$$

2.2 *Wirelength of Embedding Folded Hypercube into Windmill Graph*

A windmill graph, labeled as WM_s, is a graph having a total of $2s$ vertices. It is characterized by the presence of $s - 1$ triangles and a single pendent vertex, all of which share a common hub vertex [1].

Embedding Algorithm B

Input The s-dimensional folded hypercube FQ^s and WM_{2^s-1}, $s \geq 2$.

Algorithm Vertices in FQ^s are labeled with a Gray code sequence from 1 to $2^s - 1$. The vertices of the host graph WM_{2^s-1} are labeled accordingly:

1. The hub vertex is denoted by the label 2^s.
2. The pendent vertex is designated with the label $2^s - 1$.
3. The remaining vertices are sequentially labeled from 1 to $2^s - 2$, starting from any vertex where $(i, i + 1)$ is adjacent, i is odd, and $1 \leq i \leq 2^s - 3$.

For each vertex $x \in V(A)$, let $g(x) = x$. For any edge $(uv) \in E(A)$, $P_g(uv)$ denotes the shortest path from $g(u)$ to $g(v)$ in the windmill graph WM_{2^s-1}.

Output The embedding g of FQ^{2^s} into WM_{2^s-1} with $g(x) = x$ achieves the minimum wirelength.

Proof of Correctness For $1 \leq j \leq 2^{s-1} - 1$, let S_j be an edge cut in WM_{2^s-1} defined as $S_j = \{(2j - 1, 2^s), (2j, 2^s)\}$. For $1 \leq j \leq 2^s - 2$, let S'_j be an edge cut given by $S'_j = \{(j, j + 1)(j, 2^s)\}$ for odd j and $S'_j = \{(j - 1, j)(j, 2^s)\}$ for even j. When $j = 2^s - 1$, the edge cut is $S''_j = \{(j, j + 1)\}$, and for $j = 2^s$, the edge cut is $S''_j = \{(j - 1, j)\}$. The combined edge set $\{(2j - 1, 2^s), (2j, 2^s) : 1 \leq j \leq 2^{s-1} - 1\} \cup \{(j, j + 1)(j, 2^s) : 1 \leq j \leq 2^s - 2 \text{ and } j \text{ is odd}\} \cup \{(j - 1, 1)(j, 2^s) : 1 \leq j \leq 2^s - 2 \text{ and } j \text{ is even}\} \cup \{(j, j + 1) : j =$

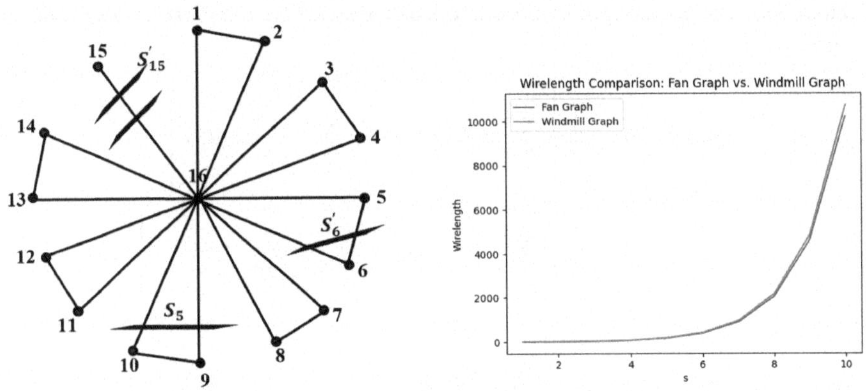

Fig. 1 (**a**) Edge cut of WM_8 (**b**) comparative analysis representation

$2^s - 1\} \cup \{(j-1, j), j = 2^s\}$ encompasses all edges of $WM_{2^{s-1}}$ twice, forming a partition of $[2E(WM_{2^{s-1}})]$; see Fig. 1a.

For $1 \leq j \leq 2^{s-1} - 1$, the removal of S_j from $E(WM_{2^{s-1}})$ results in the formation of components B_j and $\overline{B_j}$. Define $V(B_j) = \{2j-1, 2j\}$. Let $A_j = G[V(g^{-1}(B_j))]$ and $\overline{A_j} = G[V(g^{-1}(\overline{B_j}))]$. It is apparent that B_j and $\overline{B_j}$ constitute convex components, and in accordance with Lemma 1, A_j and $\overline{A_j}$ represent optimal sets with cardinalities 2 and $2^s - 2$, respectively. Thus, $EC_g(S_j)$ achieves its minimum value.

For $1 \leq j \leq 2^s - 1$, the removal of S'_j from $E(WM_{2^{s-1}})$ results in the emergence of components B'_j and $\overline{B'_j}$. Define $V(B'_j) = \{j\}$. Let $A'_j = G[V(g^{-1}(B'_j))]$ and $\overline{A'_j} = G[V(g^{-1}(\overline{B'_j}))]$. It is established that B'_j and $\overline{B'_j}$ function as convex components. In accordance with Lemma 1, A'_j and $\overline{A'_j}$ represent optimal sets with cardinalities 1 and $2^s - 1$, respectively. Hence, $EC_g(S'_j)$ achieves its minimum value.

For $j = 2^s$, excluding S''_j from $E(WM_{2^{s-1}})$ yields components B''_j and $\overline{B''_j}$. Define $V(B''_j) = \{j-1\}$. Let $A''_j = G[V(g^{-1}(B''_j))]$ and $\overline{A''_j} = G[V(g^{-1}(\overline{B''_j}))]$. Furthermore, it is evident that B''_j and $\overline{B''_j}$ serve as convex components, and according to Lemma 1, A''_j and $\overline{A''_j}$ are optimal sets with cardinalities 1 and $2^s - 1$, respectively. Therefore, $EC_g(S''_j)$ achieves its minimum value.

Theorem 2 *Consider A as the graph representing the folded hypercube in s dimensions FQ^s, where $s \geq 2$, and let B be the windmill graph on 2^s vertices, denoted as $WM_{2^{s-1}}$. Then,*

$$WL(A, B) = s \cdot 2^s + 2^{s-1} - s. \tag{9}$$

Proof By using Lemma 1, for $1 \leq j \leq 2^{s-1} - 1$, $EC_g(S_j) = 2(s+1) - 2$ and for $j = 2^s - 1,\ 2^s$, $EC_g(S'_j) = (s+1)$ and for $j = 2^s$, $EC_g(S''_j) = (s+1)$.

Hence, the wirelength is minimized, in accordance with the Partition Lemma, and we have

$$WL(A, B) = \frac{1}{2}\left[\sum_{j=1}^{2^{s-1}-1} EC_g(S_j) + \sum_{j=1}^{2^s-1}(EC_g(S_j)) + (s+1)\right]$$

$$= \frac{1}{2}\left[\sum_{j=1}^{2^{s-1}-1} (2(s+1)-2) + \sum_{j=1}^{2^s}(s+1)\right]$$

$$= s \cdot 2^s + 2^{s-1} - s.$$

3 Comparative Analysis

In this study, we focus on the wirelengths incurred when embedding the folded hypercube FQ^s into the fan graph F_{2^s} and the windmill graph on 2^s vertices. Wirelength is a critical metric, as it directly impacts communication delays and resource utilization.

3.1 Fan Graph Embedding

1. The wirelength in the fan graph embedding increases linearly with both the dimensionality of the folded hypercube and the size of the host graph.
2. The term $(1 - s)$ in the wirelength formula suggests a potential slight reduction compared to straightforward multiplication, indicating some efficiency gains.
3. The simplicity of the formula might lead to easier implementation and management of the network.

3.2 Windmill Graph Embedding

1. The additional term 2^{s-1} introduces a higher-order increase in wirelength, potentially providing more connectivity.
2. The higher-order term may result in increased wirelength, which could be a trade-off between complexity and wirelength efficiency.

In summary, folding hypercubes into fan and windmill graphs shows different wirelength characteristics. The fan graph grows linearly, offering efficiency gains, while the windmill graph grows with higher complexity, suggesting a trade-off. Designers should consider application needs: The fan graph offers simplicity and wirelength efficiency, while the windmill graph provides extra connectivity but may increase complexity and wirelength, see Fig. 1b.

4 Conclusion

In conclusion, this research on graph embedding, concentrating on folding hypercube structures into wheel-like networks like fan and windmill graphs, highlights the linear growth and potential efficiency gains in the fan graph, contrasting with the higher-order growth and complexity trade-off in the windmill graph. Looking ahead, a significant open problem is the exploration of finding the maximum subgraph for grid networks and its subsequent embedding onto grids.

Acknowledgments The work of R. Sundara Rajan is supported by the Science and Engineering Research Board (SERB), Government of India for funding under the Teachers Associateship for Research Excellence (TARE) scheme with Project No. TAR/2023/000041.

References

1. Rajan, R.S., Dominic, R., Semanicová-Fenovcíková, A.: Wirelength of 3-regular Knödel graphs into certain graphs. AIP Conf. Proc. **2797**(1), 020011 (2023). https://doi.org/10.1063/5.0148971
2. Rajasingh, I., Rajan, B., Rajan, R.S.: Embedding of hypercubes into necklace, windmill and snake graphs. Inform. Proc. Lett. **112**(12), 509–515 (2012). https://doi.org/10.1016/j.ipl.2012.03.006
3. Guo, R., Wang, Y., Fan, J., Fan, W.: Embedding hierarchical cubic networks into k-rooted complete binary trees for minimum wirelength. Int. J. Found. Comput. Sci. 1–26 (2023). https://doi.org/10.1142/S0129054123500053
4. Rajan, R.S., Rajalaxmi, T.M., Liu, J.-B., Sethuraman, G.: Wirelength of embedding complete multipartite graphs into certain graphs. Discrete Appl. Math. **280**, 221–236 (2020). https://doi.org/10.1016/j.dam.2018.05.034
5. Liu, J.-B., Arockiaraj, M., Delaila, J.N.: Wirelength of enhanced hypercube into windmill and necklace graphs. Mathematics **7**(5), (2019). https://doi.org/10.3390/math7050383
6. Rajan, R.S., Reji, R.M., Rajalaxmi, T.M.: Maximum subgraph problem for 3-regular Knödel graphs and its wirelength. Algorithms and Discrete Applied Mathematics. CALDAM 2023. Lecture Notes in Computer Science vol. 13947, pp. 403–414 (2023). https://doi.org/10.1007/978-3-031-25211-2_31
7. Reji, R.M., Rajan, R.S., Rajalaxmi, T.M.: Embedding Knödel graph into cube-like architectures: dilation optimization and wirelength analysis. J. Int. Netw. (2023). https://doi.org/10.1142/S0219265923500317
8. Rajan, R.S., Dominic, R., Sadagopan, N.: Optimal layout of $(K_p - C_p)^n$ into wheel-like networks. J. Interconnect. Netw. (2024). https://doi.org/10.1142/S0219265924500075
9. Dominic, R., Rajan, R.S., Rajalaxmi, T.M., Packiaraj, L.: Optimal layout of embedding onto folded hypercubes. Int. J. Found. Comput. Sci. Submitted to the journal
10. Quadras, J., Solomon, S.: Embedding of the folded hypercubes into tori. Math. Comput. Sci. **9** (2015). https://doi.org/10.1007/s11786-015-0223-38

Connectedness and Distance Energy of Hamacher d-Fuzzy Graphs

S. Lavanya, A. Sajiya Merlin Mahizl, J. Jeba Jesintha, and R. Yogitha

1 Introduction

Graphs are an abstraction of the relationships that have arisen in nature. The way it is represented visually by vertices drawn and connected by edges provides certain applications. In 1735, when Euler tried to solve the famous Konigsberg Bridge problem, he first came up with a basic idea of graphs, and this is said to have been its basis for Graph theory. To solve the problems of the graphs using computers, the data had to be stored in the memory in the form of a matrix. There are multiple forms of matrices related to a graph. The spectral graph theory, which studies the characteristics of eigenvalues, eigenvectors, and characteristic polynomials associated with graph matrices, is a major area of study in algebraic graph theory. Energy [6] is a concept which relates to the spectrum of graphs. In 1978, Ivan Gutman has defined it mathematically for all graphs. Graph theory has found application in a variety of real-time applications. But the vagueness in describing the information in real-life situations led to the development of fuzzy set theory and fuzzy graph theory. L.A. Zadeh developed fuzzy sets [1] in 1965 as an expansion of the traditional concept of a set to address certain attribute ambiguity. Rosenfeld in 1975 [2] developed the theory of fuzzy graphs. We consider d-fuzzy graph defined in [3] with edge membership function as Hamacher product and name it as Hamacher d-fuzzy graph [4].

S. Lavanya
Department of Mathematics, Bharathi Women's College, Chennai, India

A. Sajiya Merlin Mahizl (✉) · J. Jeba Jesintha · R. Yogitha
Department of Mathematics, Women's Christian College, University of Madras, Chennai, India

I. Rajasingh et al. (eds.), *Applied Computational Mathematics*,
Springer Proceedings in Mathematics & Statistics 480,
https://doi.org/10.1007/978-3-031-77764-6_9

2 Definitions

Definition 1 ([4]) If $d = \{d_1, d_2, \ldots, d_n\}$ is a degree sequence which realizes a crisp graph G with vertex set V and edge set E, then $\widetilde{G}_{Hd}$ =(σ, μ) is called **Hamacher *d*-fuzzy graph** defined by $\sigma : V \rightarrow [0, 1]$ and $\mu : V \times V \rightarrow [0, 1]$ where $\sigma(v_i) = \frac{d(v_i)}{n}$ for all $v_i \in V$ and $\mu(v_i, v_j) = \frac{\sigma(v_i)\sigma(v_j)}{\sigma(v_i)+\sigma(v_j)-\sigma(v_i)\sigma(v_j)}$ for all $(v_i, v_j) \in E$.

Definition 2 ([5]) The **distance matrix** $D_s(\widetilde{G})$ of a fuzzy graph $\widetilde{G}$ =(σ,μ) is an $n \times n$ matrix defined by $D_s(\widetilde{G})$ =[$d_s(v_i, v_j)$] where

$$[d_s(v_i, v_j)] = \begin{cases} \min L(P_i) & \text{if } i \neq j \\ 0 & \text{if } i = j \end{cases}$$

The eigenvalues of $D_s(\widetilde{G})$ are denoted by $\lambda_1, \lambda_2, \ldots, \lambda_n$, and the distance energy $DE(\widetilde{G})$ is defined as sum of the absolute values of distance eigenvalues of $\widetilde{G}$.

Definition 3 ([3]) The **connectedness strength matrix** C of a fuzzy graph $\widetilde{G}$ =(σ,μ) is a square matrix of order n defined by C =[c_{ij}] where

$$[c_{ij}] = \begin{cases} \mu^{\infty}(v_i, v_j), & \text{if } i \neq j \\ 0, & \text{if } i = j \end{cases}$$

The eigenvalues of C are denoted by $\gamma_1, \gamma_2, \ldots, \gamma_n$, and the connectedness energy of $\widetilde{G}$ is defined as sum of the absolute values of eigenvalues of the connectedness strength matrix C of $\widetilde{G}$.

3 Main Results

3.1 Connectedness and Distance Energy of Hamacher d-Fuzzy Uniform Graph

Theorem 3 *Let $\widetilde{G}_{Hd}$ be a Hamacher d-fuzzy uniform graph with n vertices, degree sequence $d = \{n-1, n-1, \ldots, n-1\}$, and the edge membership function represented by the Hamacher product. Then, the Connectedness energy of the Hamacher d-fuzzy uniform graph is equal to the energy of the Hamacher d-fuzzy uniform graph. In notation, $CE(\widetilde{G}_{Hd})$ =$E(\widetilde{G}_{Hd})$.*

Proof The energy and connectedness energy of $\widetilde{G}_{Hd}$ are defined as sum of the absolute values of eigenvalues of the adjacency matrix and connectedness strength matrix of $\widetilde{G}_{Hd}$, respectively. The values of (v_i, v_j) in the adjacency matrix $A(\widetilde{G}_{Hd})$ are given by

$$[\tilde{a}_{ij}] = \begin{cases} \mu(v_i, v_j), & \text{when } v_i \text{ and } v_j \text{ are adjacent} \\ 0, & \text{when } v_i \text{ and } v_j \text{ are not adjacent} \end{cases}$$

The elements of the connectedness strength matrix of $\widetilde{G}_{Hd}$ are zero only in the main diagonal, whereas the other values are given as $\mu^{\infty}(v_i, v_j)$, i.e.,

$$[c_{ij}] = \begin{cases} \mu^{\infty}(v_i, v_j), & \text{if } i \neq j \\ 0, & \text{if } i = j \end{cases}$$

In the case of Hamacher d-fuzzy uniform graph $\widetilde{G}_{Hd}$, $\mu^{\infty}(v_i, v_j) = \mu(v_i, v_j) \Longrightarrow C(\widetilde{G}_{Hd}) = A(\widetilde{G}_{Hd}) \Longrightarrow \sum \chi_i = \sum \gamma_i$. Hence, $CE(\widetilde{G}_{Hd}) = E(\widetilde{G}_{Hd})$.

Theorem 4 *Let $\widetilde{G}_{Hd}$ be a Hamacher d-fuzzy uniform graph with n vertices, degree sequence $d = \{n-1, n-1, \ldots, n-1\}$, and the edge membership function represented by the Hamacher product. Then, the Distance energy of the Hamacher d-fuzzy uniform graph is equal to the energy of the Hamacher d-fuzzy uniform graph. In notation, $DE(\widetilde{G}_{Hd}) = E(\widetilde{G}_{Hd})$.*

Proof The energy and distance energy of $\widetilde{G}_{Hd}$ are defined as the sum of the absolute values of eigenvalues of the adjacency matrix and distance matrix of $\widetilde{G}_{Hd}$, respectively. The values of (v_i, v_j) in the adjacency matrix $A(\widetilde{G}_{Hd})$ are given by

$$[\tilde{a}_{ij}] = \begin{cases} \mu(v_i, v_j), & \text{when } v_i \text{ and } v_j \text{ are adjacent} \\ 0, & \text{when } v_i \text{ and } v_j \text{ are not adjacent} \end{cases}$$

The elements of the distance matrix of $\widetilde{G}_{Hd}$ are zero only in the main diagonal, whereas the other values are given as $d_s(v_i, v_j)$, the minimum distance between (v_i, v_j), i.e.,

$$[d_s(v_i, v_j)] = \begin{cases} \min L(P_i) & \text{if } i \neq j \\ 0 & \text{if } i = j \end{cases}$$

In a Hamacher d-fuzzy uniform graph, $d_s(v_i, v_j) = \mu(v_i, v_j) \Longrightarrow A(\widetilde{G}_{Hd}) = D_s(\widetilde{G}_{Hd}) \Longrightarrow \sum \gamma_i = \sum \lambda_i$. Hence, $DE(\widetilde{G}_{Hd}) = E(\widetilde{G}_{Hd})$.

The illustration for the above theorems is given in Fig. 1.

Consider a Hamacher d-fuzzy uniform graph $\widetilde{G}_{Hd}$ on six vertices with a degree sequence of the form $d = \{5, 5, 5, 5, 5, 5\}$. Then the vertex membership function is given by $\sigma(v_i) = \frac{d(v_i)}{n}$ for $i = 1, \ldots, n$, i.e., $\sigma(v_1), \sigma(v_2), \ldots, \sigma(v_6) = \frac{5}{6}$. Similarly, the edge membership function is found using the Hamacher product given by $\mu(v_i, v_j) = \frac{\sigma(v_i)\sigma(v_j)}{\sigma(v_i)+\sigma(v_j)-\sigma(v_i)\sigma(v_j)}$, i.e., $\mu(v_i, v_j) = \frac{5}{7}$ for $i, j = 1, 2, 3, 4, 5, 6$. The adjacency, distance, and connectedness strength matrix of $\widetilde{G}_{Hd}$ is

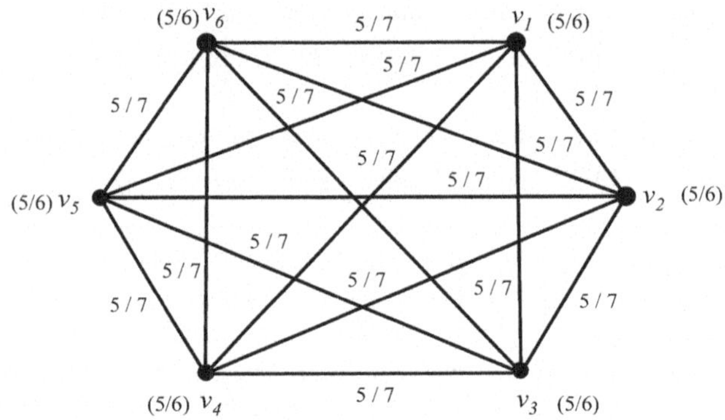

Fig. 1 Hamacher *d*-fuzzy uniform graph on six vertices

$$A(\widetilde{G}_{Hd}) = D_s(\widetilde{G}_{Hd}) = C(\widetilde{G}_{Hd}) = \begin{bmatrix} 0 & 5/7 & 5/7 & 5/7 & 5/7 & 5/7 \\ 5/7 & 0 & 5/7 & 5/7 & 5/7 & 5/7 \\ 5/7 & 5/7 & 0 & 5/7 & 5/7 & 5/7 \\ 5/7 & 5/7 & 5/7 & 0 & 5/7 & 5/7 \\ 5/7 & 5/7 & 5/7 & 5/7 & 0 & 5/7 \\ 5/7 & 5/7 & 5/7 & 5/7 & 5/7 & 0 \end{bmatrix}$$

Hence, Spec($A(\widetilde{G}_{Hd})$) = Spec($D_s(\widetilde{G}_{Hd})$) = Spec($C(\widetilde{G}_{Hd})$) = {3.5714, −0.7143, −0.7143, −0.7143, −0.7143, −0.7143}. Thus, $E(\widetilde{G}_{Hd}) = 7.143$, $DE(\widetilde{G}_{Hd}) = 7.143$ and $CE(\widetilde{G}_{Hd}) = 7.143$

Therefore, from the above, it is clear that $E(\widetilde{G}_{Hd}) = DE(\widetilde{G}_{Hd}) = CE(\widetilde{G}_{Hd})$.

3.2 Connectedness and Distance Energy of Hamacher d-Fuzzy Star Graph

Theorem 5 *Let $\widetilde{G}_{Hd}$ be a Hamacher d-fuzzy star graph with n vertices, degree sequence $d = \{n-1, 1, \ldots, 1\}$, and the edge membership function represented by the Hamacher product. Then, the Connectedness energy of the Hamacher d-fuzzy star graph is greater than the energy of the Hamacher d-fuzzy star graph. In notation, $CE(\widetilde{G}_{Hd}) > E(\widetilde{G}_{Hd})$.*

Proof The energy and connectedness energy of $\widetilde{G}_{Hd}$ are defined as the sum of the absolute values of eigenvalues of the adjacency matrix and connectedness strength matrix of $\widetilde{G}_{Hd}$, respectively. The values of (v_i, v_j) in the adjacency matrix $A(\widetilde{G}_{Hd})$ are given by

$$[\tilde{a}_{ij}] = \begin{cases} \mu(v_i, v_j), & \text{when } v_i \text{ and } v_j \text{ are adjacent} \\ 0, & \text{when } v_i \text{ and } v_j \text{ are not adjacent} \end{cases}$$

The elements of the connectedness strength matrix of $\widetilde{G}_{Hd}$ are zero only in the main diagonal, whereas the other values are given as $\mu^{\infty}(v_i, v_j)$, i.e.,

$$[c_{ij}] = \begin{cases} \mu^{\infty}(v_i, v_j), & \text{if } i \neq j \\ 0, & \text{if } i = j \end{cases}$$

Thus, for a Hamacher d-fuzzy star graph $\widetilde{G}_{Hd}$, $\mu^{\infty}(v_i, v_j) \geq \mu(v_i, v_j) \Longrightarrow C(\widetilde{G}_{Hd}) > A(\widetilde{G}_{Hd}) \Longrightarrow \sum \chi_i > \sum \gamma_i$. Hence, $CE(\widetilde{G}_{Hd}) > E(\widetilde{G}_{Hd})$.

Theorem 6 *Let $\widetilde{G}_{Hd}$ be a Hamacher d-fuzzy star graph with n vertices, degree sequence $d = \{n-1, 1, \ldots, 1\}$, and the edge membership function represented by the Hamacher product. Then, the Distance energy of the Hamacher d-fuzzy star graph is greater than the energy of the Hamacher d-fuzzy star graph. In notation, $DE(\widetilde{G}_{Hd}) > E(\widetilde{G}_{Hd})$.*

Proof The energy and distance energy of $\widetilde{G}_{Hd}$ are defined as the sum of the absolute values of eigenvalues of the adjacency matrix and distance matrix of $\widetilde{G}_{Hd}$, respectively. The values of (v_i, v_j) in the adjacency matrix $A(\widetilde{G}_{Hd})$ are given by

$$[\tilde{a}_{ij}] = \begin{cases} \mu(v_i, v_j), & \text{when } v_i \text{ and } v_j \text{ are adjacent} \\ 0, & \text{when } v_i \text{ and } v_j \text{ are not adjacent} \end{cases}$$

The elements of the distance matrix of $\widetilde{G}_{Hd}$ are zero only in the main diagonal, whereas the other values are given as $d_s(v_i, v_j)$, i.e.,

$$[d_s(v_i, v_j)] = \begin{cases} \min\{L(P_i)\}, & \text{if } i \neq j \\ 0, & \text{if } i = j \end{cases}$$

where $d_s(v_i, v_j)$ is the minimum distance between (v_i, v_j).

In a Hamacher d-fuzzy star graph, $d_s(v_i, v_j) > \mu(v_i, v_j)$, i.e., the distance between any two vertices is greater than the edge membership function $\Longrightarrow D_s(\widetilde{G}_{Hd}) > A(\widetilde{G}_{Hd}) \Longrightarrow \sum \lambda_i > \sum \gamma_i$. Hence, $DE(\widetilde{G}_{Hd}) > E(\widetilde{G}_{Hd})$.

The illustration for the above theorems is given in Fig. 2. Consider a Hamacher d-fuzzy star graph on six vertices with a degree sequence of the form $d = \{5, 1, 1, 1, 1, 1\}$. Then the vertex membership function is given by $\sigma(v_i) = \frac{d(v_i)}{n}$ for $i = 1, \ldots, n$, i.e., $\sigma(v_1) = \frac{5}{6}, \sigma(v_2) = \frac{1}{6}, \ldots, \sigma(v_6) = \frac{1}{6}$. Similarly, the edge membership function is found using the Hamacher product given by $\mu(v_i, v_j) = \frac{\sigma(v_i)\sigma(v_j)}{\sigma(v_i)+\sigma(v_j)-\sigma(v_i)\sigma(v_j)}$, i.e., $\mu(v_i, v_j) = \frac{5}{31}$ for $i, j = 1, 2, 3, 4, 5, 6$. The adjacency matrix of $\widetilde{G}_{Hd}$ is

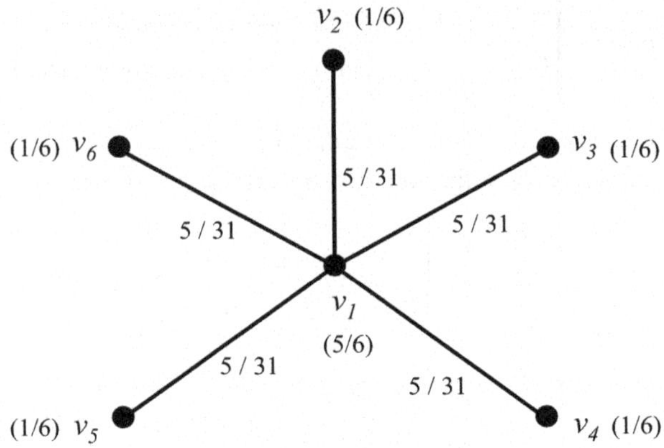

Fig. 2 Hamacher *d*-fuzzy star graph on six vertices

$$A(\widetilde{G}_{Hd}) = \begin{bmatrix} 0 & 5/31 & 5/31 & 5/31 & 5/31 & 5/31 \\ 5/31 & 0 & 0 & 0 & 0 & 0 \\ 5/31 & 0 & 0 & 0 & 0 & 0 \\ 5/31 & 0 & 0 & 0 & 0 & 0 \\ 5/31 & 0 & 0 & 0 & 0 & 0 \\ 5/31 & 0 & 0 & 0 & 0 & 0 \end{bmatrix}$$

Then, Spec($A(\widetilde{G}_{Hd})$) = {0.3607, -0.3607, 0, 0, 0, 0} $\Longrightarrow E(\widetilde{G}_{Hd}) = 0.7214$

The distance energy of $\widetilde{G}_{Hd}$ is

$$D_s(\widetilde{G}_{Hd}) = \begin{bmatrix} 0 & 5/31 & 5/31 & 5/31 & 5/31 & 5/31 \\ 5/31 & 0 & 10/31 & 10/31 & 10/31 & 10/31 \\ 5/31 & 10/31 & 0 & 10/31 & 10/31 & 10/31 \\ 5/31 & 10/31 & 10/31 & 0 & 10/31 & 10/31 \\ 5/31 & 10/31 & 10/31 & 10/31 & 0 & 10/31 \\ 5/31 & 10/31 & 10/31 & 10/31 & 10/31 & 0 \end{bmatrix}$$

Then, Spec($A(\widetilde{G}_{Hd})$) $= \{1.3843, -0.3226, -0.3226, -0.3226, -0.3226, -0.094\}$
$\Longrightarrow DE(\widetilde{G}_{Hd}) = 2.769$

The connectedness strength matrix of $\widetilde{G}_{Hd}$ is

$$C(\widetilde{G}_{Hd}) = \begin{bmatrix} 0 & 5/31 & 5/31 & 5/31 & 5/31 & 5/31 \\ 5/31 & 0 & 5/31 & 5/31 & 5/31 & 5/31 \\ 5/31 & 5/31 & 0 & 5/31 & 5/31 & 5/31 \\ 5/31 & 5/31 & 5/31 & 0 & 5/31 & 5/31 \\ 5/31 & 5/31 & 5/31 & 5/31 & 0 & 5/31 \\ 5/31 & 5/31 & 5/31 & 5/31 & 5/31 & 0 \end{bmatrix}$$

Then, $\text{Spec}(C(\widetilde{G}_{Hd})) = \{0.8065, -0.1613, -0.1613, -0.1613, -0.1613, -0.1613\}$ $\Longrightarrow CE(\widetilde{G}_{Hd}) = 1.613$. Therefore, from the above, it is clear that $CE(\widetilde{G}_{Hd}) > E(\widetilde{G}_{Hd})$ and $DE(\widetilde{G}_{Hd}) > E(\widetilde{G}_{Hd})$.

4 Conclusion

In this paper, the connectedness and distance energy for Hamacher d-fuzzy uniform graph and Hamacher d-fuzzy star graph are computed. Further work can be carried out by extending the notion of energy to other types of standard Hamacher d-fuzzy graphs and examining their properties.

References

1. Zadeh. L.A.: Fuzzy sets. Inform. Control **8**(3), 338–353 (1965)
2. Rosenfeld, A.: Fuzzy graphs. In: Zadeh, L.A., Fu, K.S., Shimura, M. (eds.) Fuzzy Sets and Their Applications, vol. 8, pp. 77–95. Academic Press, New York (1975)
3. Kalpana, K., Lavanya, S., Davvaz, B.: On defining fuzzy graph from degree sequence. J. Adv. Res. Dyn. Control Syst. **10**(1), 271–278 (2018)
4. Lavanya, S., Jeba Jesintha, J., Mahalakshmi, S.: On defining d-fuzzy graph using Hamacher product. In: National Conference on Mathematics and Computer Applications (2020)
5. Lavanya, S., Sharon Philomena, V., Malini, G.: Distance energy of a d-fuzzy graph. In: National Conference on Strategic Use of Technology in Mathematics and Computer Applications (2023)
6. Narayanan, A., Mathew, S.: Energy of a fuzzy graph. Ann. Fuzzy Math. Inform. **6**(3), 455–465 (2013)

Minimum Vertex Cover in Extended Fibonacci Cubes

R. Stalin Mary and Indra Rajasingh

1 Introduction

Fibonacci cubes are a classes of interconnection networks that are essential components of a distributed system for large-scale parallel computing. Based on the Fibonacci numbers, Hsu [7] proposed the Fibonacci cube of order n having F_n nodes, where F_n is the nth Fibonacci number and the labels for the nodes are binary strings of length $n - 2$ with no consecutive 1's. If the labels of two nodes differ in exactly one position, they are linked. Fibonacci cube may be considered to have been obtained from a hypercube after a few nodes of the hypercube become faulty.

Wu [8] proposed a series of extended Fibonacci cubes $\Gamma_n^k, k \geq 1$ using a recursive definition. The number of vertices in Γ_n^k is $2^k F_{n-k}$, where $F_1 = 1, F_2 = 1, F_3 = 2, \ldots, F_n = F_{n-1} + F_{n-2}$ [9]. The initial conditions of Γ_n^k are Q^k and Q^{k+1}, where Q^k is a hypercube of dimension k. Each vertex of Γ_n^k has an $(n-2)$-binary tuple representation where $k+1$ consecutive 1's can exist only in the extreme entries of the binary representations. These are also the classes of incomplete hypercubes. They are available in a wide range of sizes [10].

A vertex cover is a set S of vertices of G that includes at least one end vertex of each edge of G. The minimum cardinality of all the vertex covers of G is the vertex covering number of G, denoted by $\beta(G)$. Finding a minimum vertex cover is the vertex cover problem and is an NP-complete problem [3]. In coding theory, computer science, operations research, switching circuits, and electrical networks, the covering and inverse covering parameters are crucial [1]. Wireless sensor networks and network security are two areas where this idea is widely

R. Stalin Mary (✉) · I. Rajasingh
School of Advanced Sciences, Vellore Institute of Technology, Chennai, Tamil Nadu, India

I. Rajasingh et al. (eds.), *Applied Computational Mathematics*,
Springer Proceedings in Mathematics & Statistics 480,
https://doi.org/10.1007/978-3-031-77764-6_10

applied. In this paper, we have obtained the vertex covering number of Γ_n^k by studying certain aspects of the structure of Γ_n^k.

2 Preliminaries

In this section, we give the definitions necessary for our discussion.

Definition 1 ([3]) Let $G = (V, E)$ be a graph. A set $S \subset V$ is a vertex cover of E if any edge of G is incident to a vertex in S. The cardinality of a minimum vertex cover among all the vertex covers of G is the vertex covering number of G and is denoted by $\beta(G)$.

Definition 2 ([5]) An edge cover is a set of edges of G that contain as endpoints every vertex of G. The cardinality of a minimum edge cover of a graph G is the edge covering number of G and is denoted by $\beta^{'}(G)$.

Definition 3 ([2]) Matching is a collection of independent edges in G denoted by $M(G)$. A matching is said to be perfect if every vertex of G is an end vertex of exactly one edge in $M(G)$.

Definition 4 ([2]) A matching $M(G)$ is said to be maximum if for any other matching $M^{'}(G)$, $|M(G)| \geq M^{'}(G)$. $M(G)$ is referred to as a maximum-cardinality matching.

Theorem 1 (König's-Egerváry Theorem [4]) *If G is a bipartite graph, then the cardinality of a maximum matching is equal to the vertex covering number of G.*

Theorem 2 ([5]) *A perfect matching in a graph G is a minimum-size edge cover of the graph.*

Definition 5 ([8]) The series of extended Fibonacci cubes $\Gamma_n^k, k \geq 1$ is defined as $\Gamma_n^k = (V_n^k,\ E_n^k),\ n > k+1$, where $V_n^k = 0 \parallel V_{n-1}^k \cup 10 \parallel V_{n-2}^k, n \geq k+4$ and two nodes are connected by an edge in Γ_n^k if their binary representations differ in exactly one position. The initial conditions are $V_{k+2}^k = \{0, 1\}^k$, $V_{k+3}^k = \{0, 1\}^{k+1}$, and $\{0, 1\}^k$ denotes the set of binary strings of length k.

Theorem 3 ([8, 10]) *Let Γ_n^k be the n dimensional extended Fibonacci cube $k \geq 1$. Then the following results hold:*

- *The number of vertices in Γ_n^k is $2^k F_{n-k}$.*
- *The diameter of Γ_n^k is $n-2$.*
- *The degree $g(u)$ of a node u in Γ_n^k satisfies $\lceil \frac{n-(k-1)}{3} \rceil + (k-1) \leq g(u) \leq n-2$.*
- *$\Gamma_n^{n-2} = Q^{n-2}$ for $n > 2$.*
- *$\Gamma_n = \Gamma_n^0 \subset \Gamma_n^1 \subset \Gamma_n^2 \subset \ldots \subset \Gamma_n^{n-2} = Q^{n-2}$*
- *Γ_n^k is bipartite.*
- *$\Gamma_n^k, k \geq 1,\ n \geq k+4$ is hamiltonian.*

Remark 1 Γ_n^k contains Γ_{n-1}^k and Γ_{n-2}^k as subgraphs. Further Γ_{n-1}^k contains a copy of Γ_{n-2}^k, which we shall denote by $\Gamma_{n-2}^{k^*}$. The edges between Γ_{n-2}^k and $\Gamma_{n-2}^{k^*}$ in Γ_n^k constitute a perfect matching.

3 Ladder Is a Spanning Subgraph of Γ_n^k

We can redraw Γ_n^k leading to different representations such as the ladder representation, grid representation, and the cube representation. The different representations of Γ_7^1 are shown in Fig. 1. We use these different representations as per the requirement of the various problems. In this section, we denote Γ_n^1 as $G(n)$.

Definition 6 ([6]) The Ladder graph on $2n$ vertices is denoted by L_n and is defined as $P_2 \times P_n$.

Theorem 4 *$G(n)$ contains the ladder $L_{F_{n-2}}$ as a spanning subgraph.*

Proof Let $n = 5$. Then $G(5)$ itself is a ladder on six vertices. Assume the result to be true for all $G(i)$, $i \leq r - 1$. Consider $G(r)$. $G(r)$ has $G(r-1)$ and $G(r-2)$ as subgraphs. By induction hypothesis, $G(r-1)$ contains a ladder on $2F_{r-2}$ vertices which we denote by $L(r-1)$, and $G(r-2)$ contains a ladder on $2F_{r-3}$ vertices which we denote by $L(r-2)$. Now $L(r-2)$ is also a subgraph of $G^*(r-2)$ which we denote by $L^*(r-2)$. Further, $L^*(r-2)$ is isomorphic to $L(r-2)$, and the corresponding vertices of $L^*(r-2)$ and $L(r-2)$ form a perfect matching. Hence, one rung of $L^*(r-2)$ with both its end vertices a, b of degree 2 each in $L^*(r-2)$ will be adjacent to their corresponding vertices x, and y of degree 2 in $L(r-2)$ which are the end vertices of a rung in $L(r-2)$. See Fig. 2. Add the edges (a, x) and (b, y) to get a spanning ladder $L(r)$ of $G(r)$. The edges (a, x) and (b, y) combine $L(r-1)$ and $L(r-2)$ to get $L(r)$ on $2F_{n-1}$ vertices isomorphic to $L_{F_{n-1}}$.

Similarly, the following result holds good for Γ_n^k, $k \geq 1$, $n \geq k+4$. See Table 1.

Lemma 1 *Γ_n^k contains the ladder $L_{2^k F_{n-k}}$ as a spanning subgraph*

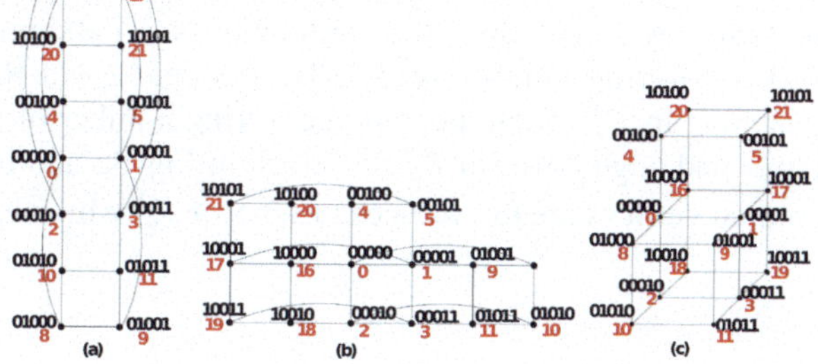

Fig. 1 Γ_7^1-(**a**) Ladder representation; (**b**) grid representation; (**c**) cube representation

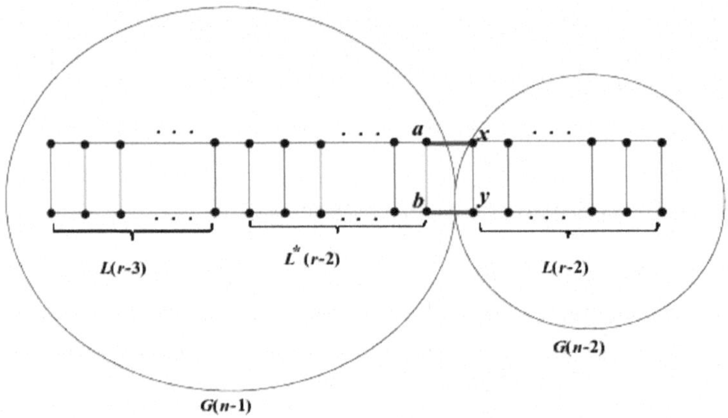

Fig. 2 Ladder is a spanning subgraph of $G(n)$

Table 1 Ladder $L_{F_{n-2}}$ as subgraphs of Γ_n^k

kn	1	2	3	4	5	6	7	8	9	10	11
Γ_n^1	–	–	L_1	L_2	L_3	L_5	L_8	L_{13}	L_{21}	L_{34}	L_{55}
Γ_n^2	–	–	–	L_2	L_4	L_6	L_{10}	L_{16}	L_{26}	L_{42}	L_{68}
Γ_n^3	–	–	–	–	L_4	L_8	L_{12}	L_{20}	L_{32}	L_{52}	L_{84}
Γ_n^4	–	–	–	–	–	L_8	L_{16}	L_{24}	L_{40}	L_{64}	L_{104}
Γ_n^5	–	–	–	–	–	–	L_{16}	L_{32}	L_{48}	L_{80}	L_{128}
Γ_n^6	–	–	–	–	–	–	–	L_{32}	L_{64}	L_{96}	L_{160}
Γ_n^7	–	–	–	–	–	–	–	–	L_{64}	L_{128}	L_{192}

The following results are easy consequences of Theorem 4.

Theorem 5 *Γ_n^k is bipancyclic for all $n \geq k + 4$.*

Corollary 1 ([8]) *Γ_n^k is hamiltonian for all $n \geq k + 4$.*

4 Vertex Covering in Γ_n^k

In this section, we determine the vertex covering number of Γ_n^k. Since Γ_n^k is bipartite, by Theorem 1, the cardinality of a maximum matching is equal to the vertex covering number of Γ_n^k. By Lemma 1 and Remark 1, there is a perfect matching in Γ_n^k. Then the vertex covering number of Γ_n^k is $2^{k-1}F_{n-k}$. The vertex cover and edge cover in Γ_8^1 are given in Fig. 3a and b, respectively, and the blue-colored vertices are the vertices in set S of Γ_8^1 which is a vertex cover of Γ_8^1.

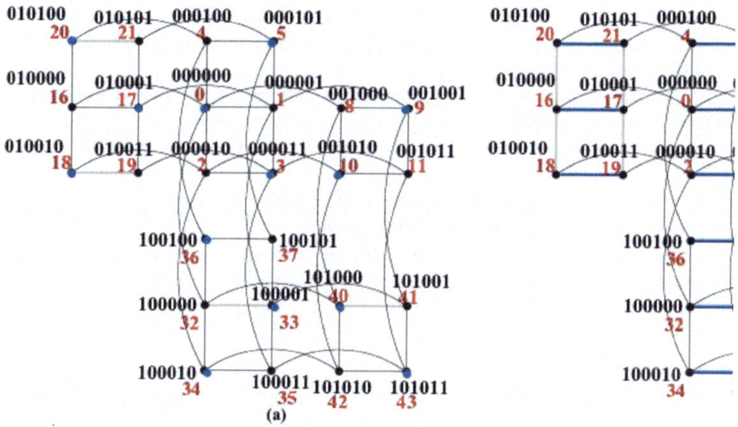

Fig. 3 (**a**) Circled vertices form a vertex cover in Γ_8^1; (**b**) edges in blue form an edge cover in Γ_8^1

Theorem 6 *The vertex covering number of Γ_n^k is given by*

$$\beta(\Gamma_n^k) = 2^{k-1}F_{n-k}$$

Theorem 7 *The edge covering number of Γ_n^k is given by*

$$\beta'(\Gamma_n^k) = 2^{k-1}F_{n-k}$$

5 Conclusion

In this paper, we find vertex covering number and edge covering number of Γ_n^k. Further, several other graph parameters like forcing number, burning number, and domination number are yet to be explored for Γ_n^k.

References

1. Bhat, P.G., Bhat, S.R.: Inverse independence number of a graph. Int. J. Comput. Appl. **42**(5), 9–13 (2012)
2. Bondy, J.A., Murty, U.S.R., et al.: Graph Theory with Applications, vol. 290. Macmillan, London (1976)
3. Cormen, T.H., Leiserson, C.E., Rivest, R.L., Stein, C.: 15.2: Matrix-chain multiplication. In: Introduction to Algorithms, pp. 331–338, 2nd edn. MIT Press/McGraw-Hill, Cambridge/New York (2001). ISBN 978-0-262032-93-3
4. Deming, R.W.: Independence numbers of graphs-an extension of the König-Egerváry theorem. Discrete Math. **27**(1), 23–33 (1979)

5. Gibbons, A.: Algorithmic Graph Theory. Cambridge University Press, Cambridge (1985)
6. Hosoya, H., Harary, F.: On the matching properties of three fence graphs. J. Math. Chem. **12**(1), 211–218 (1993)
7. Hsu, W.J.: Fibonacci cubes-a new interconnection topology. IEEE Trans. Parall. Distribut. Syst. **4**(1), 3–12 (1993)
8. Wu, J.: Extended Fibonacci cubes. IEEE Trans. Parall. Distribut. Syst. **8**(12), 1203–1210 (1997)
9. Zelina, I., Hajdu-Macelaru, M.A., Ticala, C.R.: About the cube polynomial of extended fibonacci cubes. Creat. Math. Inf. **1**, 95–100 (2018)
10. Zelina, I., Moldovan, G., Tascu, I.: On embeddings of hamiltonian paths and cycles in extended fibonacci cubes. Am. J. Appl. Sci. **5**(11), 1605–1610 (2008)

Power Dominator Sum Coloring of Graphs

Swati Kharub, Sharmila Mary Arul, and Diana Grace Thomas

1 Introduction

This paper presents the concept of power dominator sum coloring, which is a type of proper vertex coloring in graphs. This coloring scheme guarantees that each vertex power dominates every other vertex within a color class, while also minimizing the sum of labels assigned to all vertices.

Power dominator sum coloring finds applications in various fields where efficient graph coloring is crucial. In network design, particularly in wireless sensor networks, power dominator sum coloring helps minimize interference and optimize communication, ensuring that each node can effectively communicate with its neighbors. Resource allocation in systems where resources are distributed based on connectivity benefits from this coloring scheme, ensuring fair and efficient resource utilization. Task scheduling in parallel or distributed computing environments benefits from power dominator sum coloring by minimizing conflicts and optimizing task assignments, leading to improved overall system performance. Additionally, in VLSI design, this coloring technique can help reduce power consumption and enhance the efficiency of the design. These applications demonstrate the versatility and effectiveness of power dominator sum coloring in various practical scenarios.

S. Kharub · S. M. Arul (✉) · D. G. Thomas
Department of Mathematics, Saveetha School of Engineering, SIMATS, Chennai, India

Faculty of Mathematics, Saveetha Institute of Medical and Technical Sciences, Chennai, India
e-mail: sharmilamaryarul.sse@saveetha.com

I. Rajasingh et al. (eds.), *Applied Computational Mathematics*,
Springer Proceedings in Mathematics & Statistics 480,
https://doi.org/10.1007/978-3-031-77764-6_11

2 Overview of the Paper

Power domination, first introduced by Cockayne, Hedetniemi, and Mynhardt in 1998 [1], originally stemmed from the need to monitor electric power systems [2]. Harary and Haynes expanded on this idea in 2003. The concept of power domination has found application in modeling trash collection routes to reduce scheduling conflicts. This application involves the challenge of six-coloring the graph to assign workdays to routes [3–5]. Chromatic sums and sum coloring were first introduced by Supowit in 1987 using non-graph-theoretic terminology [6]. They were later independently studied in graph theoretic terms by Ewa Kubicka in her 1989 doctoral thesis [7]. Combining power dominator coloring and sum coloring, we introduce the concept of power dominator sum coloring. The power dominator sum chromatic number $\chi_{pdsum}(G)$ of a graph G is the minimum number of colors needed for this coloring and vertices are labeled in such a way that the sum denoted by $\sum_{pdsum}(G)$ is minimum.

3 Preliminaries

Before proceeding to the topic, we first go over the terminology under discussion: We now give a formal definition of power dominator sum coloring of a graph.

Definition 1 Let $v \in V(G)$; the monitoring set $M(v)$ is constructed as follows:

Step 1: $M(v) = N(v) \cup \{v\}$ where $N(v)$ is the open neighborhood of the vertex v.
Step 2: Let $u \in V(G) \backslash M(v)$. Add the vertex u to $M(v)$ whenever u has a neighbor $w \in M(v)$ such that all the neighbors of w other than u are already in $M(v)$.
Step 3: Repeat Step 2 till no more vertices be added to $M(v)$.

We say that the vertex v power dominates the vertices in $M(v)$.

Definition 2 A power dominator coloring of a graph G is a proper coloring, such that each vertex of a graph G power dominates all the vertices of at least one-color class.

Definition 3 The sum coloring problem involves finding a way to color the vertices of a given graph G using $C_1, C_2, C_3, \ldots C_k$ in such a way that the total sum of the labels for the colors of all vertices is as small as possible. An optimal coloring that achieves this minimum total sum is called an optimum coloring. The minimum number of colors required in any optimum coloring of the graph is referred to as the strength of the graph and sometimes called the chromatic number denoted by $\chi_{sum}(G)$. This smallest possible total sum of colors in a coloring of G is known as the chromatic sum of G, and is denoted by $\sum_{sum}(G)$.

Definition 4 The power dominator sum coloring is a proper coloring satisfying power dominator coloring as well as minimizing the sum of labels for the colors of all vertices. The power dominator sum chromatic number is denoted by $\chi_{pdsum}(G)$ and the smallest possible total sum of colors in a coloring of G known as the chromatic sum is denoted by $\sum_{pdsum}(G)$.

Theorem 1 *We have the following results:*

$$(a)\ \chi_{pdsum}(P_n) = 2, \sum_{pdsum}(P_n) = \begin{cases} \frac{3n}{2}, & if\ n\ is\ even,\ for\ n \geq 3 \\ \frac{3n-1}{2}, & if\ n\ is\ odd,\ for\ n \geq 4 \end{cases}$$

$$(b)\ \chi_{pdsum}(C_n) = \begin{cases} 3, & if\ n\ is\ odd,\ for\ n \geq 3 \\ 2, & if\ n\ is\ even,\ for\ n \geq 4 \end{cases}, \sum_{pdsum}(C_n) = \lceil \tfrac{n}{2} \rceil$$

$$(c)\ \chi_{pdsum}(K_n) = n, \sum_{pdsum}(K_n) = \frac{n(n+1)}{2}\ for\ n \geq 3$$

$$(d)\ \chi_{pdsum}(W_n) = \begin{cases} 3, & if\ n\ is\ odd,\ for\ n \geq 3 \\ 4, & if\ n\ is\ even,\ for\ n \geq 4 \end{cases},$$

$$\sum_{pdsum}(W_n) = \begin{cases} \frac{3(n+1)}{2}, & if\ n\ is\ odd,\ for\ n \geq 3 \\ \frac{3n}{2} + 4, & if\ n\ is\ even,\ for\ n \geq 4 \end{cases}$$

$$(e)\ \chi_{pdsum}(S_n) = 2, \sum_{pdsum}(S_n) = n + 1$$

$$(f)\ \chi_{pdsum}(ST(m,n)) = 3, \sum_{pdsum}(ST(m,n)) = m + n + 5\ for\ m, n \geq 3.$$

$$(g)\ \chi_{pdsum}(H_n) = n + 1, \sum_{pdsum}(H_n) = \frac{n^2+5n+2}{2}.\ for\ n \geq 3$$

$$(h)\ \chi_{pdsum}(B_n) = 3, \sum_{pdsum}(B_n) = 3n + 4.\ for\ n \geq 3$$

$$(i)\ \chi_{pdsum}(PN_n) = 3,\ for\ n \geq 3, \sum_{pdsum}(PN_n) = \begin{cases} \frac{3n+4}{2}, & if\ n\ is\ even,\ for\ n \geq 3 \\ \frac{3n+5}{2}, & if\ n\ is\ odd,\ for\ n \geq 3 \end{cases}$$

Proof

(i) Consider G be a path graph P_n with vertices $v_1, v_2, \ldots, v_n$. Assign colors 1 and 2 alternately to the vertices of G. Each vertex v_i (odd index) power dominates the color class 2, and each vertex v_i (even index) power dominates the color class 1. Thus the result.

(ii) Consider G be a cycle graph C_n with vertices $v_1, v_2, \ldots, v_n$. Assign colors 1 and 2 alternately to the vertices of G. Each vertex v_i(odd index) power dominates all the vertices of color class 2, and each vertex v_i (even index) power dominates all the vertices of the color class 1. Hence the proof.

(iii) Consider G be a complete graph K_n with vertices $v_1, v_2, \ldots, v_n$. Assign distinct colors 1, 2, ..., n to the vertices of G. Each vertex v_i power dominates its own color class, ensuring a power dominator sum coloring with $\chi_{pdsum}(K_n) = n$ and hence the sum of labels.

(iv) Consider G be a wheel graph W_n with $n \geq 4$. The vertices of W_n consist of a cycle with $(n - 1)$ vertices $(v_1, v_2, .., v_{(n-1)})$ and a central vertex (hub) denoted as v_n. When n is even, assign color 4 to the central vertex v_n, color 3 to v_1, and colors 1,2,1,2... alternatively to remaining $(n - 2)$ vertices respectively. Since v_n is in the monitoring set for all vertices, hence every vertex power dominates color class 4 and by adding labels we can find the sum. When n is odd, assign color 3 to the central vertex v_n and give color 1,2,1,2, ... alternatively to $v_1, v_2, .., v_{(n-1)}$; this assignment ensure that each vertex power dominates color class 3 and sum can be calculated by adding the given labels. Hence the proof.

(v) Consider G be a star graph S_n with $n \geq 4$. The vertices of S_n consist of a central vertex v_n which is adjacent to remaining $(n - 1)$ vertices $(v_1, v_2, .., v_{(n-1)})$. Now, put vertex v_n in single color class 2 and put the remaining $(n - 1)$ vertices in color class 1. Since v_n is in the monitoring set for all vertices hence all the vertices power dominates color class 2 and adding the labels we get the sum. Hence the proof.

(vi) Consider G be a Bistar graph $ST(m, n)$ with $m, n \geq 4$. The vertices of $ST(m, n)$ consist of a two adjacent central vertex v, u, $(v_1, v_2, .., v_m)$ adjacent to v and $(u_1, u_2, .., u_n)$ adjacent to u. Assign color class 1 to $(v_1, v_2, .., v_m, u_1, u_2, .., u_n)$, color class 2 to v, and color class 3 to u and adding the labels we get the sum. Hence the proof.

(vii) Consider G be a helm graph H_n with $n \geq 3$ described as the graph formed by adding a pendant edge to each node of an n-wheel graph. Assign color 1 to the central and the pendent vertices, and then assign colors $2, 3, 4, \ldots, n + 1$ to the remaining n vertices on the nodes. Adding the labels for each vertex we can find the sum. Since, every vertex is having at least one node in their monitoring set and we have given single color classes to the nodes so, the condition for power dominator coloring is satisfied. Hence the proof.

(viii) Consider G be a book graph B_n with $n \geq 3$ with spine vertices as v_1, v_2 and remaining $2n$ vertices on n leaves. We give single color class 3 to v_1, color class 1 to v_2, and the remaining are given v_1, v_2 to the adjacent vertices for each leaf. Since both v_1, v_2 are in the monitoring set of all vertices, hence the power dominator coloring is satisfied and finding the sum of labels will give us the value of $\sum_{pdsum}$.

(ix) Consider G be a pan graph PN_n with $n \geq 3$ with $v_{(n+1)}$ as the pendent vertex adjacent vertex v_n of the cycle graph C_n. Assign single color class 3 to v_n; put $v_{(n+1)}$ in color class 1; and starting with v_1, assign alternate color class 1 and 2 to remaining $(n - 1)$ vertices. Since, v_n is there in the monitoring set for all the vertices and giving labels in this manner will minimize the sum so the condition of power dominator sum coloring is satisfied. Hence the proof.

4 Proof of Correctness

In the context of power dominator sum coloring, each vertex must power dominate all the vertices within at least one-color class. Reducing the number of color classes would violate this condition. Hence, the power dominator sum chromatic number $\chi_{pdsum}(G)$ is the minimum number of colors required to satisfy this condition and the vertices are given labels in such a way that the sum of labels $\sum_{pdsum}$ is minimum for the given graph.

5 Conclusion

The power dominator sum coloring algorithm presents a methodical way to color graphs, ensuring that both the power dominator and sum coloring conditions are satisfied. The coloring technique described in this paper can also be applied to solve a broader range of graphs.

References

1. Cockayne, E.J., Hedetniemi, S.T.: Towards a theory of domination in graphs. Networks **7**, 247–261 (1977)
2. Haynes, T.W., Hedetniemi, S.M., Hedetniemi, S.T., Henning, M.A.: Domination in graphs applied to electric power networks, SIAM J. Discrete Math. **15**(4), 519–529 (2002)
3. Kumar, K.S., Gnanamalar, D.N., Subramanian, K.G.: Graphs and power dominator colorings. Ann. Pure Appl. Math. **11**(2), 67–71 (2016)
4. Kumar, K.S., Gnanamalar, D.N., Subramanian, K.G.: Power dominator coloring of corona graphs. Int. J. Pure Appl. Math. **118**(23), 59–67 (2018)
5. Uma Maheswari, A., Bala Samuvel, J., Azhagarasi, S.: Power dominator coloring of special kind of graphs. Kala Sarovar **23**(04)(X1), 58–62 (2020)
6. Supowit. K. J.: Finding a maximum planar subset of a set of nets in a channel. IEEE Trans. Comput.-Aided Design Integr. Circ. Syst. **6**(1), 93–94 (1987)
7. Kubicka, E.M.: The Chromatic Sum and efficient tree algorithms. Ph.D. Thesis, Western Michigan University (1989)

Roman Domination Number for Fractal Cubic Networks

S. Arulanand, R. Sundara Rajan, and T. M. Rajalaxmi

1 Introduction

When the Roman Empire came under siege in the nineteenth century AD, Emperor Constantine had to decide where to position his four field army divisions to protect eight districts. His plan was to arrange the army units, so that, in the event of a battle, one of them could be sent straight to the unguarded area, allowing each zone to be either secured by its own army or by an adjacent army consisting of two troops. Constantine made the decision to post two army battalions in Rome and another two in his new city, Constantinople. Therefore, reaching Britain by yourself in a single step was not doable. In actuality, Constantine's heirs lost authority over Britain. The foremost goal is to deploy a minimal number of legions overall. This challenge is highlighted in Cockayne et al. [1], where they introduced a Roman Dominating mapping based on events that transpired in the nineteenth century AD. A mapping f from the node set to the values 0, 1, 2 that satisfies the requirement that each node u for having value zero is connected to at least one node v for which having the value two is considered a Roman domination on a network G. A Roman dominating function's weight can be expressed as the addition of all the node's weights. The Roman domination number of a graph G is the lowest weight of a function on Roman domination of G. Cockayne et al. perceived that the Roman domination always lies between the domination and twice the domination number for any network G, where $\gamma(G)$ symbolizes the domination number, and further

S. Arulanand (✉) · R. S. Rajan
Department of Mathematics, Hindustan Institute of Technology and Science, Chennai, Tamil Nadu, India

T. M. Rajalaxmi
Department of Mathematics, Sri Sivasubramaniya Nadar College of Engineering, Chennai, Tamil Nadu, India

I. Rajasingh et al. (eds.), *Applied Computational Mathematics*,
Springer Proceedings in Mathematics & Statistics 480,
https://doi.org/10.1007/978-3-031-77764-6_12

they have characterized trees for which $\gamma_R(G) \leq \gamma(G)+2$, and if a graph is Roman graph (or Roman), then $\gamma_R(G) = 2\gamma(G)$.

Let $G = (V, E)$ be a network, a mapping $\psi : V \to \{0, 1, 2\}$. Let V_0, V_1, and V_2 be the set of nodes under ψ that have been given the values 0, 1, and 2, respectively. The ordered partitions of (V_0, V_1, V_2) and the mapping $\psi : V \to \{0, 1, 2\}$ have a one-to-one correspondence. Thus, $\psi = (V_0, V_1, V_2)$. The formulation of this RDF definition was inspired by an article written by Ian Stewart "Defend the Roman Empire!" [2]. The results were achieved using a type of zero-one integer programming as described by ReVelle and Rosing [3]. Similar challenges are beneficial in deciding the optimal placement for a new hospital, gas station, or fast-food outlet in a community, as well as determining the strategic positioning of security in a specific area. Roman domination or one of its derivatives is often applied to represent and address such optimization problems. The exploration of Roman domination within network contexts has been examined in [4–8, 14, 15].

The names $V(G)$ and $E(G)$ called as lines and nodes of an undirected simple network G, $N(c) = \{d \in V(G)|cd \in E(G)\}$ represents the open neighbor of node c in $V(G)$, and $N[c] = N(C) \cup \{c\}$ is the closed neighbor of the node c. *link* in a graph is an edge characterized by having distinct endpoints, contrasting with a *loop* where the ends are identical. When two or more linkages share the same pair of endpoints, they are referred to as *parallel edges*. In a graph G, the *degree* (or valence) of a vertex is denoted by $deg_G(v)$. If every node v in $V(G) \setminus D$ is adjacent to at least one node in D, then a set D of nodes in a graph G is considered a dominating set of G. For more results on domination, see [9].

2 Fractal Cubic Networks

Enabling communication among a significant quantity of consistently duplicated computer memory sets, commonly referred to as nodes or processors, frequently necessitates the use of multiprocessor interconnection networks. Message passing is a prevalent method employed to achieve comprehensive synchronization and transmission among processors, enabling programmed execution instead of relying on shared memory. There has been a discernible enthusiasm for the creation and execution of multiprocessor interconnection networks, driven by the availability of affordable yet potent microchips and remembrance pieces. The advancement of central processing units (CPUs), interconnection networks, and transmitting procedures stands as the focal areas of research in mainframes and extremely equivalent computing structures. The interconnection network, which involves the links connecting lots of computers, holds a crucial role in the evolution of a supercomputer [10]. A commonly used interconnection network topology, the hypercube, exhibits several noteworthy traits, containing orderliness, stability, straightforward routing, strong connectivity, and a continuous architecture. In current spans, thorough research has been undertaken on diverse properties of hypercubes [11].

While numerous studies have explored variations of the hypercube mentioned earlier, one of the abovementioned hypercube variants is called the fractal cubic network introduced in [12]. For this network, so far the Roman domination problem has not been investigated. This inspiration and motivation prompted us to determine the Roman domination problem on FCN.

The formation of FCN(r) given in [12] is as follows: For $r \geq 0$, the fractal cubic network FCN(r) $= (V_1(r), E_1(r))$ has the node and line set, for FCN(0) is the four-cycle with the nodes 00, 10, 11, 01.

More specifically, FCN(0) is equal to Q_2. If $r \geq 1$, FCN(r) = 11 concatenation of FCN($r-1$) union 01 concatenation of FCN^{-1}($r-1$) union 00 concatenation of FCN($r-1$) union 10 concatenation of FCN^{-1}($r-1$), where FCN^{-1}($r-1$) is reverse order labeling of FCN($r-1$), then $V_1(r)$ = 11 concatenation of $V_1(r-1)$ union 01 concatenation of $V_1(r-1)$ union 10 concatenation of $V_1(r-1)$ union 00 concatenation of $V_1(r-1)$, and $E_1(r)$ = 11 concatenation of $E_1(r-1)$ union 01 concatenation of $E_1(r-1)$ union 10 concatenation of $E_1(r-1)$ union 00 concatenation of $E_1(r-1)$ union E' where $E' = \{(e_i, e_j) | e_i = ab$ and $e_j = cd$ and $(ab) \oplus (cd) = 1$, the label of e_i and e_j is same except ab of e_i and cd of $e_j\}$. The eternal edges of the ith order can be acquired through any of the following conditions.

(a) $a_{2i} \oplus b_{2i} = 1$ and $a_{2i+1} \oplus b_{2i+1} = 0, 1 \leq i \leq r$
(b) $a_{2i} \oplus b_{2i} = 0$ and $a_{2i+1} \oplus b_{2i+1} = 1, 1 \leq i \leq r$.

The fractal cubic network FCN(0) and FCN(1) is shown in Fig. 1. The Eulerian and Hamiltonian routing properties of the fractal cubic network are studied in [12]. The discussions on the metric dimension and defect-accepting metric dimension for FCN can be found in [13]. In this paper, we calculate the Roman domination for FCN.

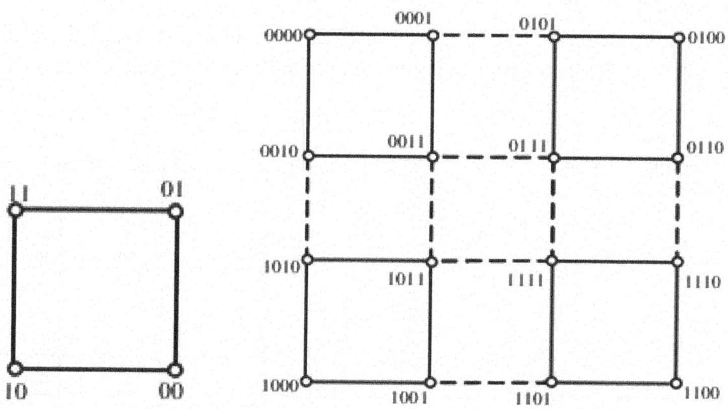

Fig. 1 (**a**) FCN(0); (**b**) FCN(1)

3 Main Results

To comprehend the proof technique, we require the upcoming results.

Theorem 1 *Let G be any network, then* $\gamma(G) \geq \lceil \frac{n}{1+\Delta} \rceil$.

Theorem 2 *Let G be a network of order n with highest degree* Δ*; then we have the following:*

(*i*) $\gamma_R(G) \geq \lceil \frac{2n}{\Delta+1} \rceil$.
(*ii*) $\gamma(G) \leq \gamma_R(G) \leq 2\gamma(G)$.
(*iii*) *Let G be any graph G, then* $\gamma(G) = \gamma_R(G)$ *if and only if* $G = \overline{K_n}$, *where* $\overline{K_n}$ *is a complement graph of* K_n.
(*iv*) *G is Roman if and only if* $\gamma_R(G) = 2\gamma(G)$.

Theorem 3 *Let G be the r-dimensional fractal cubic network FCN(r),* $r \geq 1$, *then* $\gamma(G) = 4^r$.

Proof We establish this by mathematical induction on r. When $r = 1$, by Theorem 1, $\gamma(\text{FCN}(1)) \geq \lceil \frac{16}{5} \rceil = 4$. To prove the other inequality, let $D_1 = \{u_1, u_2, u_3, u_4\} = \{0001, 0110, 1101, 1010\}$ be the dominating set of FCN(1), then by Hamming distance, $\bigcup_{i=1}^{4} N[u_i] = V(\text{FCN}(1))$ and $N[u_i] \cap N[u_j] = \emptyset$ for all i, j and $i \neq j$. Thus, $\gamma(\text{FCN}(1)) \leq 4$ and hence $\gamma(\text{FCN}(1)) = 4$. In Fig. 2, the dominating vertices of FCN(1) are colored in black.

By rotational symmetry, we get another minimum dominating set $D'_1 = \{u'_1, u'_2, u'_3, u'_4\} = \{0101, 1110, 1001, 0010\}$ in FCN(1), but it is the translation of D_1. Now to prove the uniqueness, we suppose consider a minimum dominating set $D''_1 = \{u''_1, u''_2, u''_3, u''_4\}$, then we have the following four cases. **Case** (*i*): If any one node in D''_1 is of degree 4, then at least one of the two-degree vertex (vertices) will not be dominated by D''_1.
Case (*ii*): If any one node in D''_1 is of degree 2, then at least one of the two-degree vertex (vertices) or four-degree vertex (vertices) will not be dominated by D''_1.

Fig. 2 Dominating vertices in FCN(1)

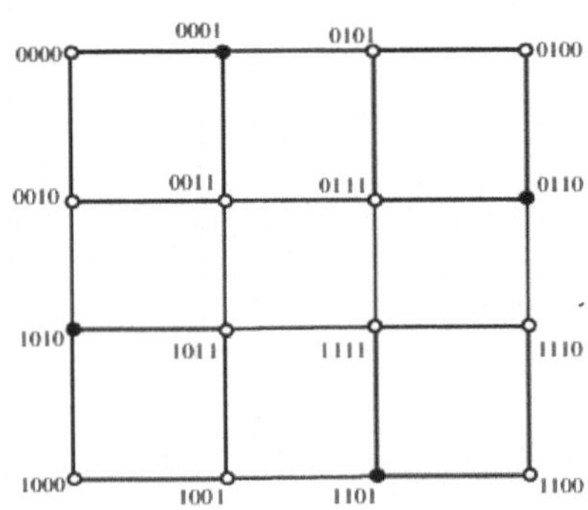

Case (iii): If any one node in D_1'' will be in D_1' and the remaining nodes will be in D_1, then at least one of the two-degree vertex (vertices) will not be dominated by D_1''. If all the four vertices in D_1'' will be in D_1', then $D_1'' = D_1'$.
Case (iv): If any one node in D_1'' will be in D_1' and the remaining nodes will not be in D_1, then at least one of the two-degree vertex (vertices) will not be dominated by D_1''.

By the above four cases, we get a contradiction to our assumption. Hence, the dominating set of FCN(1) is unique.

Now, let us assume that the result is true for $r = k$. That is, $\gamma(\text{FCN}(k)) = |D_k| = |D_k'| = 4^k$. Consider the fractal cubic network of dimension $k+1$. Since FCN$(k+1)$ contains four copies of FCN(k), say FCN$^1(k)$, FCN$^2(k)$, FCN$^3(k)$, and FCN$^4(k)$. Since the dominating set of FCN$^i(k)$ is unique for all $i \in [4]$ and by the construction of FCN$(k+1)$ from FCN(k), the dominating vertices in FCN$^i(k)$ will not dominate any vertex in FCN$^j(k)$, for all $i, j \in [4]$ and $i \neq j$. Hence, the dominating set of FCN$(k+1)$ is four times of FCN(k). Thus, $\gamma(\text{FCN}(k+1)) = 4(\gamma(\text{FCN}(k))) = 4 \times 4^k = 4^{k+1}$. Hence the theorem.

Corollary 1 *Let G be the r-dimensional fractal cubic network FCN(r), $r \geq 1$; then the Roman domination of G is given by $\gamma_R(G) = 2(4^r)$.*

Proof Since the dominating set D of FCN(r) is unique and $N[u_i] \cap N[u_j] = \emptyset$ for all $u_i, u_j \in D, i \neq j, 1 \leq i, j \leq 4^r$. Thus, all vertices in D will receive the weight two. Hence, $\gamma_R(\text{FCN}(r)) = 2(4^r)$.

For illustration, the Roman dominating vertices of FCN(2) are given in Fig. 3.

Remark Since the graph FCN(r), $r \geq 1$ satisfies the condition (iv) in Theorem 2, we say that FCN(r) is a Roman graph.

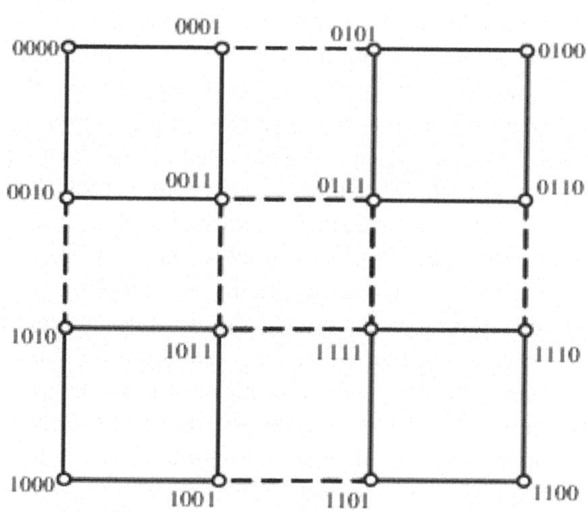

Fig. 3 The Roman dominating vertices in FCN(2)

4 Implementation

Roman domination, a concept in graph theory, involves the exploration of domination within graphs. A method for assigning labels or tokens to the nodes of a network is employed, ensuring that each node is either labeled (dominated) or adjacent to another labeled vertex. This results in the formation of a dominating set, where the labeled vertices collectively dominate the graph. The function responsible for this labeling or tokenizing process is known as Roman dominating mapping. The Roman domination parameter of a network is defined as the lowest value required for such a dominating set to be established.

Although Roman domination is a mathematical abstraction without direct, tangible applications in real life, its underlying principles can be employed in practical scenarios where the notions of domination or covering are pertinent. Here are a few instances where the concept of Roman domination can find practical utility.

Surveillance and Security Imagine a scenario involving a network of security cameras strategically placed in a building or public space. The objective is to optimize the coverage of every point in the area while minimizing the number of required cameras. This scenario can be represented as a network, where nodes symbolize potential camera locations and edges denote the coverage range of each camera. By applying principles of Roman domination, it becomes possible to determine the smallest number of cameras necessary to achieve comprehensive coverage.

Wireless Communication In the realm of wireless communication networks, it is essential to guarantee that every region falls within the coverage range of at least one access point. By representing this scenario as a graph, where vertices denote potential access point locations and edges indicate their signal range, the principles of Roman domination can be applied to determine the minimum number of access points needed for comprehensive coverage.

Resource Allocation In scenarios involving resource allocation, such as the distribution of medical supplies during a crisis, the goal is to diminish the number of distribution centers while ensuring comprehensive coverage of every affected area. This challenge can be translated into a graph representation, where nodes signify potential distribution centers and edges denote the reach or capacity of each center. Leveraging Roman domination principles can aid in devising an efficient distribution strategy for optimal resource utilization.

Network Design In the realm of computer or communication network design, there may be a preference to minimize the number of active nodes, such as routers, while ensuring that all devices maintain the ability to connect. The application of Roman domination concepts allows for the determination of the smallest number of active nodes required to achieve complete network connectivity.

In our scenario, the various regions, districts, or specific locations can be likened to a fractal cubic network. In this context, our identified dominating set nodes, strategically positioned as nodes in the network, play a crucial role in efficiently allocating facilities to those individuals who genuinely seek to benefit.

In the context of Roman domination, the regions or states within a country can be viewed as a fractal cubic network as well. The selected dominating set nodes, analogous to key strategic positions, can be utilized by the country's emperor to position army units strategically. This ensures the safety and protection of all regions in the country during times of war or any adverse situations.

5 Concluding Remarks

Interconnection networks play a pivotal role in the execution of parallel computing systems as they significantly influence their scalability. Communication efficiency stands out as a critical performance indicator in parallel computing, with the diameter of an interconnection network being a key metric for assessing communication effectiveness. Various interconnection network designs have been proposed to date. Hypercubes are widely favored in the realm of connectivity networks because of their advantageous features. Numerous hypercube variations exist, offering the possibility of modifying certain linkages. One such variation is the fractal cubic network. In this study, we have found the domination and Roman domination numbers for fractal cubic network $\mathrm{FCN}(r)$, $r \geq 1$, and also we have shown that it is a Roman graph. A similar problem for other cube-like networks is under investigation.

References

1. Cockayne, E.J., Dreyer, P.A., Hedetniemi, S.M., Hedetniemi, S.T.: Roman domination in graphs. Discrete Math. **278**(1), 11–22 (2004)
2. Stewart, I.: Defend the roman empite! Sci. Am. **281**(6), 136–138 (1999)
3. Revelle, C.S., Rosing, K.E.: Defendens imperium romanum: a classical problem in military strategy. Am. Math. Monthly **107**, 585–594 (2000)
4. Henning, M.A., Hedetniemi, S.T.: Defending the roman empire—a new strategy. Discrete Math. **266**(1), 239–251 (2003). The 18th British Combinatorial Conference
5. Xing, H.-M., Chen, X., Chen, X.-G.: A note on roman domination in graphs. Discrete Math. **306**(24), 3338–3340 (2006)
6. Arulanand, S., Sundara Rajan, R., Prabhu, S., Stephen, S.: Certain domination numbers for Cartesian product of graphs. J. Discrete Math. Sci. Cryptogr. **27**(3), 1045–1058 (2024)
7. Rajan, R.S., Arulanand, S., Prabhu, S., Rajasingh, I.: 2-power domi- nation number for knodel graphs and its application in communication networks. RAIRO-Oper. Res. **57**(6), 3157–3168 (2023)
8. Arulanand, S., Rajan, R.S., Prabhu, S.: 2-domination number for spe- cial classes of hypercubes, enhanced hypercubes and knodel graphs. International J. Netw. Virtual Organis. **29**(2), 168–183 (2023)
9. Arulanand, S., Rajan, R.S., Prabhu, S.: On 2-domination number in certain octahedral networks. Int. J. Netw. Virtual Organis. **27**(3), 251–266 (2022)

10. Akers, S.B., Krishnamurthy, B.: A group-theoretic model for symmetric intercon- nection networks. IEEE Trans. Comput. **38**(4), 555–566 (1989)
11. Harary, F., Hayes, J.P., Wu, H.-J.: A survey of the theory of hypercube graphs. Comput. Math. Appl. **15**(4), 277–289 (1988)
12. Karci, A., Selcuk, B.: A new hypercube variant: Fractal cubic network graph. Eng. Sci. Technol. Int. J. **18**(1), 32–41 (2015)
13. Arulperumjothi, M., Klavzar, S., Prabhu, S.: Redefining fractal cubic networks and determining their metric dimension and fault-tolerant metric dimension. Appl. Math. Comput. **452**, 128037 (2023)
14. Henning, M.A.: A characterization of roman trees. Discuss. Math. Graph Theory **22**(2), 325–334 (2002)
15. Henning, M.A.: Defending the roman empire from multiple attacks. Discrete Math. **271**(1), 101–115 (2003)

Analyzing the Factors Associated with the Severity of Nonalcoholic Fatty Liver Disease (NAFLD) Using Graph Theory and Matrix Approach Method

S. Teresa Arockiamary and E. Virgin Christilda

1 Introduction

The liver is a large organ located on the right side of the body, above the abdominal cavity, and directly beneath the diaphragm, top of the stomach, and ribcage. The human liver is enormous and has a reddish-brown hue. The liver is responsible for sustaining the body's health and efficiency. The liver performs several vital functions such as detoxifying foreign substances and secreting various enzymes required for different biochemical processes including protein synthesis, beta-oxidation of fatty acids, glycolysis, gluconeogenesis, and the tricarboxylic acid cycle [1]. Alcoholic liver disease (ALD) and nonalcoholic fatty liver disease (NAFLD) are both significant causes of chronic liver disease, and NAFLD is prevalent in at least 25% populations throughout the world. NAFLD can progress and lead to fibrosis of the liver, cirrhosis, hepatocellular carcinoma (HCC), and liver transplantation [2].

In the wider population, NAFLD is exceedingly typical, and in the years ahead, it is expected to grow increasingly prevalent. In Asia, NAFLD typically occurs in 30% of cases. Turkey has the greatest prevalence of NAFLD in Europe (48.4%), followed by Italy (38.2%). The widespread presence of NAFLD has been 25–27% in Germany, Portugal, and Spain. In North America, NAFLD is 35.3% prominent. In South America, 59% of people have NAFLD. 13.5% is considered to be the estimated average prevalence of NAFLD in Africa. It is reasonable and fair to anticipate an increase in the incidence of NAFLD given the rising rates of obesity, diabetes, and insulin resistance that have been witnessed in India over the last two decades. However, there is a dearth of awareness about the incidence of NAFLD in India.

S. T. Arockiamary · E. V. Christilda (✉)
Department of Mathematics, Stella Maris College, Affliated to the University of Madras, Chennai, India

I. Rajasingh et al. (eds.), *Applied Computational Mathematics*,
Springer Proceedings in Mathematics & Statistics 480,
https://doi.org/10.1007/978-3-031-77764-6_13

1.1 Literature Review

Graph theory matrix approach (GTMA) is a methodical and logical decision-making process. The matrix technique is effective in assessing directional graphs, especially with several nodes and complicated visualizations. The permanent function of a matrix encompasses all attribute structural components and their relative relevance, providing a comprehensive understanding of the selection problem [3]. The GTMA approach provides a simplified computing procedure as it incorporates interdependencies along with assessing the attributes and offers a graphical representation of the numerous criteria and their interrelations which makes it more advantageous than the other methods [4].

Diagnosing the patient using CART(classification and regression tree) and identifying the type of liver disease using CBR(case-based reasoning) paved the way for raising the precision of liver disease [5]. The strong association between NAFLD and CVD (cardiovascular disease) and the factors responsible were discussed. Some of the established risk factors of NAFLD along the Asia Pacific region are investigated. An efficient technique for determining fibrosis and cirrhosis is the C-methacetin breath test (MBT). It has been explored how epigenetic factors, environmental factors, certain risk factors, and genetic correlations with the development of NAFLD interact with medical problems and monetary burden.

1.2 Research Gaps

The identification, diagnosis, and treatment of NAFLD were the main priorities. There is no quantitative modeling of the risk variables and their interdependencies, and the risk factors that contributed to it were also limited to a handful of variables. To rate or quantify the variables linked to the severity of the NAFLD, none of the MCDM techniques were used. The GTMA, which is used in other disciplines, has yet to be utilized in any health-related issues. To alleviate these constraints, we provided the following:

(a) The proposed framework provides an enhanced context for examining the variables influencing liver disease and using triangular fuzzy numbers for data received in linguistic terms and defuzzifying them to achieve better outcomes.
(b) The rank of the relative importance of the factors taken into consideration is determined using the FAHP (fuzzy analytic hierarchy process).

1.3 Objectives and Motivation

Several key facts motivated us to undertake this current research work, and the objectives are described as follows.

(a) The widespread diseases due to lifestyle changes and post pandemic situations throughout the world.
(b) Applicability in several other diseases which can be also modeled similarly or using any of the MCDM methods.
(c) To understand the nature of the factors influencing liver diseases and to make necessary lifestyle changes.

2 Research Methodology

The initial step was to consider the factors based on the literature review. Numerous reputable search engines were employed to investigate the research database. We employed numerous search terms for perusing these research databases. The flowchart Fig. 1 which depicts the methodology is described as follows.

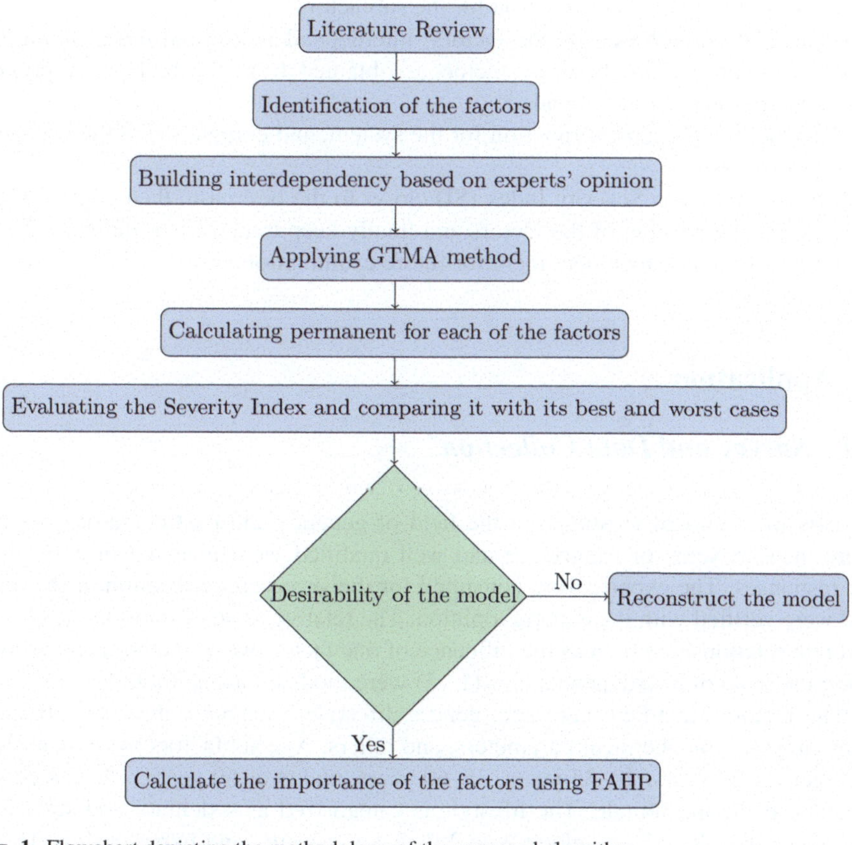

Fig. 1 Flowchart depicting the methodology of the proposed algorithm

The next step is to determine the appropriate factors and their subfactors. They are visualized using digraphs, and GTMA method is performed. If the model works better, we use FAHP to rate the significance of the factors.

2.1 Proposed Algorithm

The algorithm is briefly elaborated as follows.

- Identify the various factors leading to the severity of NAFLD and broadly group these factors.
- Construct a digraph of the subfactors depending on how they influence one another.
- Create a matrix of subfactors. Use tables at the subsystem level to transform linguistic terms into triangular fuzzy numbers. It estimates the numerical values for factor interactions and their relative importance.
- Calculate the permanent function for the subfactors.
- Create a digraph based on the factors' interdependencies, and the quantitative value of interactions between factors is obtained from the table after proper interpretation by professionals.
- Calculate the permanent function for the system, and compare with the best and the worst cases.
- If the model has a Severity Index (SI) closer to the best case, then apply FAHP to obtain the ranking of the factors and finally keep track of the findings of this investigation and document them for future examination.

3 Application

3.1 Survey and Data Collection

Experts from Global Hospitals in the field of genetics and gastroenterology with more than 25 years of experience and well qualified were involved in collecting the responses. The experts were consulted for the successful collection of the data and were verified with the experts' opinion. The relative scale of importance (1−9) and interrelationships such as the influence of one factor over the other is measured using the scale of interdependencies (1−5) were modeled using Table 1.

The factors identified were age, gender, lifestyle, metabolic diseases, climate, liver enzymes, biochemical parameters, and others. Age subfactors were identified as less than 20 years, 20–35 years, 35–60 years, and over 60 years. There are two genders: male and female. The lifestyle is categorized as sedentary and stressful. The metabolic disorders include type 2 diabetes, obesity, and hypertension. There are three climate zones: tropical, temperate, and cold. The liver enzymes were

Table 1 Relative scale of importance

Lingusitic terms	TFN's	Linguistic terms	TFN's
Extremely Unimportant (EUI)	(1, 1, 1)	No Influence (NI)	(1, 1, 1)
Very Very Unimportant (VVUI)	(1, 2, 3)	Weak Influence (WI)	(2, 3, 4)
Very Unimportant (VUI)	(2, 3, 4)	Medium Influence (MI)	(4, 5, 6)
Unimportant (UI)	(3, 4, 5)	Strong Influence(SI)	(6, 7, 8)
Medium Importance (MI)	(4, 5, 6)	Extremely Strong Influence (ESI)	(9, 9, 9)
Important (I)	(5, 6, 7)		
Very Important (VI)	(6, 7, 8)		
Very Very Important (VVI)	(7, 8.9)		
Extremely Important (VI)	(9, 9, 9)		

Fig. 2 Severity of NAFLD

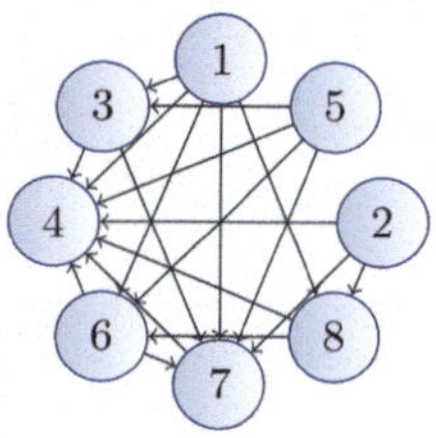

identified as AST, ALT, and GGT. The biochemical parameters include high triglycerides, low HDL, high LDL, high total cholesterol, elevated fasting sugar, HbA1c, T-bilirubin, T-protein, and globulin. Other influences included dietary habits and the PNPLA3 allele [2, 6, 7].

3.2 *Construction of the Mathematical Model*

The term A_i denotes the importance of the factor i, while a_{ij} represents the impact of factor i over factor j and the weights of the edges are obtained using experts' opinions. The subfactors of the factors are calculated using the same approach, and we obtain the permanent of each matrix A_k for $k = 1, 2, \ldots, 8$. The digraph is depicted below in Fig. 2. The factors age, gender, lifestyle, metabolic diseases, climate, liver enzymes, biochemical parameters, and others are numbered as $1, 2, \ldots, 8$, respectively. The permanent values for each of the factors are given below

1. Age: $Per(A_1) = 635$
2. Gender: $Per(A_2) = 19$
3. Lifestyle: $Per(A_3) = 53$
4. Metabolic disorders: $Per(A_4) = 498$
5. Climate: $Per(A_5) = 138$

6. Liver enzymes: $Per(A_6) = 450$
7. Biochemical Parameters: $Per(A_7) = 55{,}138{,}668$
8. Other factors: $Per(A_8) = 82$

The variable permanent matrix containing all the 8 factors is given as:

$$A = \begin{pmatrix} A_1 & a_{12} & a_{13} & a_{14} & a_{15} & a_{16} & a_{17} & a_{18} \\ a_{21} & A_2 & a_{23} & a_{24} & a_{25} & a_{26} & a_{27} & a_{28} \\ a_{31} & a_{32} & A_3 & a_{34} & a_{35} & a_{36} & a_{37} & a_{38} \\ a_{41} & a_{42} & a_{43} & A_4 & a_{45} & a_{46} & a_{47} & a_{48} \\ a_{51} & a_{52} & a_{53} & a_{54} & A_5 & a_{56} & a_{57} & a_{58} \\ a_{61} & a_{62} & a_{63} & a_{64} & a_{65} & A_6 & a_{67} & a_{68} \\ a_{71} & a_{72} & a_{73} & a_{74} & a_{75} & a_{76} & A_7 & a_{78} \\ a_{81} & a_{82} & a_{83} & a_{84} & a_{85} & a_{86} & a_{87} & A_8 \end{pmatrix}$$

$$= \begin{pmatrix} 635 & 1 & 2 & 4 & 1 & 3 & 3 & 5 \\ 1 & 19 & 1 & 3 & 1 & 1 & 2 & 4 \\ 1 & 1 & 53 & 5 & 1 & 1 & 3 & 1 \\ 1 & 1 & 1 & 498 & 1 & 1 & 1 & 1 \\ 1 & 1 & 3 & 3 & 138 & 2 & 2 & 1 \\ 1 & 1 & 1 & 5 & 1 & 450 & 4 & 1 \\ 1 & 1 & 1 & 5 & 1 & 1 & 55{,}138{,}668 & 1 \\ 1 & 1 & 1 & 5 & 1 & 2 & 1 & 82 \end{pmatrix}$$

$Per(A) = 8.9960 \times 10^{22}$

Severity Index $= \text{Log}_{10}(8.9960 \times 10^{22}) = 22.9540$

The best case of the Severity Index (SI) is found to be 27.3822, and the worst case of the Severity Index(SI) is found to be 13.28. Hence, our model is desirable, and we calculate the importance of the factors using FAHP.

3.3 Categorizing the Importance of the Factors Using Fuzzy Analytical Hierarchy Process

FAHP is employed as a part of the approach, since rating the factors using GTDM is desired when the number of subfactors for each factor is more or less the same. As a result, we use FAHP to rank each factor in relative order. The importance of the factors is listed based on the ratings of the decision-makers. The factors are rated based on the linguistic scale of importance using Table 2. The factors are normalized and then ranked whose final results are discussed.

Table 2 Importance and influence of the factors

Lingusitic terms	Triangular fuzzy numbers	Lingusitic terms	Triangular fuzzy numbers
Exceptionally Low (EL)	(1, 1, 1)	Extremely Weak Influence (EWI)	(1, 1, 1)
Very Low (VL)	(1, 2, 3)	Weak Influence	(2, 3, 4)
Low (L)	(2, 3, 4)	Weakly Uninfluential	$(\frac{1}{4}, \frac{1}{3}, \frac{1}{2})$
Below Average (BA)	(3, 4, 5)	Medium Influence (MI)	(4, 5, 6)
Average (A)	(4, 5, 6)	Medium Unifluential (MUI)	$(\frac{1}{6}, \frac{1}{5}, \frac{1}{4})$
Above Average (AA)	(5, 6, 7)	Strong Influence (SI)	(6, 7, 8)
High (H)	(6, 7, 8)	Strong Uninfluential (SUI)	$(\frac{1}{8}, \frac{1}{7}, \frac{1}{6})$
Very High (VH)	(7, 8.9)	Extremely Strong Influence (VSI)	(9, 9, 9)
Exceptionally High (EH)	(9, 9, 9)	Extremely Strong Uninfluential (VSUI)	$(\frac{1}{9}, \frac{1}{9}, \frac{1}{9})$

4 Results and Discussion

4.1 *Range of the Factors and the Severity Index*

The range of the SI is elaborated as follows. Maximum value of the Severity Index occurs when the elements in the main diagonal have a maximum value. The values of the best case include $A_1 = 7128$, $A_2 = 82$, $A_3 = 89$, $A_4 = 893$, $A_5 = 758$, $A_6 = 806$, $A_7 = 1.0347 \times 10^9$, $A_8 = 82$. The Severity Index of the best case is 2.4110×10^{27}. The logarithmic value of the Severity Index calculated for the best case is $\text{Log}_{10}(2.4110 \times 10^{27}) = 27.3822$

Similarly, minimum value of the Severity Index occurs when the elements in the main diagonal have a minimum value. The values of the worst case include $A_1 = 24$, $A_2 = 2$, $A_3 = 9$, $A_4 = 37$, $A_5 = 6$, $A_6 = 14$, $A_7 = 1{,}303{,}200$, $A_8 = 2$. The Severity Index of the worst case is 1.8359×10^{13}. The logarithmic value of the Severity Index calculated for the worst case is $\text{Log}_{10}(1.8359 \times 10^{13}) = 13.2638$.

4.2 *Inference*

The results obtained using FAHP are explained as follows. The biochemical parameters fairly contribute to the severity of NAFLD. It is followed by the dietary patterns and PNPLA3 allele (other factors). These two factors are influenced to a greater percentage than the other factors, so that they lead to the severity. The metabolic disorders and the lifetsyle changes today play a significant role in contributing to the sickness of the liver followed by the liver enzymes, age, gender, and the climate. The factors are ranked below in Table 3 using the FAHP methodology elaborated in [8]. The factors are normalized to obtain accurate results.

Table 3 Ranking the factors

Factors	Values	Ranking
Age	0.0648574	6
Gender	0.053355	7
Lifetsyle	0.144067	3
Metabolic disorders	0.117597	4
Climate	0.02425	8
Liver enzymes	0.096676	5
Biochemical parameters	0.322612	1
Other factors	0.176569	2

5 Conclusion, Limitations, and Future Works

The proposed technique modeled the Severity Index (SI), along with ranking the factors. It can be applied to a variety of real-world challenges. It provides an insight based on the interrelationships of the factors, which are illustrated by digraphs. It provides an enhanced view of the model, because it includes numerous factors and its subfactors.

However, the article does have limitations. It will be time-consuming to list all of the probable components at once, and errors are inevitable. Furthermore, the methodology utilized is solely reliant on the experts' input to allow for various computations. Inadequate data recording may also result in incorrect outcomes.

The factors can be evaluated using any other decision-making approach. The present case study focuses solely on the severity of NAFLD. Alcoholic fatty liver disease can also be modeled similarly.

References

1. Veeranki, S.R., Varshney, M.: Intelligent techniques and comparative performance analysis of liver disease prediction. Int. J. Mech. Eng. **7**(1), 489–503 (2022)
2. Younossi, Z.M.: Non-alcoholic fatty liver disease—a global public health perspective. J. Hepatol. **70**(3), 531–544 (2019). https://doi.org/10.1016/j.jhep.2018.10.033
3. Agrawal, S., Singh, R.K., Murtaza, Q.: Disposition decisions in reverse logistics: graph theory and matrix approach. J. Cleaner Prod. **137**, 93–104 (2016). https://doi.org/10.1016/j.jclepro.2016.07.045
4. Baykasoglu, A.: A review and analysis of graph theoretical-matrix permanent approach to decision making with example applications. Artif. Intell. Rev. **42**, 573–605(2014). https://doi.org/10.1007/s10462-012-9354-y
5. Lin, R.-H.: An intelligent model for liver disease diagnosis. Artif. Intell. Med. **47**(1), 53–62 (2009). https://doi.org/10.1016/j.artmed.2009.05.005
6. Stachowska, E., Portincasa, P., Jamiol-Milc, D., Maciejewska-Markiewicz, D., Skonieczna-Zydecka, K.: The relationship between prebiotic supplementation and anthropometric and biochemical parameters in patients with NAFLD—a systematic review and meta-analysis of randomized controlled trials. Nutrients **12**(11), 3460 (2020). https://doi.org/10.3390/nu12113460

7. Ghouri, N., Preiss, D., Sattar, N.: Liver enzymes, nonalcoholic fatty liver disease, and incident cardiovascular disease: a narrative review and clinical perspective of prospective data. Hepatology **52**(3), 1156–1161 (2010). https://doi.org/10.1002/hep.23789
8. Pehlivan, N. Y., Turan, P., Ahmet Ç.: Comparison of methods in FAHP with application in supplier selection. Fuzzy analytic hierarchy process, pp. 45–76. Chapman and Hall/CRC (2017). https://dx.doi.org/10.1201/9781315369884

Some Results on Variants of Lucky Labeling of Extended Polytopes

S. Teresa Arockiamary and Rini Angeline Sahayamary

1 Introduction

The research in graph theory took its wings in different areas like graph coloring, graph labeling, etc., Graph labeling is a technique in which vertices, edges, or both can be labeled in a graph. One of the graph labeling techniques is lucky labeling. The concept of lucky labeling was introduced by S. Czerwinski et al. [1]. Later this concept was studied by Ashwini J. et al. [2]. Some of its variants are proper lucky labeling, d-lucky labeling, proper d-lucky labeling, e-lucky labeling, and d-lucky face labeling. We are interested in studying the behavior of d-lucky labeling and d-lucky face labeling for some extended polytopes. The concept of d-lucky labeling was introduced by Mirka Miller et al. for certain networks [3], and several other authors have continued the study for different graphs [4–6]. Later the concept of d-lucky face labeling was introduced by Teresa Arockiamary and Rini Angeline Sahayamary and studied for some graphs [7, 8]. We have considered the graphs of the polytopes and extended them in a particular manner. The construction of these extended polytopes is briefed in the following section.

2 Construction of Extended Polytopes

In this section, we study the construction of the extended polytopes R_n^*, A_n^* and T_n^*. The graphs of these polytopes are obtained by adding edges to the graph of the polytopes R_n, A_n and T_n, and their graphs are exhibited as follows:

S. T. Arockiamary · R. A. Sahayamary (✉)
Department of Mathematics, Stella Maris College (Autonomous), Affiliated to University of Madras, Chennai, Tamil Nadu, India

I. Rajasingh et al. (eds.), *Applied Computational Mathematics*,
Springer Proceedings in Mathematics & Statistics 480,
https://doi.org/10.1007/978-3-031-77764-6_14

2.1 Graph of Convex Polytope R_n^*

The convex polytope R_n is extended by adding a layer of pentagons and adding edges to join the pentagons to form the extended polytope R_n^*. The graph of R_n^* is shown in Fig. 1.

2.2 Graph of Convex Polytope A_n^*

The convex polytope A_n is extended by adding edges and a circle. The graph of A_n^* is shown in Fig. 2.

2.3 Graph of Convex Polytope T_n^*

The convex polytope T_n is extended by adding a T_n to form the extended polytope T_n^*. The graph of T_n^* is shown in Fig. 3.

3 Results and Discussion

Definition 1 Let l be a labeling of the vertices of a graph G by positive integers. Define $c(u) = d(u) + \Sigma_{v \in N(u)} l(v)$ where $d(u)$ denotes the number of edges incident

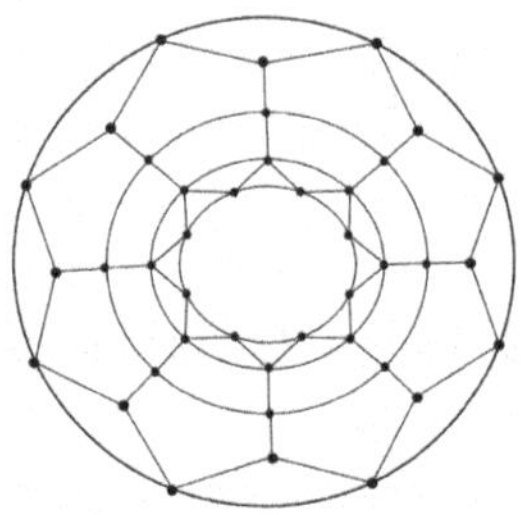

Fig. 1 Extended polytope of R_n^*

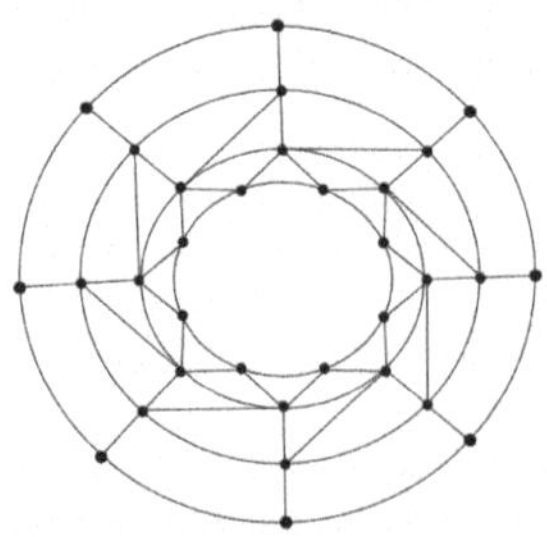

Fig. 2 Extended polytope of A_n^*

Fig. 3 Extended polytope of T_n^*

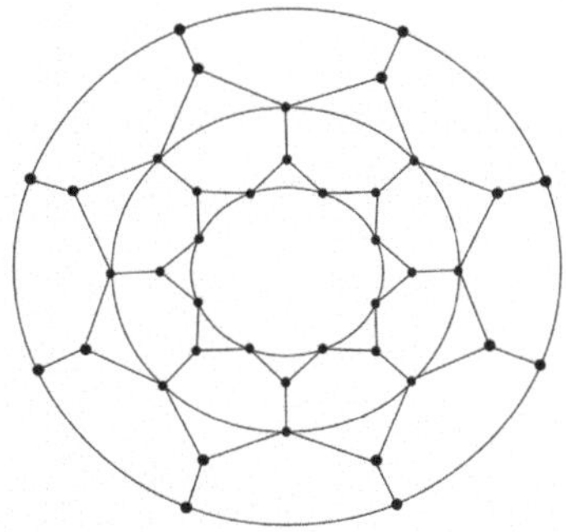

Fig. 4 Extended polytope of R_8^*

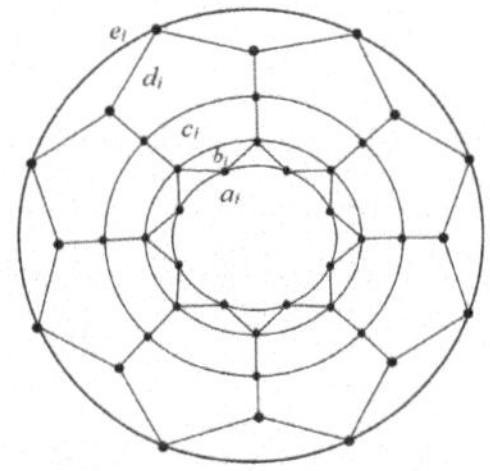

to u. For each pair of vertices u and v adjacent in G, we define a labeling l as d-lucky if $c(u) \neq c(v)$. The least positive integer k such that a graph G has a d-lucky labeling with the set of labels $\{1, 2, \ldots, k\}$ is the d-lucky number for that graph, indicated by η_{dl}(G).

Definition 2 A labeling $f : F(G) \rightarrow \{1, 2, \ldots, r\}$ is said to be d-lucky face labeling if $f(f_i) = d(f_i) + l(f_i) + \sum_{f_j \in N(f_i)} l(f_j) + |v_i|$, where f_i denotes the faces of the graph G, $d(f_i)$ denotes the degree of the faces, $l(f_i)$ denotes the label of the face considered, $l(f_j)$ denotes the labels of the adjacent faces of f_i and f_j belongs to the open neighbourhood of f_i, and $|v_i|$ denotes the cardinality of the vertices forming the face f_i, satisfying the condition $f(f_i) \neq f(f_j)$ for every pair of adjacent faces. The least positive integer satisfied by the faces of the graph is the d-lucky face number denoted by $\eta_{df}(G)$".

3.1 Results on d-Lucky Labeling

Theorem 1 *The extended polytope admits d-lucky labeling and the* $\eta_{dl}(R_n^*) = 2$.

Proof By definition, the condition for d-lucky labeling is $c(u) \neq c(v)$. To prove this condition, the graph of extended polytope R_8^* is under discussion. See Fig. 4. The graph has five layers, namely, a_i, b_i, c_i, d_i and e_i as Fig. 4.

We label the graph as follows:

$$l\left(a_i\right) = \begin{cases} 1 & 1 \leq i \leq 2n-1, \quad for\ i\ odd \\ 2 & 2 \leq i \leq 2n \quad\quad\quad for\ i\ even \end{cases}$$

$$l\left(b_i\right) = 1\ for\ all\ i$$

$$l\left(c_i\right) = \begin{cases} 1 & 1 \leq i \leq 2n-1, \quad for\ i\ odd \\ 2 & 2 \leq i \leq 2n \quad\quad\quad for\ i\ even \end{cases}$$

$$l\left(d_i\right) = \begin{cases} 1 & 1 \leq i \leq 2n-1, \quad for\ i\ odd \\ 2 & 2 \leq i \leq 2n \quad\quad\quad for\ i\ even \end{cases}$$

$$l\left(e_i\right) = 1\ for\ all\ i$$

We take $c(u) \neq c(v)$. Consider the pair of adjacent vertices a_i with label 1 and b_i with label 1 which are adjacent in G. Degrees of a_i and b_i are 4 and 5, respectively. By definition $c(a_i) = 4 + 6 = 10$ and $c(b_i) = 5 + 6 = 11$. Therefore, $c(a_i) \neq c(b_i)$. See Fig. 5. The same type of argument holds perfect for all the pair of vertices adjacent in G. This proves that $c(u) \neq c(v)$. Therefore, R_n^* accepts d-lucky labeling and $\eta_{dl}\left(R_n^*\right) = 2$. Since 2 is an Euler lucky number, the graph R_n^* is an Euler lucky graph.

Theorem 2 *The extended polytope* A_n^* *accepts d-lucky labeling and the* $\eta_{dl}\left(A_n^*\right) = 2$.

Proof The graph of extended polytope A_8^* is considered. We begin the vertex labeling from a_i to the outermost layer d_i. See Fig. 6. We label the vertices as follows:

$$l\left(a_i\right) = \begin{cases} 1 & 1 \leq i \leq 2n-1, \quad for\ i\ odd \\ 2 & 2 \leq i \leq 2n \quad\quad\quad for\ i\ even \end{cases}$$

$$l\left(b_i\right) = \begin{cases} 1 & 1 \leq i \leq 2n-1, \quad for\ i\ odd \\ 2 & 2 \leq i \leq 2n \quad\quad\quad for\ i\ even \end{cases}$$

$$l\left(c_i\right) = \begin{cases} 1 & 1 \leq i \leq 2n-1, \quad for\ i\ odd \\ 2 & 2 \leq i \leq 2n \quad\quad\quad for\ i\ even \end{cases}$$

$$l\left(d_i\right) = 1\ for\ all\ i$$

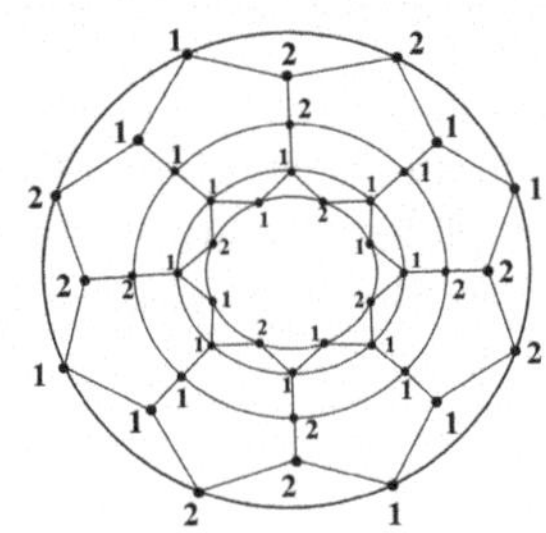

Fig. 5 d-lucky labeling of R_8^*

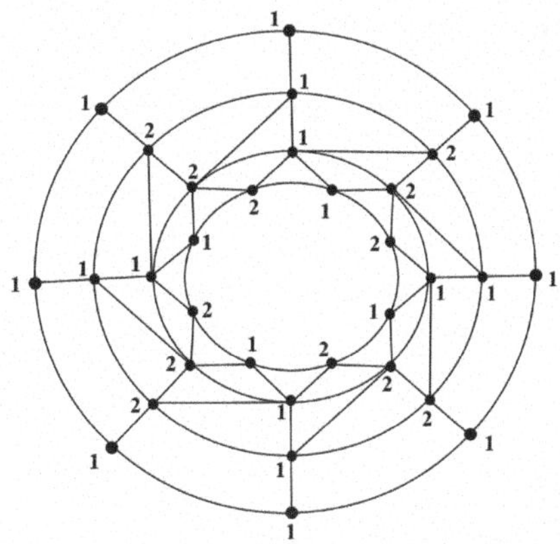

Fig. 6 d-lucky labeling of A_8^*

We take $c(u) \neq c(v)$. Consider the pair of adjacent vertices a_i with label 1 and b_i with label 2 which are adjacent in$G$. Degrees of a_i and b_i are 4 and 6, respectively. By definition $c(a_i) = 4 + 7 = 11$ and $c(b_i) = 6 + 8 = 14$. Therefore, $c(a_i) \neq c(b_i)$. The same type of argument holds for all the pairs of vertices which are adjacent in G. This implies that $c(u) \neq c(v)$ and the graph A_n^* accepts d-lucky labeling. The $\eta_{dl}(G) = 2$. Since 2 is an Euler lucky number, the graph A_n^* is an Euler lucky graph.

Theorem 3 *The extended polytope* T_n^* *accepts d-lucky labeling and the* $\eta_{dl}\left(T_n^*\right) = 2$.

Proof We aim to show that the graph of extended polytope T_n^* is d-lucky. Then we claim that $c(u) \neq c(v)$. To prove this condition, we consider T_8^*. See Fig. 7. The graph has five layers, namely, the innermost a_i, b_i, c_i, d_i and the outermost e_i, and the labels are as follows:

$$l(a_i) = \begin{cases} 1 & 1 \leq i \leq 2n-1, \quad for\ i\ odd \\ 2 & 2 \leq i \leq 2n \quad\quad\quad for\ i\ even \end{cases}$$

$$l(b_i) = \begin{cases} 1 & 1 \leq i \leq 2n-1, \quad for\ i\ odd \\ 2 & 2 \leq i \leq 2n \quad\quad\quad for\ i\ even \end{cases}$$

$$l(c_i) = 1\ for\ all\ i$$

$$l(d_i) = \begin{cases} 1 & 1 \leq i \leq 2n-1, \quad for\ i\ odd \\ 2 & 2 \leq i \leq 2n \quad\quad\quad for\ i\ even \end{cases}$$

$$l(e_i) = \begin{cases} 2 & 1 \leq i \leq 2n-1, \quad for\ i\ odd \\ 1 & 2 \leq i \leq 2n \quad\quad\quad for\ i\ even \end{cases}$$

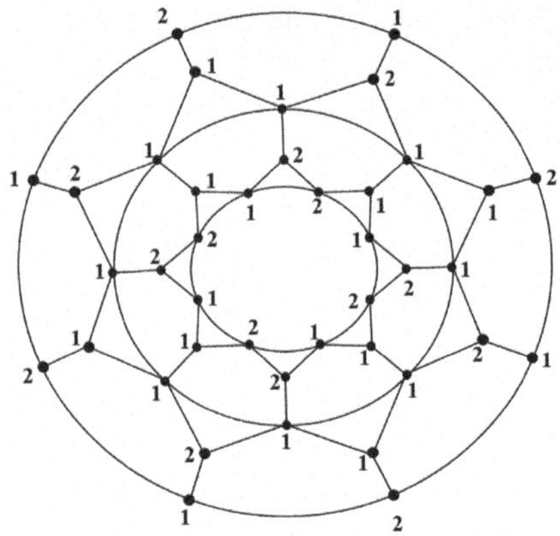

Fig. 7 *d*-lucky labeling of T_8^*

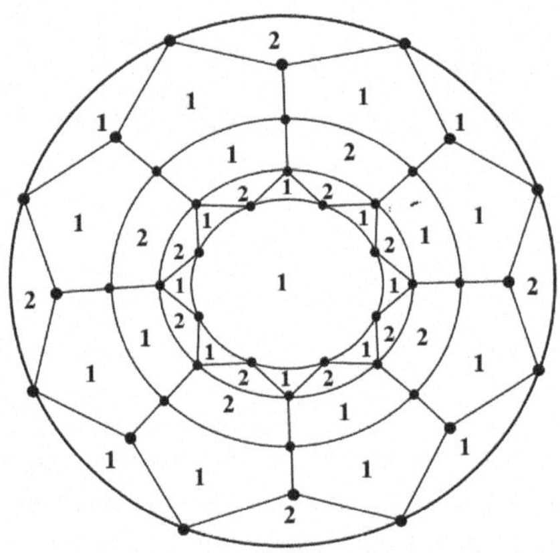

Fig. 8 *d*-lucky face labeling of R_8^*

From definition of 3.1, it is clear that $c(u) \neq c(v)$. Therefore, the graph T_n^* is *d*-lucky. The $\eta_{dl}(G) = 2$. Since 2 is an Euler lucky number, the graph T_n^* is an Euler lucky graph.

3.2 Results on d-Lucky Face Labeling

Theorem 4 *The extended polytope R_n^* accepts d-lucky face labeling. $\eta_{df}(G) = 2$.*

Proof As in Theorem 3.1.1, we consider the graph of R_n^*. We label the faces in such a way it satisfies *d*-lucky face labeling. See Fig. 8. The face labels are:

$$f_i = 1 \begin{cases} for\ i = 1 \\ for\ 2 \leq i \leq 5n, \quad i\ even \end{cases}$$

$$f_i = 2, \quad 3 \leq i \leq 5n + 1, \quad i\ odd$$

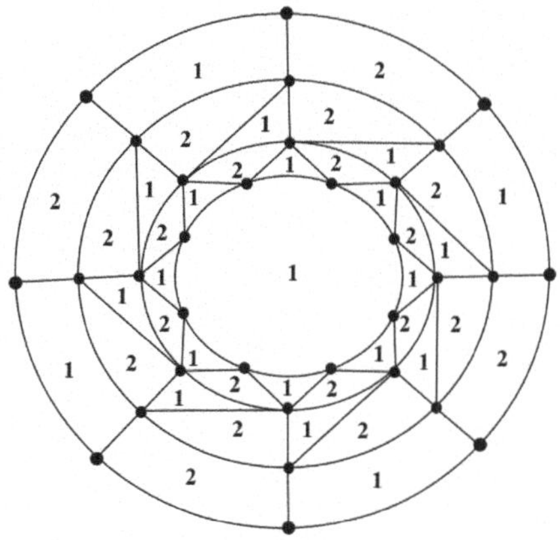

Fig. 9 *d*-lucky face labeling of A_8^*

This shows that the condition $f(f_i) \neq f(f_j)$ holds for all pairs of adjacent faces. Therefore, the graph satisfies *d*-lucky face labeling. Therefore, $\eta_{df}(G) = 2$. Since $\eta_{df}(G)$ is an Euler lucky number, the graph R_n^* is an Euler lucky graph.

Theorem 5 *The extended polytope* A_n^* *accepts d-lucky face labeling and the* $\eta_{df}(G) = 2$.

Proof The graph of extended polytope A_8^* is considered as in Theorem 3.1.2. We begin the face labeling from the innermost face. See Fig. 9. We label the faces as follows:

$$f_i = 1 \begin{cases} for\ i = 1 \\ for\ 2 \leq i \leq 5n, \quad\quad i\ even \end{cases}$$

$$f_i = 2, \quad 3 \leq i \leq 5n+1, \quad i\ odd$$

This satisfies the condition $f(f_i) \neq f(f_j)$ for all pairs of faces which are adjacent in G. Therefore, $\eta_{df}(G) = 2$ and since $\eta_{df}(G)$ is an Euler lucky number, the graph A_n^* is an Euler lucky graph.

Theorem 6 *The extended polytope* T_n^* *accepts d-lucky face labeling.* $\eta_{df}(G) = 2$.

Proof We aim to prove that the graph of the extended polytope T_n^* is *d*-lucky. We label the faces as follows:

$$f_i = 1 \begin{cases} for\ i = 1, \\ for\ all\ i,\ 2 \leq i \leq n+1, \\ \quad\quad 2n+2 \leq i \leq 3n+1 \\ for\ i\ even,\ n+2 \leq i \leq 2n, \\ \quad\quad 3n+2 \leq i \leq 4n \end{cases}$$

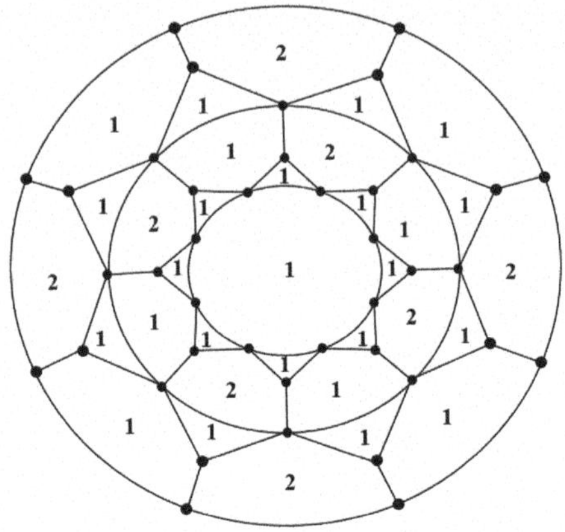

Fig. 10 *d*-lucky face labeling of T_8^*

$$f_i = 2\begin{cases} for\ n+3 \le i \le 2n+1, \\ \quad 3n+3 \le i \le 4n+1, \quad i\ odd \end{cases}$$

See Fig. 10. This condition $f(f_i) \neq f(f_j)$ holds for every pair of adjacent faces. Therefore, the graph T_n^* satisfies *d*-lucky face labeling. Therefore, $\eta_{df}(G) = 2$. Since $\eta_{df}(G) = 2$ is an Euler lucky number, the graph T_n^* is an Euler lucky graph.

4 Conclusion

We identified the $\eta_{dl}(G)$ and $\eta_{df}(G)$ for the abovementioned graphs and have also proved that the graphs are Euler lucky. We would like to extend our work to study the behavior of Projection graphs.

References

1. Czerwinski, S., Grytczuk, J., Zelazny, W.: Lucky labeling of graphs. Inf. Process. Lett. **109**, 1078–1081 (2009)
2. Ashwini, J., Selvam, S.P., Gnanajothi, R.B.: Some new results on lucky labeling. Sci. J. **20**(1), 0365–0370 (2023)
3. Miller, M., Rajasingh, I., Emilet, D.A., Jemilet, D.A.: *d*-lucky labeling of graphs. Procedia Comput. Sci. **57**, 766–771 (2015)
4. Emilet, D.A., Rajasingh, I.: *d*-lucky labeling of cycle of ladder, *n*-sunlet and Helm graphs. Int. J. Pure Appl. Math. **109**(10), 219–227 (2016)
5. Sahayamary, R.A.A., Arockiamary, S.T.: *d*-lucky labeling of honeycomb network. Int. J. Comput. Sci. Eng. **7**(5), 35–39 (2019)
6. Sahayamary, R.A.A., Arockiamary, T.: *d*-Lucky Labeling of certain graphical structures. Adv. Appl. Math. Sci. **21**(3), 6687–6694 (2022)

7. Sahayamary, R.A.A., Arockiamary, T.: *d*-lucky face labeling of graphs. Purakala **32**, 304–309 (2023). (Print only)
8. Arockiamary, T., Sahayamary, R.A.A.: *d*-lucky face labeling of polytopes. Indian J. Sci. Technol. **16**(3), 39–43 (2023)

Topological Indices of Vemurafenib for the Cure of Melanoma Cancer

Arul Jeya Shalini and R. Ramya

1 Introduction

Melanoma cancer stands as the leading cause of death attributed to skin cancer and tends to be more frequently reported and accurately diagnosed compared to nonmelanoma skin cancers. For instance, melanoma of the skin is experiencing a notable rise within the white population in the United States [1] albeit with a diminishing growth rate in its incidence. The likelihood of receiving a melanoma diagnosis over a lifetime is estimated at 1.91% for white men and 1.37% for white women [2]. Key environmental risk factors implicated in this trend include sun exposure during childhood [3] and episodes of intense intermittent sun exposure [4].

In the field of medical science, the study of molecular structure, topological indices, and their correlation with pharmaceutical, chemical [5, 6], medical, and biological properties of drugs involves the calculation of these indices. A molecular structure is essentially a graph, where edges represent bonds and vertices represent atoms. The molecular graphs discussed are simple graphs without cycle formations but may have multiple edges. The chemical graph of the drug vemurafenib is shown in Fig. 1. Amidst current technological advancements, pharmaceutical and chemical techniques have rapidly evolved, resulting in the development of numerous crystalline compounds, new medicines, and nano-materials each year. Defining the chemical properties of these compounds and drugs requires extensive chemical experiments, expanding the scope for analysis by pharmaceutical and chemical researchers. The topological index of a molecule's structure serves as a nonempirical numerical quantity that quantifies the structure and its divergent arrangement. These

A. J. Shalini (✉) · R. Ramya
PG Department of Mathematics, Women's Christian College, University of Madras, Chennai, India

I. Rajasingh et al. (eds.), *Applied Computational Mathematics*,
Springer Proceedings in Mathematics & Statistics 480,
https://doi.org/10.1007/978-3-031-77764-6_15

indices act as total functions, assigning an actual number to the molecular structure and are considered as descriptors for analysis. According to the IUPAC definition [7], a topological index (or molecular structure descriptor) is a numerical value associated with chemical constitution for correlation of chemical structure with various physical properties, chemical reactivity, or biological activity.

2 Methodology

The primary focus of this study is the application of an algebraic polynomial approach to assess the topological indices of specific antiviral medication designed for melanoma cancer treatment. The study involves gathering the chemical structure of vemurafenib. The chemical graph of vemurafenib is shown in Fig. 1. As hydrogen atoms (vertices) do not impact graph isomorphism, the analysis is conducted on hydrogen-suppressed molecular graphs. The methodology employs degree counting, analytical techniques, and graph theoretical software to derive the outcomes.

Results are formulated by utilizing separations, extracting closed forms of NM-polynomial and M-polynomials. A three-dimensional polynomial surface is generated using Maple 2023 software which is shown in (a) and (b). The application of NM-polynomial and M-polynomial aids in calculating neighborhood degree-based and degree sum-based indices through mathematical operators, as presented in Tables 1 and 2 [8]. Edge separation forms of hydrogen-removed molecular graph for vemurafenib compound are established based on the neighborhood degree and degree sum of end vertices.

Definition 1 ([9, 10]) If $G = (V, E)$ is a graph and $v \in V$, then $d_v(G)$ denotes the degree of v. Let m_{ij} denote the no of edges uv in $E(G) \ni d(u) = i, d(v) = j$. The ***M*-polynomial** of G is

$$M(G; p, q) = \sum_{i \le j} m_{ij}(G) p^i q^j. \tag{1}$$

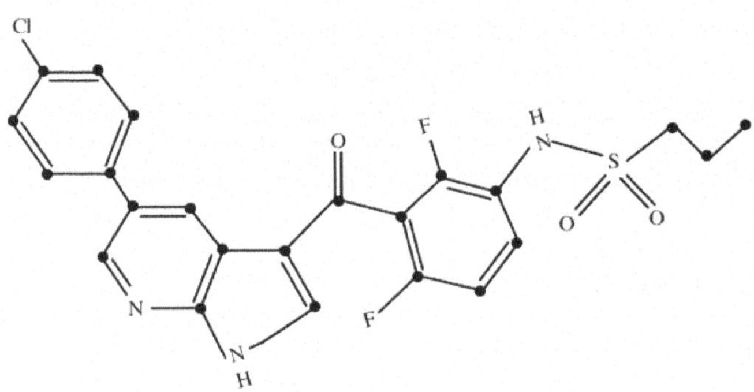

Fig. 1 Molecular graph of vemurafenib

Table 1 Degree-based formulae

Topological indices	Deduction from $M(G)$
1st Zagreb Index(M_1)	$(D_p + D_q)(M(G))_{p=q=1}$
2nd Zagreb Index(M_2)	$(D_p D_q)(M(G))_{p=q=1}$
Forgotten Index(F)	$(D_p^2 + D_q^2)(M(G))_{p=q=1}$
2nd Modified Zagreb Index($^m M_2$)	$(I_p I_q)(M(G))_{p=q=1}$
General Randic Index(R_α)	$(D_p^\alpha D_q^\alpha)(M(G))_{p=q=1}$
Redifined 3rd Zagreb Index($ReZG_3$)	$D_p D_q (D_p + D_q)(M(G))_{p=q=1}$
Symmetric Division Index(SDD)	$(D_p I_q + I_p D_q)(M(G))_{p=q=1}$
Harmonic Index(H)	$2I_p J(M(G))_{p=q=1}$
Inverse Sum Index(I)	$I_p J D_p D_q M(G)_{p=q=1}$
Augmented Zagreb Index (A)	$I_p^3 Q_{-2} J D_p^3 D_q^3 (M(G))_{p=q=1}$
General Randic Index (RR_α)	$I_p^\alpha I_q^\alpha (M(G))_{p=q=1}$

Table 2 Neighborhood degree-based formulae

Topological indices	Deduction from $NM(G)$
1st Zagreb Index(M_1^1)	$(D_p + D_q)(NM(G))_{p=q=1}$
2nd Zagreb Index(M_2^*)	$(D_p D_q)(NM(G))_{p=q=1}$
Forgotten Index(F_N^*)	$(D_p^2 + D_q^2)(NM(G))_{p=q=1}$
2nd Modified Zagreb Index($^{nm} M_2$)	$(I_p I_q)(NM(G))_{p=q=1}$
General Randic Index(NR_α)	$(D_p^\alpha D_q^\alpha)(NM(G))_{p=q=1}$
Redifined 3rd Zagreb Index(ND_3)	$D_p D_q (D_p + D_q)(NM(G))_{p=q=1}$
Symmetric Division Index(ND_5)	$(D_p I_q + I_p D_q)(NM(G))_{p=q=1}$
Harmonic Index(NH)	$2I_p J(NM(G))_{p=q=1}$
Inverse Sum Index(NI)	$I_p J D_p D_q NM(G)_{p=q=1}$
Augmented Zagreb Index (S)	$I_p^3 Q_{-2} J D_p^3 D_q^3 (NM(G))_{p=q=1}$
General Randic Index (NRR_α)	$I_p^\alpha I_q^\alpha (NM(G))_{p=q=1}$

Definition 2 ([9, 10]) Let neighborhood degree sum of u in G be denoted by $d_u^*(G)$ which is the sum of degree of all adjacent vertices of u in G. Let the no of edges uv in $E(G) \ni d^*(u) = i, d^*(v) = j$ be denoted by m_{ij}^*. The **Neighbourhood M-polynomial** of G is defined as

$$NM(G; p, q) = \sum_{i \leq j} m_{ij}^*(G) p^i q^j. \tag{2}$$

where
$D_p(M(G)) = p\frac{\partial M(G)}{\partial p}$, $D_q(M(G)) = q\frac{\partial M(G)}{\partial q}$, $I_p(m(G)) = \int_0^p \frac{(M(t,q))}{t}dt$,
$I_q(M(G)) = \int_0^q \frac{(M(p,t))}{t}dt$, $J(M(G)) = M(p, p)$, $Q_\alpha(M(G)) = p^\alpha M(G)$

3 Results

Theorem 1 *Let the molecular graph of vemurafenib be denoted by G. Then we have*

$$M(G) = pq^2 + 4pq^3 + 2pq^4 + 6p^2q^2 + 13p^2q^3 + 2p^2q^4 + 8p^3q^3$$

$$NM(G) = p^2q^3 + p^3q^5 + 2p^3q^6 + 2p^3q^7 + 2p^4q^6 + 7p^5q^5 + p^5q^6 + 6p^5q^7 + p^5q^8 + p^6q^6 +$$

$$2p^6q^7 + p^6q^8 + p^6q^9 + 3p^7q^7 + 2p^7q^8 + 2p^7q^9 + p^8q^8.$$

Proof Let the molecular graph of vemurafenib be denoted by G. From Fig. 1, $|E(G)| = 36$. Let

$\Re_{(r,s)} = \{uv \in E(G) : d(u) = r, d(v) = s\}$ and $|\Re_{(r,s)}| = \eta_{(r,s)}$. In Fig. 2a, we observe that

$\eta_{(1,2)} = 1$, $\eta_{(1,3)} = 4$, $\eta_{(1,4)} = 2$, $\eta_{(2,2)} = 6$, $\eta_{(2,3)} = 13$, $\eta_{(2,4)} = 2$ and $\eta_{(3,3)} = 8$.

The M-polynomial of G can be resulted as

$$M(G; p, q) = \sum_{r \leq s} \eta_{rs}(G) p^r q^s$$

$$= \eta_{(1,2)} pq^2 + \eta_{(1,3)} pq^3 + \eta_{(1,4)} pq^4 + \eta_{(2,2)} p^2q^2 + \eta_{(2,3)} p^2q^3 + \eta_{(2,4)} p^2q^4 + \eta_{(3,3)} p^3q^3$$

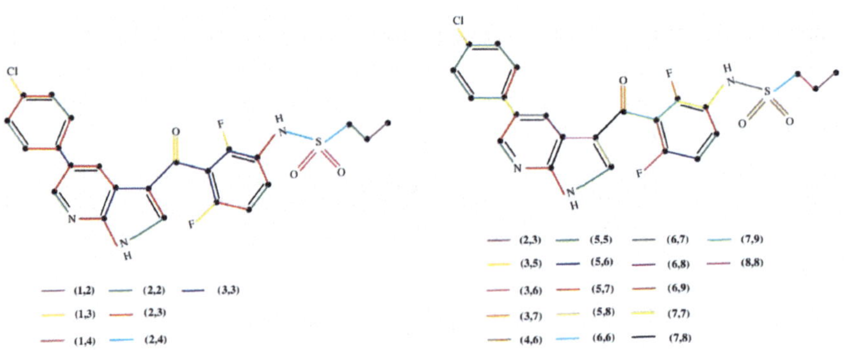

(a) (b)

Fig. 2 Edge partition classes of (**a**) degree and (**b**) neighborhood degree

By the substitution of $\eta_{(r,s)}$ values,

$$M(G) = pq^2 + 4pq^3 + 2pq^4 + 6p^2q^2 + 13p^2q^3 + 2p^2q^4 + 8p^3q^3$$

Let $\Re^*_{(r,s)} = \{uv \in E(G) : d^*(u) = r, d^*(v) = s\}$. Let $|\Re^*_{(r,s)}| = \eta^*_{(r,s)}$

From Fig. 2b, we observe that $\eta^*_{(2,3)} = 1$, $\eta^*_{(3,5)} = 1$, $\eta^*_{(3,6)} = 2$, $\eta^*_{(3,7)} = 2$, $\eta^*_{(4,6)} = 2$, $\eta^*_{(5,5)} = 1$, $\eta^*_{(5,6)} = 1$, $\eta^*_{(5,7)} = 6$, $\eta^*_{(5,8)} = 1$, $\eta^*_{(6,6)} = 1$, $\eta^*_{(6,7)} = 2$, $\eta^*_{(6,8)} = 1$, $\eta^*_{(6,9)} = 1$, $\eta^*_{(7,7)} = 3$, $\eta^*_{(7,8)} = 2$, $\eta^*_{(7,9)} = 2$ and $\eta^*_{(8,8)} = 1$.

The NM-polynomial of G can be resulted as

$$NM(G; p, q) = \sum_{r \le s} \eta^*_{rs}(G) p^r q^s$$

$$= \eta^*_{(2,3)}p^2q^3 + \eta^*_{(3,5)}p^3q^5 + \eta^*_{(3,6)}p^3q^6 + \eta^*_{(3,7)}p^3q^7 + \eta^*_{(4,6)}p^4q^6 + \eta^*_{(5,5)}p^5q^5 +$$

$$\eta^*_{(5,6)}p^5q^6 + \eta^*_{(5,7)}p^5q^7 + \eta^*_{(5,8)}p^5q^8 + \eta^*_{(6,6)}p^6q^6 + \eta^*_{(6,7)}p^6q^7 + \eta^*_{(6,8)}p^6q^8 +$$

$$\eta^*_{(6,9)}p^6q^9 + \eta^*_{(7,7)}p^7q^7 + \eta^*_{(7,8)}p^7q^8 + \eta^*_{(7,9)}p^7q^9 + \eta^*_{(8,8)}p^8q^8$$

By the substitution of $\eta^*_{(r,s)}$ values,

$$= p^2q^3 + p^3q^5 + 2p^3q^6 + 2p^3q^7 + 2p^4q^6 + 7p^5q^5 + p^5q^6 + 6p^5q^7 + p^5q^8 + p^6q^6 +$$

$$2p^6q^7 + p^6q^8 + p^6q^9 + 3p^7q^7 + 2p^7q^8 + 2p^7q^9 + p^8q^8$$

Theorem 2 *Let the molecular graph of vemurafenib be denoted by G. Then,*

1. $M_1(G) = 178$, $M_1^1(G) = 424$

2. $M_2(G) = 212$, $M_2^*(G) = 1273$

3. $F(G) = 480$, $F_N^*(G) = 2668$

4. ${}^{m}M_2(G) = 7.139$, ${}^{nm}.M_2(G) = 1.096$

5. $R_\alpha(G) = (2)^\alpha + 4(3)^\alpha + 2(4)^\alpha + 6(4)^\alpha + 13(6)^\alpha + 2(8)^\alpha + 8(9)^\alpha$,

$$NR_\alpha(G) = (6)^\alpha + (15)^\alpha + 2(18)^\alpha + 2(21)^\alpha + 2(36)^\alpha + 7(25)^\alpha$$
$$+ (30)^\alpha + 6(35)^\alpha + (40)^\alpha +$$

$$(36)^\alpha + 2(42)^\alpha + (48)^\alpha + (54)^\alpha + 3(49)^\alpha + 2(56)^\alpha + 2(63)^\alpha + (64)^\alpha$$

$$6.\ ReZG_3(G) = 1108,\ ND_3(G) = 16{,}278$$

$$7.\ SDD(G) = 85.5,\ ND_5(G) = 187.171$$

$$8.\ H(G) = 15,\ NH(G) = 6.4509$$

$$9.\ I(G) = 41.53,\ NI(G) = 103.211$$

$$10.\ A(G) = 285.366,\ S(G) = 1612.4$$

$$11.\ RR_\alpha M(G) = \frac{1}{(2)^\alpha} + \frac{4}{(3)^\alpha} + \frac{2}{(4)^\alpha} + \frac{6}{(4)^\alpha} + \frac{13}{(6)^\alpha} + \frac{2}{(8)^\alpha} + \frac{8}{(9)^\alpha},$$

$$NRR_\alpha(NM(G)) = \frac{1}{(6)^\alpha} + \frac{1}{(15)^\alpha} + \frac{2}{(18)^\alpha} + \frac{2}{(21)^\alpha} + \frac{2}{(36)^\alpha} + \frac{7}{(25)^\alpha} + \frac{1}{(30)^\alpha} + \frac{6}{(35)^\alpha} +$$

$$\frac{1}{(40)^\alpha} + \frac{1}{(36)^\alpha} + \frac{2}{(42)^\alpha} + \frac{1}{(48)^\alpha} + \frac{1}{(54)^\alpha} + \frac{3}{(49)^\alpha} + \frac{2}{(56)^\alpha} + \frac{2}{(63)^\alpha} + \frac{1}{(64)^\alpha}$$

Proof Initially, we will calculate the degree-based indices. From Theorem 1,

$$M(G) = pq^2 + 4pq^3 + 2pq^4 + 6p^2q^2 + 13p^2q^3 + 2p^2q^4 + 8p^3q^3$$

From this,

$$(D_p + D_q)(M(G)) = pq^2(48p^2q + 12pq^2 + 65pq + 24p + 10q^2 + 16q + 3)$$

$$(D_pD_q)(M(G)) = 2pq^2(36p^2q + 8pq^2 + 39pq + 12p + 4q^2 + 6q + 1)$$

$$(D_p^2 + D_q^2)(M(G)) = pq^2(144p^2q + 24pq^2 + 169pq + 48p + 34q^2 + 40q + 5)$$

$$(I_pI_q)(M(G)) = \frac{8}{9}p^3q^3 + p^2(\frac{1}{4}q^4 + \frac{13}{6}q^3 + \frac{3}{2}q^2) + p(\frac{1}{2}q^4 + \frac{4}{3}q^3 + \frac{1}{2}q^2)$$

$$D_pD_q(D_p{+}D_q)(M(G)) = 2pq^2(216p^2q{+}48pq^2{+}195pq{+}48p{+}20p^2{+}24q{+}6)$$

$$(D_pI_q{+}I_pD_q)M(G) = \frac{1}{6}[pq^2(96p^2q{+}30pq^2{+}169pq{+}72p{+}51q^2{+}80q{+}15)]$$

$$(D_p^\alpha D_q^\alpha) = (2)^\alpha pq^2 + 4(3)^\alpha pq^3 + 2(4)^\alpha pq^4 + 6(4)^\alpha p^2q^2 + 13(6)^\alpha p^2q^3$$

$$+ 2(8)^{\alpha} p^2 q^4 + 8(9)^{\alpha} p^3 q^3$$

$$2I_p J(M(G)) = \frac{1}{3} p^3 [10p^3 + 18p^2 + 15p + 2]$$

$$I_p J D_p D_q M(G) = \frac{1}{15} p^3 [220p^3 + 258p^2 + 135p + 10]$$

$$I_p^3 Q_{-2} J D_p^3 D_q^3 M(G) = \frac{p(23{,}139p^3 + 23{,}488p^2 + 13{,}284p + 1728)}{216}$$

$$I_p^{\alpha} I_q^{\alpha} M(G) = \frac{1}{(2)^{\alpha}} pq^2 + \frac{4}{(3)^{\alpha}} pq^3 + \frac{2}{(4)^{\alpha}} pq^4 + \frac{6}{(4)^{\alpha}} p^2 q^2 + \frac{13}{(6)^{\alpha}} p^2 q^3 + \frac{2}{(8)^{\alpha}} p^2 q^4 + \frac{8}{(9)^{\alpha}} p^3 q^3$$

Thus,

$$M_1(G) = [pq^2(48p^2 q + 12pq^2 + 65pq + 24p + 10q^2 + 16q + 3)]_{p=q=1} = 178$$

$$M_2(G) = [2pq^2(36p^2 q + 8pq^2 + 39pq + 12p + 4q^2 + 6q + 1)]_{p=q=1} = 212$$

$$F(G) = [pq^2(144p^2 q + 24pq^2 + 169pq + 48p + 34q^2 + 40q + 5)]_{p=q=1} = 480$$

$${}^{m}M_2(G) = \left[\frac{8}{9} p^3 q^3 + p^2(\frac{1}{4} q^4 + \frac{13}{6} q^3 + \frac{3}{2} q^2) + p(\frac{1}{2} q^4 + \frac{4}{3} q^3 + \frac{1}{2} q^2)\right]_{p=q=1} = 7.139$$

$$R_{\alpha}(G) = [(2)^{\alpha} pq^2 + 4(3)^{\alpha} pq^3 + 2(4)^{\alpha} pq^4 + 6(4)^{\alpha} p^2 q^2 + 13(6)^{\alpha} p^2 q^3 + 2(8)^{\alpha} p^2 q^4 + 8(9)^{\alpha} p^3 q^3]_{p=q=1}$$
$$= (2)^{\alpha} + 4(3)^{\alpha} + 2(4)^{\alpha} + 6(4)^{\alpha} + 13(6)^{\alpha} + 2(8)^{\alpha} + 8(9)^{\alpha}$$

$$ReZG_3(G) = [2pq^2(216p^2 q + 48pq^2 + 195pq + 48p + 20p^2 + 24q + 6)]_{p=q=1} = 1108$$

$$SDD(G) = \left[\frac{1}{6}[pq^2(96p^2q + 30pq^2 + 169pq + 72p + 51q^2 + 80q + 15)]\right]_{p=q=1} = 85.5$$

$$H(G) = \left[\frac{1}{3}p^3[10p^3 + 18p^2 + 15p + 2]\right]_{p=q=1} = 15$$

$$I(G) = \left[\frac{1}{15}p^3(220p^3 + 258p^2 + 135p + 10)\right]_{p=q=1} = 41.53$$

$$A(G) = \left[\frac{p(23,139p^3 + 23,488p^2 + 13,284p + 1728)}{216}\right]_{p=q=1} = 285.366$$

$$RR_\alpha M(G) = \left[\frac{1}{(2)^\alpha}pq^2 + \frac{4}{(3)^\alpha}pq^3 + \frac{2}{(4)^\alpha}pq^4 + \frac{6}{(4)^\alpha}p^2q^2 + \frac{13}{(6)^\alpha}p^2q^3 + \frac{2}{(8)^\alpha}p^2q^4 + \frac{8}{(9)^\alpha}p^3q^3\right]_{p=q=1}$$

$$= \frac{1}{(2)^\alpha} + \frac{4}{(3)^\alpha} + \frac{2}{(4)^\alpha} + \frac{6}{(4)^\alpha} + \frac{13}{(6)^\alpha} + \frac{2}{(8)^\alpha} + \frac{8}{(9)^\alpha}$$

To evaluate the neighbourhood degree sum-based indices, let us consider

$$NM(G) = p^2q^3 + p^3q^5 + 2p^3q^6 + 2p^3q^7 + 2p^4q^6 + 7p^5q^5 + p^5q^6 + 6p^5q^7 + p^5q^8 + p^6q^6 + 2p^6q^7 + p^6q^8 + p^6q^9 + 3p^7q^7 + 2p^7q^8 + 2p^7q^9 + p^8q^8$$

Later, the neighborhood degree sum-based indices are obtained by applying the above operations.

Hereby the proof is complete. The 3D M-polynomial and NM-polynomial of vemurafenib are shown in Fig. 3.

4 Conclusion

The investigation focuses on examining specific topological characteristics of the molecular structure of vemurafenib, a substance used to prevent the onset and

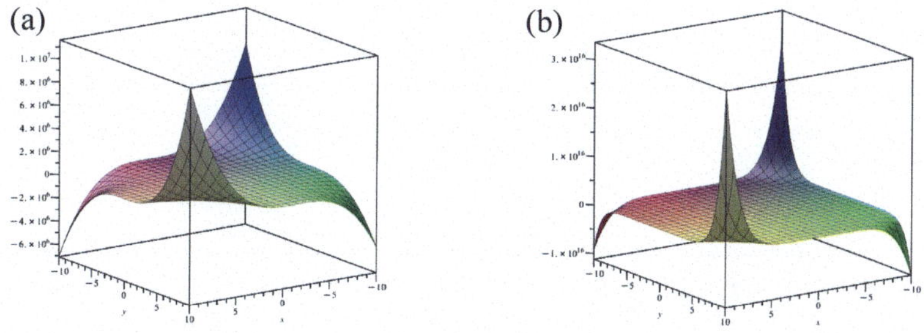

Fig. 3 3D M-polynomial (**a**) and NM-polynomial (**b**) representation of vemurafenib

spread of melanoma cancer. The analysis involves the computation of various neighborhood-based and degree-based indices. Scientists rely on these topological indices to predict a range of properties, including boiling point (B.P), enthalpy, and entropy. This research particularly explores the expressions of specific topological indices for vemurafenib. In the realm of medical science, comprehending the medical, chemical, pharmaceutical, and biological attributes of molecular structures is essential for drug design, a goal achieved through the calculation of topological indices.

References

1. Ries, L.A., Wingo, P.A., Miller, D.S., Howe, H.L., Weir, H.K., Rosenberg, H.M., et al.: The annual report to the nation on the status of cancer, 1973–1997, with a special section on colorectal cancer. Cancer **88**, 2398–424 (2000)
2. Ries, L.A., Eisner, M.P., Kosary, C.L., Hankey, B.F., Miller, B.A., Clegg, L.: SEER Cancer Statistics Review, 1973–1997. National Cancer Institute, Bethesda (2000)
3. Mack, T.M., Floderus, B.: Malignant melanoma risk by nativity, place of residence at diagnosis, and age at migration. Cancer Causes Control **2**, 401–11 (1991)
4. Gilchrest, B.A., Eller, M.S., Geller, A.C., Yaar, M.: The pathogenesis of melanoma induced by ultraviolet radiation. N Engl. J. Med. **340**, 1341–8 (1999)
5. Chen, Z., Dehmer, M., Emmert-Streib, F., Shi, Y.: Entropy bounds for dendrimers. Appl. Math. Comput. **242**, 462–472 (2014)
6. Gutman, I., Trinajstic, N.: Graph theory and molecular orbitals total Φ-electron energy of alternant hydrocarbons. Chem. Phys. Lett. **17**, 535–538 (1972)
7. Gutman, I.: Degree-based topological indices. Croatica Chem. Acta **86**(4), 351–361 (2013)
8. Divya Shree, B.K., Jagadeesh, R., Siddabasappa: Application of topological indices of tenofovir chemical structures for the cure of HIV/AIDS patients. Turkish J. Comput. Math. Edu. **12**, 1693–1706 (2021)
9. Deutsch, E.: M-Polynomial and Degree-Based Topological Indices. Polytechnic Institute of New York University, New York (2014)
10. Mondal, S., De, N., Pal, A.: Topological indices of some chemical structures applied for the treatment of COVID-19 patients. Polycyclic Aromatic Compounds **42**, 1220–1234 (2022)

Antimagic Labeling of Network Graphs

N. K. Vinodhini, J. Jeba Jesintha, and Y. Princy

1 Introduction

Graph labeling [3] entails assigning integers to vertices, edges, or both, with adherence to specific conditions. Rosa [6] proposed β-valuations and various labeling methods as implements to decompose the complete graph into isomorphic subgraphs. Graph labeling finds its use in social network analysis, circuit design, bioinformatics, and so on. In the year 1990, Hartsfield and Ringel [4] introduced antimagic graphs. The name antimagic labeling comes from the fact that it is closely associated with magic square and magic labeling. Therefore, the word "“antimagic” refers to the reverse of magic or the arrangement of numbers, so that no two sums are equal. Chu and Hsu [5] pioneered generalized honeycomb torus by combining families of honeycomb torus. The *n-dimensional pancake graph* is defined in [2] and proved the star structure connectivities. The braid graph is defined in [1] and established to be felicitous. The applications of antimagic labeling span multiple domains, encompassing network security, cryptography, coding theory, and data compression. In this paper, generalized honeycomb torus, pancake graph, and braid graph are shown to be antimagic.

N. K. Vinodhini (✉) · J. J. Jesintha · Y. Princy
Department of Mathematics, Anna Adarsh College for Women, Chennai, India

PG Department of Mathematics, Women's Christian College, University of Madras, Chennai, India

I. Rajasingh et al. (eds.), *Applied Computational Mathematics*,
Springer Proceedings in Mathematics & Statistics 480,
https://doi.org/10.1007/978-3-031-77764-6_16

2 Preliminary Definition

Definition 1 ([2]) The *n-dimensional pancake graph* which is represented as P_n has the vertex set $V(P_n) = \{u = u_1u_2 \cdots u_n | u_i \in < n >, u_i = u_j \text{ for } i = j\}$, where $< n > = \{1, 2, \ldots, n\}$ and the edge set is given by $E(P_n) = \{(u, v) | u = u_1u_2 \cdots u_ju_n,\ v = u^j = u_ju_{j-1} \cdots u_2u_1u_{j+1} \cdots u_{n-1}u_n$ and $2 \leq j \leq n\}$ in such a way that it is evident that P_n is made up of $(n-1)$ types of edges for every $j \in < n >$. Here, u^j represents the distinct j-neighbor of u. The edge uu^j is referred to as a j-edge.

3 Main Result

Theorem 1 *The generalized honeycomb torus $GHT(3, n, d)$, where $n = 20k - 10$, $d = \frac{n}{2}$, and $k =$, positive integer is antimagic.*

Proof Let $GHT(3, n, d)$ be the generalized honeycomb torus. The vertex set of $GHT(3, n, d)$ is given by $V(GHT(3, n, d)) = \{(i, j) | 0 \leq i < 3,\ 0 \leq j < n\}$. The edge set $E(GHT(3, n, d))$, where $n = 20k - 10$, $d = \frac{n}{2}$ and k=positive integer, is given by
$E(GHT) = \{E_1\} \cup \{E_2\} \cup \{E_3\} \cup \{E_4\} \cup \{E_5\} \cup \{E_6\} \cup \{E_7\}$.
The edge set joining (i, j) to any vertex (k, l) is classified as the sets given below:

$$E_1 = \{\beta_\lambda | i = k = 0,\ j = l \pm (modn)\}$$

$$E_2 = \{h_\lambda | i = k = 1,\ j = l \pm (modn)\}$$

$$E_3 = \{e_\lambda | i = k = 2,\ j = l \pm (modn)\}$$

$$E_4 = \{\alpha_\lambda | i = 0, k = 1, i + j \text{ is even}\}$$

$$E_5 = \{g_\lambda | i = 1, k = 1, i + j \text{ is even }\}$$

$$E_6 = \{f_\lambda | i = 0, k = m - 1,\ j = l \pm (modn)\}$$

$$E_7 = \{\gamma_\lambda | i, k = 0, 1, 2,\ j = l \pm (modn)\}$$

The edge labelings are defined as follows:
$\zeta : E(GHT(3, n, d)) \rightarrow \{1, 2, 3, \ldots, 3(n + d)\}$

$$\zeta(e_\lambda) = \begin{cases} \lambda + 1, & 0 \leq \lambda \leq d - 1 \\ d + \lambda + 1, & d \leq \lambda \leq 2d - 2 \end{cases}$$

$$\zeta(f_\lambda) = d + \lambda + 1, \ \ \lambda = 0, 1, 2, \ldots, d - 1$$

$$\zeta(g_\lambda) = 3d + \lambda, \ \ \lambda = 0, 1, 2, \ldots, d - 1$$

$$\zeta(h_\lambda) = 4d + \lambda, \ \ \lambda = 0, 1, 2, \ldots, 2d - 2$$

$$\zeta(\alpha_\lambda) = 6d + \lambda - 1, \ \ \lambda = 0, 1, 2, \ldots, d - 1$$

$$\zeta(\beta_\lambda) = 7d + \lambda - 1, \ \ \lambda = 0, 1, 2, \ldots, 2d - 2$$

$$\zeta(\gamma_\lambda) = \begin{cases} 9d - 1 + \lambda, & \lambda = 0, 1 \\ 9d - 4 + \lambda, & \lambda = 2 \end{cases} \tag{1}$$

The induced vertex labelings are as follows:

$$\xi(0, \lambda) = \begin{cases} 15d + \frac{5}{2}(\lambda - 1), & \lambda = 1, 3, \ldots 2d - 3 \\ 20d + \frac{1}{2}(5\lambda - 8), & \lambda = 2, 4, \ldots, 2d - 2 \end{cases}$$

$$\xi(1, \lambda) = \begin{cases} 11d + \frac{1}{2}(5\lambda - 3), & \lambda = 1, 3, \ldots 2d - 3 \\ 14d + \frac{1}{2}(5\lambda - 4), & \lambda = 2, 4, \ldots, 2d - 2 \end{cases}$$

$$\xi(2, \lambda) = \begin{cases} \frac{1}{2}[7d + 6\lambda + 1], & \lambda = d \\ 3d + \frac{1}{2}(5\lambda + 1), & \lambda = 1, 3, \ldots, d - 2 \\ \frac{1}{2}[5\lambda + 3(d + 1)], & \lambda = 2, 4, \ldots d - 1 \\ \frac{1}{2}[5(d + \lambda) + 3], & \lambda = d + 1, d + 3, \ldots 2d - 2 \\ \frac{1}{2}[5(\lambda + 2d) + 1], & \lambda = d + 2, d + 4, \ldots 2d - 3 \end{cases}$$

$$\xi(0, 0) = 22d - 3$$

$$\xi(1, 0) = 19d - 1$$

$$\xi(2, 0) = 10d + \frac{1}{2}(d - 1)$$

$$\xi(0, n - 1) = 22d - n - 4$$

$$\xi(1, n - 1) = 10n - d - 3$$

$$\xi(2, n - 1) = 8n - 4 \tag{2}$$

The $3n$ vertices are labeled and observed to be distinct. Therefore, $GHT(3, n, d)$ where $n = 20k - 10$, $d = \frac{n}{2}$, and $k =$ positive integer is antimagic.

Theorem 2 *The pancake graph P_4 is antimagic.*

Proof Let P_n be a pancake graph, where $n = 4$. Let H_1, H_2, H_3 and H_4 denote the hexagons in P_4 in clockwise direction. Let the vertices of H_1, H_2, H_3, and H_4 be denoted by $x_1, x_2, \ldots, x_6$, $u_1, u_2, \ldots, u_6$, $w_1, w_2, \ldots, w_6$, and $v_1, v_2, \ldots, v_6$ in anticlockwise direction, respectively. Let the edges of H_1, H_2, H_3, and H_4 be denoted by $e_1, e_2, \ldots, e_6$, $f_1, f_2, \ldots, f_6$, $h_1, h_2, \ldots, h_6$, and $g_1, g_2, \ldots, g_6$ in anticlockwise direction, respectively. Denote the edges x_4u_4, v_1w_1 by β_1, and β_2, respectively. The edges u_3v_6, x_5w_2, x_6v_5, and u_2w_3 are denoted by α_1,α_2,α_3, and α_4, respectively. The edges v_4w_4, w_6u_5, x_1u_1, and x_3v_2 are denoted by γ_1,γ_2,γ_3,γ_4, respectively. The edges x_2v_3 and u_6w_5 are denoted by y_1, y_2, respectively.

The edge labeling $\zeta : E(P_n) \rightarrow \{1, 2, \ldots, n!\frac{(n-1)}{2}\}$ where $n = 4$ is as follows:

$$\zeta(e_\lambda) = \begin{cases} \lambda, & \lambda = 1, 2 \\ 9 - \lambda, & \lambda = 3, 4, 5, 6 \end{cases}$$

$$\zeta(f_\lambda) = \begin{cases} 6 + \lambda, & \lambda = 1, 2 \\ 15 - \lambda, & \lambda = 3, 4, 5, 6 \end{cases}$$

$$\zeta(g_\lambda) = \begin{cases} 3n + \lambda, & \lambda = 1, 2 \\ (5n + 1) - \lambda, & \lambda = 3, 4, 5, 6 \end{cases}$$

$$\zeta(h_\lambda) = \begin{cases} 4n + 2 + \lambda, & \lambda = 1, 2 \\ 3(2n + 1) - \lambda, & \lambda = 3, 4, 5, 6 \end{cases}$$

$$\zeta(\alpha_\lambda) = 6n + \lambda, \quad \lambda = 1, 2, 3, 4$$

$$\zeta(\beta_\lambda) = 7n + 3\lambda, \lambda = 1, 2$$

$$\zeta(\gamma_\lambda) = \begin{cases} 7n + \lambda, & \lambda = 1, 2 \\ 9n - \lambda, & \lambda = 3, 4 \end{cases}$$

$$\zeta(y_\lambda) = 9n - 2 + \lambda, \quad \lambda = 1, 2 \tag{3}$$

The induced vertex labelings are as follows:

$$\xi(x_\lambda) = \begin{cases} 9n + \lambda, & \lambda = 1, 2 \\ 9n + 2(\lambda - 1), & \lambda = 3, 4 \\ 9n + 4 - \lambda, & \lambda = 5, 6 \end{cases}$$

$$\xi(u_\lambda) = \begin{cases} 10n + 9 + \frac{1}{2}(\lambda - 1), & \lambda = 1, 5 \\ 10(n + 1) + \frac{1}{2}(4 + \lambda), & \lambda = 4, 6 \\ 10n - 1 + 2\lambda, & \lambda = 2, 3 \end{cases}$$

$$\xi(v_\lambda) = \begin{cases} 13n + 9 + \frac{1}{2}(5\lambda - 3), & \lambda = 1, 3 \\ 13n + 5 + \frac{1}{2}(5\lambda - 6), & \lambda = 2, 4 \\ 12n + 7 + \lambda, & \lambda = 5 \\ 12n + 2 + \lambda, & \lambda = 6 \end{cases}$$

$$\xi(w_\lambda) = \begin{cases} 10(n + 3) + 3 + \lambda, & \lambda = 1 \\ 10(n + 3) + \frac{1}{3}(\lambda + 3), & \lambda = 3, 6 \\ 13n + 11 + \lambda, & \lambda = 2 \\ 15n + 5\lambda - 4, & \lambda = 4, 5 \end{cases} \tag{4}$$

The 4! vertices are labeled and observed to be distinct. Hence, pancake graph P_4 is antimagic.

Theorem 3 *The braid graph $B(n)$ is antimagic.*

Proof Let the braid graph consists of two paths represented as P'_n and P''_n. Let u_λ and $1 \leq \lambda \leq n$ denote the vertices of P'_n, and let v_λ , $1 \leq \lambda \leq n$ denote the vertices of P''_n. Let e_λ and $1 \leq \lambda \leq n - 1$ denote $u_\lambda u_{\lambda+1}$. The edges $v_\lambda v_{\lambda+1}$ are denoted by f_λ and $1 \leq \lambda \leq n-1$. The edges $u_\lambda v_{\lambda+1}$ are denoted by $x_\lambda and$ $1 \leq \lambda \leq n-1$, and the edges $v_\lambda u_{\lambda+2}$ are denoted by y_λ, $1 \leq \lambda \leq n-2$. The braid graph $B(n)$ consists of $4n - 5$ edges and $2n$ vertices.

The edge labelings $\zeta : E(B(n)) \to \{1, 2, 3, \ldots, 4n - 5\}$ are as follows:

$$\zeta(e_\lambda) = \lambda, \quad \lambda = 1, 2, \ldots, n - 1$$

$$\zeta(x_\lambda) = n - 1 + \lambda, \quad \lambda = 1, 2, \ldots, n - 1$$

$$\zeta(y_\lambda) = 3n + \lambda - 3, \quad \lambda = 1, 2, \ldots, n - 2$$

$$\zeta(f_\lambda) = 2n + \lambda - 2, \quad \lambda = 1, 2, \ldots, n - 1 \tag{5}$$

The induced vertex labeling $\xi : V(B(n)) \rightarrow Z^+$ is as follows:

$$\xi(v_\lambda) = 8n + 4\lambda - 10, \quad \lambda = 2, 3, 4, \ldots, n-2$$

$$\xi(v_1) = 5n - 3$$

$$\xi(v_{n-1}) = 8n - 10$$

$$\xi(v_n) = 5n - 5$$

$$\xi(u_\lambda) = 3\lambda + n - 2, \quad \lambda = 1, 2$$

$$\xi(u_\lambda) = 4\lambda + 4n - 7, \quad \lambda = 3, 4, \ldots, n-1$$

$$\xi(u_n) = 5n - 6 \tag{6}$$

The $2n$ vertices are labeled and observed to be distinct. Hence, the braid graph $B(n)$ is antimagic.

4 Conclusion

In this paper, $GHT(3, n, d)$, where $n = 20k - 10$, $d = \frac{n}{2}$, and $k =$ positive integer, pancake graph P_4, and braid graph are shown to be antimagic.

References

1. Devakirubanithi, D., Jesintha, J.J., Abigail, M.V.: Felicitious labeling of varied bistar, braid and globe graph. Ind. Eng. J. **52**, 155–160 (2023)
2. Dilixiati, S., Sabir, E., Meng, J.: Star structure connectivities of pancake graphs and burnt pancake graphs. Int. J. Parall. Emerg. Distrib. Syst. **36**, 440–448 (2021)
3. Gallian, J.A.: A dynamic survey of graph labeling. Electr. J. Combinat. 1, DS6 (2023)
4. Hartsfield, N., Ringel, G.: Pearls in Graph Theory, pp.108–110. Academic, San Diego (1990)
5. Hsu, L.Y., Chu, H.J.: Generalized honeycomb torus. Inf. Process. Lett. **86**, 185–190 (2003)
6. Rosa, A.: On certain valuations of the vertices of a graph. In: Theory of Graphs (International Symposium, Rome, July 1966), Gordon and Breach, N. Y. and Dunod Paris , pp.349–355 (1967)

Edge Vertex Prime Labeling (EVPL) of Certain Trees and Graphs

Sharon Philomena V and M. Gomathi

1 Introduction

A great deal of graph labeling may be traced back to Rosa [4] labeling in 1967. Edge-vertex prime labeling is a specialized concept within graph theory, a field of mathematics that studies the properties and relationships of networks composed of nodes and edges. In this context, a graph is a collection of vertices (also known as nodes) and edges (connections between vertices). The main idea behind edge-vertex prime labeling is to create a unique numerical representation for each vertex and edge within a graph, utilizing prime numbers to ensure distinctiveness. This labeling scheme often has applications in areas such as network optimization, communication networks, and computer science. Roger Entringer first proposed the idea of prime labeling and was introduced by Tout, Dabboucy and Howalla [8] in 1982. In 2017, Parmar [2] proved that wheel, fan and friendship graphs admits edge vertex prime labeling. Also Parmar (2018) [3] proved that the graphs $K_{2,n}$ and $K_{3,n}$ is an edge vertex prime graph. Shirmali and Parmar (2019) [7] proved that bistar, centipede trees, coconut, double coconut and banana tress are edge verted prime graph. Simaringa and Muthukumaran (2019) [5] proved triangular and rectangular book are EVPL. Also Simaringa and Muthukumaran [6] investigated that crown, union of cycles and wheel related graphs is an edge verted prime graph. The process of assigning prime numbers to vertices and edges typically involves specific rules or constraints to ensure that the resulting labeling adheres to certain properties or objectives. These rules may vary depending on the specific context or problem being addressed. Overall, edge-vertex prime labeling represents a fascinating aspect

Sharon Philomena V (✉) · M. Gomathi
PG department of Mathematics, Women's Christian College, University of Madras, Chennai, India
e-mail: sharonphilomena@wcc.edu.in

I. Rajasingh et al. (eds.), *Applied Computational Mathematics*,
Springer Proceedings in Mathematics & Statistics 480,
https://doi.org/10.1007/978-3-031-77764-6_17

of graph theory, offering a structured approach to represent and analyze complex networks using prime numbers. Its applications span across diverse fields, making it a valuable tool for researchers and practitioners alike in understanding and solving problems related to graph structures.

2 Results

In this heading, we deduce four theorems which are the results developed in this paper.

Theorem 1 *For any integer $k \geq 2$, the olive tree T_k permits prime labeling of edge vertex.*

Proof Let $G(V, E) = T_k, k \geq 2$. Let l_1 be the length of the 1st branch, l_2 be the length of the 2nd branch, and $\ldots$, l_k be the length of the kth branch. Then G has $l_1 + l_2 + \ldots + l_k + 1$ vertices and $l_1 + l_2 + \ldots + l_k$ edges.

The vertex set for the graph is given by

$$V = l_1 + l_2 + \ldots + l_k + 1$$

The edge set is given by,

$$E = l_1 + l_2 + \ldots + l_k$$

Let $g : V \cup E \to \{1, 2, \ldots, 2(l_1 + l_2 + \ldots + l_k)\}$ be a bijective function that is defined in this way.

$$g(v_m) = 2m - 1, \qquad 1 \leq m \leq l_1 + l_2 + \ldots + l_k + 1 \tag{1}$$

$$\begin{aligned} g(v_m v_{m+1}) &= 2m, \quad m = l_2 + 1, l_3 + 2, l_3 + 3, \ldots \\ g(v_1 v_m) &= 2m - 2, \quad m = 2, l_1 + 2, l_1 + l_2 + 2, \ldots, l_1 + l_2 + \ldots + l_{k-1} + 2 \end{aligned} \tag{2}$$

The conditions for the graph are as follows
$g(v_1), g(v_m)$ and $g(v_1 v_m)$
$g(v_m), g(v_{m+1})$ and $g(v_m v_{m+1})$ are each prime relative to one another in pairs.

Finding greatest common divisor for the following
For each edge $v_1 v_m \in E$,

$$\begin{aligned} \{g(v_1), g(v_m)\} &= \{1, 2m - 1\} = 1 \\ \{g(v_1), g(v_1 v_m)\} &= \{1, 2m - 2\} = 1 \\ \{g(v_m), g(v_1 v_m)\} &= \{2m - 1, 2m - 2\} = 1 \end{aligned} \tag{3}$$

For each edge $v_m v_{m+1} \in E$,

$$\begin{aligned}\{g(v_m), g(v_{m+1})\} &= \{2m-1, 2m+1\} = 1 \\ \{g(v_m), g(v_m v_{m+1})\} &= \{2m-1, 2m\} = 1 \\ \{g(v_{m+1}), g(v_m v_{m+1})\} &= \{2m+1, 2m\} = 1\end{aligned} \tag{4}$$

Therefore,the Olive tree permits EVPL .

Theorem 2 *For any integer $m \geq 3$, the F tree FP_m permits prime labeling of edge vertex.*

Proof Let $G(V, E) = FP_m, m \geq 3$. Then G has $m+2$ vertices and $m+1$ edges.

Consider $u_1, u_2, \ldots, u_n$ as the vertices of the path, with v_1 attached to u_n and v_2 attached to u_1 as pendant vertices.

The vertex set is given by

$$V = \{u_1 \cup u_l \cup v_m\}, l = 2, \ldots, m, m = 1, 2$$

The edge set is given by

$$\begin{aligned}E &= E_1 \cup E_2 \cup E_3 \cup B_4 \\ E_1 &= u_1 u_l, l = 2, n \\ E_2 &= u_1 v_m, m = 2 \\ E_3 &= u_n v_m, m = 1 \\ E_4 &= u_l u_{l+1}, l = 2, \ldots, n-2\end{aligned}$$

Define a bijective function $g : V \cup E \to \{1, 2, \ldots, 2m+3\}$ as follows:

$$\begin{aligned}&g(u_1) = 1 \\ &g(u_l) - 2l - 1, l - 2, \ldots, n \\ &g(v_m) = 2m + 2n - 1, m = 1, 2\end{aligned} \tag{5}$$

$$\begin{aligned}g(u_1 u_l) &= 2l - 2, l = 2, n \\ g(u_1 v_m) &= 2m + 2n - 2, m = 2 \\ g(u_n v_m) &= 2m + 2n - 2, m = 1 \\ g(u_l u_{l+1}) &= 2l, l = 2, \ldots, n-2\end{aligned} \tag{6}$$

The conditions for the graph are as follows:
$g(u_1), g(u_l)$, and $g(u_1 u_l)$ for $l = 2, n$.
$g(u_1), g(v_m)$, and $g(u_1 v_m)$ for $m = 2$.
$g(u_n), g(v_m)$ and $g(u_n v_m)$ for $m = 1$.

$g(u_l)$, $g(u_{l+1})$, and $g(u_l u_{l+1})$ for $l = 2, \ldots, n-2$.
are each prime relative to one another in pairs.

Finding greatest common divisor for the following:

For $l = 2, n$

$$\begin{aligned} \{g(u_1), f(u_l)\} &= \{1, 2l-1\} = 1 \\ \{g(u_1), f(u_1 u_l)\} &= \{1, 2l-2\} = 1 \\ \{g(u_l), f(u_1 u_l)\} &= \{2l-1, 2l-2\} = 1 \end{aligned} \tag{7}$$

For $m = 2$

$$\begin{aligned} \{g(u_1), g(v_m)\} &= \{1, 2m+2n-1\} = 1 \\ \{g(u_1), g(u_1 v_m)\} &= \{1, 2m+2n-2\} = 1 \\ \{g(v_m), g(u_1 v_m)\} &= \{2m+2n-1, 2m+2n-2\} = 1 \end{aligned} \tag{8}$$

For $m = 1$

$$\begin{aligned} \{g(u_n), g(v_m)\} &= \{2n-1, 2m+2n-1\} = 1 \\ \{g(u_n), g(u_n v_m)\} &= \{2n-1, 2m+2n-2\} = 1 \\ \{g(v_m), g(u_n v_m)\} &= \{2m+2n-1, 2m+2n-2\} = 1 \end{aligned} \tag{9}$$

For $l = 2, \ldots, n-2$

$$\begin{aligned} \{g(u_l), g(u_{l+1})\} &= \{2l-1, 2l+1\} = 1 \\ \{g(u_l), g(u_l u_{l+1})\} &= \{2l-1, 2l\} = 1 \\ \{g(u_{l+1}), g(u_1 u_{l+1})\} &= \{2l+1, 2l\} = 1 \end{aligned} \tag{10}$$

Therefore, the F tree permits EVPL.

Theorem 3 *For any integer $m \geq 3$, the Y tree Y_m permits prime labeling of edge vertex.*

Proof Let $G(V, E) = Y_m, m \geq 3$. Then G has $m+1$ vertices and m edges.

Suppose $u_1, u_2, \ldots, u_n$ represent the vertices of the path, with v_1 serving as the pendant vertex connected to the vertex u_1 in P_n.

The vertex set is given by $V = \{u_1 \cup u_l \cup v_1\}, l = 2, \ldots, n$

The edge set is given by $E = E_1 \cup E_2 \cup B_3$

$$\begin{aligned} E_1 &= u_1 u_{l+1}, l = 2, \ldots, n-2 \\ E_2 &= u_1 u_l, l = 2, n \\ E_3 &= u_1 v_1, l = 2, n \end{aligned}$$

Define a bijective function $g : V \cup E \to \{1, 2, \ldots, 2n + 1\}$ as follows:

$$\begin{aligned} g(u_1) &= 1 \\ g(u_l) &= 2l - 1, l = 2, \ldots, n \\ g(v_1) &= 2n \end{aligned} \tag{11}$$

$$\begin{aligned} g(u_1 u_l) &= 2l - 2, l = 2, n \\ g(u_1 v_1) &= 2n \\ g(u_l u_{l+1}) &= 2l, l = 2, \ldots, n - 2 \end{aligned} \tag{12}$$

The conditions for this graph are given as follows:
$g(u_l)$, $g(u_{l+1})$, and $g(u_l u_{l+1})$ for $l = 2, \ldots, n - 2$.
$g(u_1)$, $g(u_l)$ and $g(u_1 u_l)$ for $l = 2, n$.
$g(u_1)$, $g(v_1)$ and $g(u_1 v_1)$ are each prime relative to one another in pairs.

Finding greatest common divisor for the following:

For $l = 2, \ldots, n - 2$

$$\begin{aligned} \{g(u_l), g(u_{l+1})\} &= \{2l-, 2l + 1\} = 1 \\ \{g(u_l), g(u_l u_{l+1})\} &= \{2l - 1, 2l\} = 1 \\ \{g(u_{l+1}), g(u_1 u_{l+1})\} &= \{2l + 1, 2l\} = 1 \end{aligned} \tag{13}$$

For $l = 2, n$

$$\begin{aligned} \{g(u_1), g(u_l)\} &= \{1, 2l - 1\} = 1 \\ \{g(u_1), g(u_1 u_l)\} &= \{1, 2l - 2\} = 1 \\ \{g(u_l), g(u_1 u_l)\} &= \{2l - 1, 2l - 2\} = 1 \end{aligned} \tag{14}$$

For $m = 2$

$$\begin{aligned} \{g(u_1), g(v_1)\} &= \{1, 2n + 1\} = 1 \\ \{g(u_1), g(u_1 v_1)\} &= \{1, 2n\} = 1 \\ \{g(v_1), g(u_1 v_1)\} &= \{2n + 1, 2n\} = 1 \end{aligned} \tag{15}$$

Therefore, the Y tree permits EVPL.

Theorem 4 *An (n, t)- Kite graph permits prime labeling of edge vertex, where $n \geq 3$ and $t \geq 1$*

Proof Let $G(V, E)$ be an (n, t)- Kite graph, where $n \geq 3$, $t \geq 1$ Then G has $n + t$ vertices and $n + t$ edges. The graph G is described as follows: Let u_1 be the apex of the cycle of length n and the t—edge path. Let $u_1, u_2, \ldots, u_n$ be the vertices of C_n labeled in the clockwise direction from u_1, which is the apex vertex.

Let $v_1, v_2, \ldots, v_m$ be the vertices of the t—edge path attached to the apex vertex u_1 which is also a vertex of the t—edge path.

The vertex set for this graph is given by

$$V = \{u_1 \cup u_m \cup v_l\};\ 1 \leq l \leq t, 2 \leq m \leq s$$

The edge set for this graph is given by

$$E = B_1 \cup E_2 \cup E_3$$

$$E_1 = u_1 v_1$$

$$E_2 = \begin{cases} 2m - 2t - 2, & m = 2 \\ 2m + 2t, & m = s \end{cases}$$

$$E_3 = 2l + 2, 1 \leq l \leq t - 1$$

Define a bijective function $g : V \cup E \rightarrow \{1, 2, \ldots, 2(n + t)\}$ as follows:

$$\begin{aligned} g(u_1) &= 1 \\ g(v_l) &= 2l + 1, 1 \leq l \leq t \\ g(u_m) &= 2m + 2t, m = 2, \ldots, n \end{aligned} \tag{16}$$

$$\begin{aligned} g(u_1 v_l) &= 2l, l = 1 \\ g(u_1 u_m) &= \begin{cases} 2m + 2t - 2, & m = 2 \\ 2m + 2t, & m = n \end{cases} \\ g(v_l v_{m+1}) &= 2m + 2, 1 \leq l \leq t - 1 \\ g(u_m u_{m+1}) &= 2m + 2t, 2 \leq m \leq n - 1 \end{aligned} \tag{17}$$

The conditions for the graph are as follows:
$g(u_1)$, $g(v_l)$ and $g(u_1 v_l)$
$g(u_1)$, $g(u_m)$ and $g(u_1 u_m)$
$g(u_m)$, $g(u_{m+1})$ and $g(u_m u_{m+1})$
are each prime relative to one another in pairs.

Finding greatest common divisor for the following:
For $l = 1$

$$\begin{aligned} \{g(u_1), g(v_l)\} &= \{1, 2l + 1\} = 1 \\ \{g(u_1), g(u_1 v_l)\} &= \{1, 2l\} = 1 \\ \{g(v_l), g(u_l v_l)\} &= \{2l + 1, 2l\} = 1 \end{aligned} \tag{18}$$

For $m = 2, n$

$$\{g(u_1), g(u_m)\} = \{1, 2m + 2t - 1\} = 1$$

$$\{g(u_1), g(u_1 u_m)\} = \begin{cases} (1, 2m + 2t - 2), & m = 2 \\ (1, 2m + 2t), & m = n \end{cases} = 1 \tag{19}$$

$$\{g(u_m), g(u_1 u_m)\} = \begin{cases} (2m + 2t - 1, 2m + 2t - 2), & m = 2 \\ (2m + 2t - 1, 2m + 2t), & m = n \end{cases} = 1$$

For $1 \leq i \leq t - 1$

$$\begin{aligned} \{g(v_l), g(v_{l+1})\} &= \{2l + 1, 2l + 3\} = 1 \\ \{g(v_l), g(v_l v_{l+1})\} &= \{2l + 1, 2l + 2\} = 1 \\ \{g(v_{l+1}), g(v_l v_{l+1})\} &= \{2l + 3, 2l + 2\} = 1 \end{aligned} \tag{20}$$

For $2 \leq m \leq n - 1$

$$\begin{aligned} \{g(u_m), g(u_{m+1})\} &= \{2m + 2t - 1, 2m + 2t + 1\} = 1 \\ \{g(u_m), g(u_m u_{m+1})\} &= \{2m + 2t - 1, 2l\} = 1 \\ \{g(u_{m+1}), g(u_m u_{m+1})\} &= \{2m + 2t + 1, 2m + 2t\} = 1 \end{aligned} \tag{21}$$

Therefore, the (n, t)—Kite graph permits EVPL.

3 Conclusion

We have demonstrated in this study that Edge vertex prime labeling is admissible for the olive tree, F-tree, Y-tree, and (n, t)—Kite graph.

References

1. Jagadesh, R., Baskar Babujee, J.: Edge vertex prime labeling for some class of graphs. In: National Conference on Recent Trends in Mathematics and its Applications, pp. 24–25 (2017)
2. Parmar, Y.: Edge vertex prime labeling for wheel, fan and friendship graph. Int. J. Math. Stat. Invent. **5**(8), 23–29 (2017)
3. Parmar, Y.: Edge vertex prime labeling for $K_{2,n}$ and $K_{3,n}$ graphs. Math. J. Interdiscipl. Sci. **6**(2), 167–180 (2018)
4. Rosa, A.: On certain valuations of the vertices of a graph, theory of graphs. In: International Symposium, Rome, July 1966, Gordon and Breach, N. Y. and Dunod Paris, pp. 349–355 (1967)

5. Simaringa, M., Muthukumaran, S.: Edge vertex prime labeling of some graphs. Malaya J. Mat. **7**(2), 264–268 (2019)
6. Simaringa, M., Muthukumaran, S.: Edge vertex prime labeling of graphs. Malaya J. Mat. **7**(3), 572–578 (2019)
7. Shrimali, N.P., Parmar, Y.M.: Edge vertex prime labeling for some trees. JASC: J. Appl. Sci. Comput. **VI**(I), 1236–1249 (2019)
8. Tout, A., Dabboucy, A.N., Howalla, K.: Prime labeling of graphs. Natl Acad. Sci. Lett. **11**, 365–368 (1982)

Domination Equitable Coloring for Family of Ladder Graphs

Rebekal Haribabu and Sharmila Mary Arul

1 Introduction

Discrete structured graphs are used in various fields as abstract models for problem-solving and process analysis of real-world issues. The relationships between the things in most domains such as the bonds in DNA, the work schedule, and the links between atoms and elements in chemistry—are better represented by graphs. This results in graph theoretic scheduling issues that are articulated through the use of graph coloring problems. Here, domination coloring along with equitable concept is explored for some ladder graphs.

2 Overview of the Paper

The famous four-color hypothesis from the nineteenth century, which states that any planar map may be colored with four colors so that neighboring nations have distinct colors, served as the impetus for the study of graph coloring [1]. In 1976, Appel and Haken managed to find a solution to the issue. It is well known that coloring a graph with the fewest colors feasible is an NP-hard issue. Due to the contributions of Berge in 1958 [1] and Ore in 1962 [2], dominance is now a formalized theoretical field in graph theory. A graph can be equitably k-colorable without also being equitably

R. Haribabu (✉)
Department of Mathematics, Ethiraj College for Women, Chennai, India

S. M. Arul
Department of Mathematics, Saveetha School of Engineering, Saveetha Institute of Medical and Technical Sciences, Chennai, India
e-mail: sharmilamaryarul.sse@saveetha.com

I. Rajasingh et al. (eds.), *Applied Computational Mathematics*,
Springer Proceedings in Mathematics & Statistics 480,
https://doi.org/10.1007/978-3-031-77764-6_18

(k + 1) colorable, in contrast to the regular appropriate coloring of the graph, which lacks monotonicity. In 1973, Walter Meyer [3] was the first to propose the idea of equitability. His motivation came from an application by Tucker [4], in which the vertices represent garbage collection routes with the caveat that two vertices are united if the relevant routes do not execute on the same day. Recent advances, most notably those of Chellali and Volkmann [5, 6], reveal relationships between dominance parameters and graph chromatic number. Gera et al.'s [7] dominator coloring concept, scrutinized by Arumugam et al. [8] and Gera [9, 10], sparked extensive research. In some graph families, this study explores a unique equitable coloring variation that combines vertex and color class domination. Our work adds to the complex interaction of coloring, dominance, and equitability, enhancing this fascinating research topic, by examining correlations among domination equitable characteristics in this context [11–13].

3 Definitions

Domination Coloring in Graphs A domination coloring of a graph G is a proper vertex coloring of G such that each vertex of G dominates at least one color class, and each color class is dominated by at least one vertex. The minimum number of colors among all domination colorings is called the domination chromatic number, denoted by $\chi_{dd}(G)$.

Equitable Coloring in Graphs A proper coloring of a graph G is said to be equitably $k-$ colorable if the number of vertices of any two color classes $C_1, C_2, \ldots, C_k$ of a graph G differs by at most one. That is, $||C_i| - |C_j|| \leq 1$ for every $1 \leq i, j \leq k$. The chromatic number of an equitable coloring is denoted by $\chi_e(G)$.

Domination Equitable Coloring in Graphs A domination equitable coloring of G is a proper vertex coloring such that

(i) each vertex v dominates at least one color class
(ii) each color class is dominated by at least one vertex and
(iii) if G has k color classes $C_1, C_2, C_3, \ldots, C_k$, the number of vertices in any two color classes C_i and C_j differs by at most one: $||C_i| - |C_j|| \leq 1$ for every $1 \leq i, j \leq k$.
The minimum number of colors required for a domination equitable coloring of G is denoted by $\chi_{dde}(G)$ and is referred to as the domination equitable chromatic number of G.

Ladder The ladder graph L_n is the product graph $P_2 X P_n$ where $n \geq 2$.

Open Ladder An open ladder graph $OL_n, n \geq 2$ is obtained from two paths of length $n - 1$ with $V(OL_n) = \{u_i, v_i / 1 \leq i \leq n\}$ and $E(OL_n) = \{u_i u_{i+1}, v_i v_{i+1} / 1 \leq i \leq n - 1\} U \{u_i v_i / 2 \leq i \leq n - 1\}$.

Slanting Ladder A slanting ladder graph SL_n is the graph obtained from two paths $u_1u_2 \ldots u_n$ and $v_1v_2 \ldots v_n$ by joining each u_i with v_{i+1}, $1 \leq i \leq n-1$.

Triangular Ladder A triangular ladder graph $TL_n, n \geq 2$ is the graph obtained from L_n by adding the edges u_iv_{i+1}, $1 \leq i \leq n-1$. The vertices of TL_n are u_i and v_i where $i = 1, 2, 3 \ldots n$.

Theorem 1 ([14]) *Let P_n be a path of length n. Then $\chi_{dde}(P_n) \geq 2\lfloor \frac{n}{3} \rfloor$*

By using Theorem 1, the domination equitable chromatic number can be determined for some ladder graphs which are discussed as follows:

Theorem 2 *Let $L_n(P_2XP_n)$ be a ladder graph with 2n vertices.*

$$\text{Then } \chi_{dde}(L_n) = \begin{cases} 2, & n = 2, \\ 2\lfloor \frac{2n}{3} \rfloor + mod(2n, 3), & n > 2 \end{cases}$$

Proof Consider the ladder graph L_n with $2n$ vertices (Fig. 1a). Name the ith vertex of L_n from the left as v_i, $1 \leq i \leq 2n$, and let $D_i = \{v_{3i-2}, v_{3i-1}, v_{3i}\}$, $1 \leq i \leq \lfloor \frac{2n}{3} \rfloor$. Partition L_n into domination equitable color classes as follows:

(*i*) Let the mid vertex of D_i be in color class C_{2i-1} and the remaining two vertices of every D_i be in C_{2i}, $1 \leq i \leq \lfloor \frac{2n}{3} \rfloor$.

(*ii*) When $2n \equiv 1 \pmod 3$ and $2n \equiv 2 \pmod 3$, color the vertices $v_{3\lfloor \frac{2n}{3} \rfloor+1}$ and $v_{3\lfloor \frac{2n}{3} \rfloor+2}$ as $C_{2\lfloor \frac{2n}{3} \rfloor+1}$ and $C_{2\lfloor \frac{2n}{3} \rfloor+2}$, respectively.

Theorem 3 *Let OL_n be an open ladder graph with 2n vertices.*

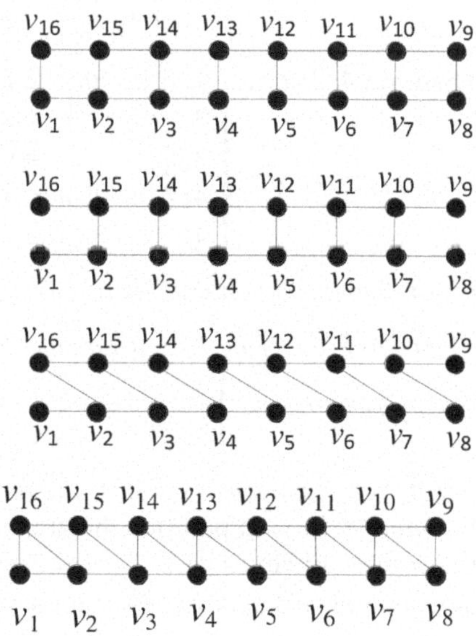

Fig. 1 (**a**) Ladder (L_8); (**b**) open ladder (OL_8); (**c**) slanting ladder (SL_8); (**d**) triangular ladder (TL_8)

$$Then\ \chi_{dde}(OL_n) = \begin{cases} 4, & n = 2, \\ 4\lfloor \frac{n}{3} \rfloor, & n \equiv 0\ mod\ 3 \\ 4\lfloor \frac{n}{3} \rfloor + 2, & n \equiv 1\ mod\ 3 \\ 4\lfloor \frac{n}{3} \rfloor + 3, & n \equiv 2\ mod\ 3 \end{cases}$$

Proof Name the vertex of degree one from the left as v_1 and all other vertices as $v_2, v_3, \ldots, v_{2n}$ in the anticlockwise sense (Fig. 1b). Partition the vertices of OL_n into color classes as follows:

Case (i) When $mod(n, 3) = 0$
Color the mid vertex in $D_i = \{v_{3i-2}, v_{3i-1}, v_{3i}\}$ as a color class C_{2i-1} and the remaining two vertices of every D_i as C_{2i}, $1 \le i \le 2\lfloor \frac{n}{3} \rfloor$.

Case (ii) When $mod(n, 3) = 1$
Color the mid vertex in $D_i = \{v_{3i-2}, v_{3i-1}, v_{3i}\}$ and $D'_j = \{v_{3j}, v_{3j+1}, v_{3j+2}\}$ as a color class C_{2i-1} and C_{2j-1}, the remaining two vertices of every D_i and D'_j as C_{2i} and C_{2j}, $1 \le i \le \lfloor \frac{n-1}{3} \rfloor$, $\lfloor \frac{n-1}{3} \rfloor + 1 \le j \le \lfloor \frac{2n}{3} \rfloor$, respectively.
Also color the vertices $v_{3\lfloor \frac{n-1}{3} \rfloor+1}$ as $C_{2\lfloor \frac{2n}{3} \rfloor+1}$ and $v_{3\lfloor \frac{n-1}{3} \rfloor+2}$ as $C_{2\lfloor \frac{2n}{3} \rfloor+2}$.

Case (iii) When $mod(n, 3) = 2$
Color the mid vertex in $D_i = \{v_{3i-2}, v_{3i-1}, v_{3i}\}$ and $D'_{j-1} = \{v_{3j-1}, v_{3j}, v_{3j+1}\}$ as a color class C_{2i-1} and C_{2j-3}, the remaining two vertices of every D_i and D'_{j-1} as C_{2i} and C_{2j-2}, $1 \le i \le \lfloor \frac{n-2}{3} \rfloor$, $\lfloor \frac{n-2}{3} \rfloor + 2 \le j \le \lfloor \frac{2n}{3} \rfloor$, respectively.
Also color the vertices $v_{3\lfloor \frac{n-2}{3} \rfloor+1}$ as $C_{2\lfloor \frac{2n}{3} \rfloor-1}$, $v_{3\lfloor \frac{n-2}{3} \rfloor+2}$ and $v_{3\lfloor \frac{n-2}{3} \rfloor+4}$ as $C_{2\lfloor \frac{2n}{3} \rfloor}$ and $v_{3\lfloor \frac{n-2}{3} \rfloor+3}$ as $C_{2\lfloor \frac{2n}{3} \rfloor+1}$, respectively.

Theorem 4 *Let SL_n be a slanting ladder graph with $2n$ vertices.*

$$Then\ \chi_{dde}(SL_n) = \begin{cases} 3, & n = 2, \\ 4\lfloor \frac{n}{3} \rfloor, & n \equiv 0\ mod\ 3 \\ 4\lfloor \frac{n}{3} \rfloor + 2, & n \equiv 1\ mod\ 3 \\ 4\lfloor \frac{n}{3} \rfloor + 3, & n \equiv 2\ mod\ 3 \end{cases}$$

Proof Same as Theorem 2 (Fig. 1c)

Theorem 5 *Let TL_n be a triangular ladder graph with $2n$ vertices.*

$$Then\ \chi_{dde}(TL_n) = \begin{cases} 3, & n = 2, \\ 2\lfloor \frac{2n}{3} \rfloor + mod(2n, 3), & n > 2 \end{cases}$$

Proof When $mod(n, 3) = 0, 1$, the proof is same as Theorem 2 (Fig. 1d).
For $mod(n, 3) = 2$, partition the graph TL_n into color classes as follows:
Color the mid vertex in $D_i = \{v_{3i-2}, v_{3i-1}, v_{3i}\}$ and $D'_{j-1} = \{v_{3j-1}, v_{3j}, v_{3j+1}\}$ as a color class C_{2i-1} and C_{2j-3}, the remaining two vertices of every D_i and D'_{j-1} as C_{2i} and C_{2j-2}, $1 \le i \le \lfloor \frac{n-2}{3} \rfloor$, $\lfloor \frac{n-2}{3} \rfloor + 2 \le j \le \lfloor \frac{2n}{3} \rfloor$, respectively.

Also color the vertices $v_{3\lfloor\frac{n-2}{3}\rfloor+1}$ and $v_{3\lfloor\frac{n-2}{3}\rfloor+3}$ as $C_{2\lfloor\frac{2n}{3}\rfloor-1}$, $v_{3\lfloor\frac{n-2}{3}\rfloor+2}$ as $C_{2\lfloor\frac{2n}{3}\rfloor}$ and $v_{3\lfloor\frac{n-2}{3}\rfloor+4}$ as $C_{2\lfloor\frac{2n}{3}\rfloor+1}$, respectively.

4 Conclusion

In this paper, the domination equitable chromatic number which is a proper coloring variant is determined for some standard ladder graphs. Our future investigation is focused on finding χ_{dde} of networks, jump graphs, and family of tadpole graphs.

References

1. Berge, C.: Theory of Graphs and Its Applications, no. 2. Collection Universitaire de Mathematiques, Dunod (1958)
2. Ore, O.: Theory of Graphs, vol. 38. American Mathematical Society, Colloq. Publications, AMS, Providence (1962)
3. Pardalos, P.M., Mavridou, T., Xue, J.: The graph coloring problem: A bibliographic survey. In: Handbook of Combinatorial Optimization, pp. 1077–1141. Springer, Boston (1998)
4. Vernold Vivin, J., Kaliraj, K., Akbar Ali, M.M.: Equitable coloring on total graphs of bigraphs and central graphs of cycles and paths. Int. J. Math. Math. Sci. **2011**, Article ID 279246, 5 pages (2011)
5. Chellali, M., Maffray, F.: Dominator colorings in some classes of graphs. Graphs Comb. **28**, 97–107 (2021)
6. Chellali, M., Volkmann, L.: Relations between the lower domination parameters and the chromatic number of a graph. Discret. Math. **274**, 1–8 (2004)
7. Gera, R.M., Rasmussen, C., Horton, S.: Dominator colorings and safe clique partitions. Congr. Numer. **181**, 19–32 (2006)
8. Arumugam, S., Bagga, J., Chandrasekar, K.R.: On dominator colorings in graphs. Proc. Math. Sci. **122**, 561–571 (2012)
9. Gera, R.M., On dominator colorings in graphs. Graph. Theory Notes N. Y. **52**, 25–30 (2007)
10. Gera, R.M.: On the dominator colorings in bipartite graphs. In: Proceedings of the 4th International Conference on Information Technology, New Generations, Las Vegas, NV, vol. 2–4, pp. 947–952 (2007)
11. Henning, M.A.: Total dominator colorings and total domination in graphs. Graph. Comb. **31**, 953–974 (2015)
12. Kazemi, A.P.: Total dominator chromatic number and Mycieleskian graphs. Util. Math. **103**, 129–137 (2017)
13. Kazemi, A.P.: Total dominator coloring in product graphs. Util. Math. **94**, 329–345 (2014)
14. Haribabu, R., Arul, S.M.: Domination equitable coloring of graphs. Int. J. Math. Comput. Sci. **19**, 1417–1420 (2024)

Large Graphs with Small Burning Numbers

Angelin Jemima Rajasingh and N. Parthiban

1 Introduction

Burning number of a graph is a sequence of vertices selected using a certain protocol called a burning sequence. Spread of data existing at a certain vertex covers the entire network (graph) in a shortest duration of time, if it follows the burning sequence. On the other hand, spread of a pandemic with the outbreak at an initial vertex spreads to the entire network at a fast rate, if it follows the burning sequence. In this case, the spread of the disease can be controlled to a great extent, if preventive measures are undertaken at the vertices of the burning sequence. In computer networks, spread of viruses can be curtailed. In wireless sensor networks, the sensors that monitor and record data placed in the positions of the vertices in a burning sequence can be used to measure the environmental conditions such as humidity and wind [1–4].

We define a burning sequence in a graph G as follows [5]:

Step 1: Assign all vertices of the graph to a status of unburnt vertices.
Step 2: Select a new vertex v and call it burned. Further, burn all the vertices adjacent to the already burnt vertices.
Step 3: Continue Step 2 till all the vertices in the network are burned.

The sequence of selecting the vertices in Step 2 is called a *burning sequence*. Selecting minimum number of vertices in a burning sequence is the *burning number problem*. This minimum number is the burning number of the graph G and is denoted by $b(G)$.

A. J. Rajasingh (✉) · N. Parthiban
Department of Data Science and Business System, School of Computing, SRM Institute of Science and Technology Kattankulathur, Chennai, India
e-mail: parthibn@srmist.edu.in

I. Rajasingh et al. (eds.), *Applied Computational Mathematics*,
Springer Proceedings in Mathematics & Statistics 480,
https://doi.org/10.1007/978-3-031-77764-6_19

A variant of burning number problem is the r-burning number problem [1]. r-burning number problem is an extension of the burning number problem. It begins with choosing r vertices at a time and continue the burning sequence by choosing r new vertices at every stage. The *r-burning number* of G is the minimum number of steps to burn all vertices and is denoted by $b_r(G)$. The 1-burning number is the same as the well-known burning number problem. The generalization to r-burning number enables us to study different worst-case contagion scenarios by varying the spread factor r.

Large graphs consist of thousands of vertices and tens of thousands of edges between these vertices. In this paper, we explore graphs that could be large, but whose burning numbers are either 2 or 3.

2 Graphs with Burning Number 2 or 3

Definition 1 A vertex in a graph G is called a *universal vertex* if it is adjacent to all other vertices in G. A universal vertex is called an apex vertex of the graph.

Theorem 1 *A graph G with a universal vertex has b(G)=2.*

Proof Let v be an apex vertex of G. Choose v in the burning sequence of G. Choosing any other arbitrary vertex will trigger the burning of all other vertices of G as they are all neighbours of the already burnt vertex v. Thus $b(G) = 2$.

Corollary 1 *Let G be a wheel graph. Then b(G)=2.*

Definition 2 Let G_1 and G_2 be graphs. The join of G_1 and G_2 denoted by $G_1 \nabla G_2$ is the graph with $V(G_1 \nabla G_2) = V(G_1) \cup V(G_2)$, $E(G_1 \nabla G_2) = E(G_1) \cup E(G_2) \cup E$ where E is the set of all edges connecting each vertex of G_1 with every vertex of G_2. It is a commutative operation for unlabelled graphs. We observe that even if G_1 or G_2 is large, $G_1 \nabla G_2$ is large.

Theorem 2 *Let G_1 and G_2 be graphs. Suppose one of the graphs say G_1 has a vertex v with degree of v equal to $|V(G_1)|-1$ or $|V(G_1)|-2$. Then $b(G_1 \nabla G_2) = 2$. Otherwise $b(G_1 \nabla G_2) = 3$.*

Proof

Case 1: deg v=$|V(G1)| - 1$. In other words, v is a universal vertex. Choosing vertex v in Round 1 and any other arbitrary vertex in Round 2 in the burning sequence which burns every vertex in $G_1 \nabla G_2$.

Case 2: deg v=$|V(G1)| - 2$. Let u be the vertex of G_1 which is not adjacent to v. Then choosing vertex v in Round 1 and u in Round 2 in the burning sequence which burns every vertex in $G_1 \nabla G_2$. Thus, $b(G_1 \nabla G_2) = 2$.

Case 3: No vertex of degree $|V(G_1)| - 1$ or $|V(G_2)| - 2$ exists in G_1 (or G_2). These two arbitrarily chosen vertices in G_1 and a third vertex in G_2 are a burning sequence. In this case $b(G_1 \nabla G_2) = 3$.

3 Conclusion

Determining burning number of graphs obtained by graph operations such as Cartesian product of graphs and tensor product of graphs would be an interesting line of research.

References

1. Mondal, D., Parthiban, N., Kavitha, V., Rajasingh, I.: APX-Hardness and Approximation for the k-Burning Number Problem. ArXiv, abs/2006.14733 (2020)
2. Bonato, A., Janssen, J., Roshanbin, E.: How to burn a graph. Int. Math. **12**(1–2), 85–100 (2016)
3. Bonato, A., Janssen, J.C.M., Roshanbin, E.: Burning a graph as a model of social contagion. In: Bonato, A., Graham, F.C., Pralat, P. (Eds.), Proceedings of the 11th International Workshop on Algorithms and Models for the Web Graph (WAW). LNCS, vol. 8882, pp.13–22. Springer, Berlin (2014)
4. Bessy, S., Bonato, A., Janssen, J., Rautenbach, D., Roshanbin, E.: Bounds on the burning number. Discrete Appl. Math. **235**, 16–22 (2018)
5. Bonato, A.: A survey of graph burning. arXiv preprint arXiv:2009.10642 (2020)

Equality of Domination Parameters in Certain Necklace Graphs

V. Shalini and Indra Rajasingh

1 Introduction

A graph is determined by its points vertices or nodes and lines (edges or links) joining certain pairs of these points. Research in domination problems originated in the 1950s, but it gained considerable momentum from 1970 onwards [1–4]. Cockayne et al. [5] first proposed the idea of total domination, which has since found extensive use in fault tolerance. It was Kulli and Sigarkanthi who first suggested the inverse dominating set [6], and this concept was examined by numerous graph theorists. Cockayne et al. [7] presented the notion of secure domination, which was later explored by Benecke et al. [8]. Additionally, several authors have undertaken further studies on secure and secure total dominating sets [9–12].

We give below the necessary definitions required for our work. A few results available in the literature have also been listed out.

Definition 1 ([3, 5]) Let D be a subset of the vertex set of graph G. If each node in $G \setminus D$ is adjacent to some vertex in D, then the set D in G yields a dominating set. The minimum cardinality of a dominating set is called the domination number of G denoted by $\gamma(G)$. Applying the condition to every vertex in G instead of $G \setminus D$, we get a total dominating set and the total domination number $\gamma_t(G)$. Cardinality of a minimum dominating set with independent set of vertices is the independent domination number denoted by $\gamma_i(G)$.

Definition 2 ([6]): Let D be the minimum dominating set of G. A dominating set of G in $G \setminus D$ is called an inverse dominating set. The cardinality of a minimum inverse dominating set is called the inverse domination number denoted by $\gamma^{'}(G)$.

V. Shalini (✉) · I. Rajasingh
School of Advanced Sciences, Vellore Institute of Technology, Chennai, Tamil Nadu, India

I. Rajasingh et al. (eds.), *Applied Computational Mathematics*,
Springer Proceedings in Mathematics & Statistics 480,
https://doi.org/10.1007/978-3-031-77764-6_20

Definition 3 ([7]) If for every graph ($G = (V, E)$) and any subset ($D \subseteq V$), such that for each node $v \in V \setminus D$, there exists a node $u \in D$ adjacent to v and resultant set still remains a dominating set even after moving u to $V \setminus D$ yields secure dominating set. The minimum cardinality of a secure dominating set is the secure domination number denoted by $\gamma_s(G)$. Applying the condition to every vertex in G instead of $G \setminus D$, we get a secure total dominating set and the secure total domination number $\gamma_{st}(G)$.

Definition 4 ([13]) Let G_m be a graph on m vertices labeled $v_1, v_2, \ldots, v_m$ and let K_{t_i} be a complete graph on t_i vertices, $1 \leq i \leq m$. Merging vertex v_i of G with any one vertex of K_{t_i}, $1 \leq i \leq m$, yields a G-necklace graph denoted by $GN(G_m; K_{t_1}, K_{t_2}, \ldots, K_{t_m})$.

Theorem 1 ([14]) *For the path-necklace graph the domination and independent domination numbers are equal.*

Theorem 2 ([14]) *Let G be the C-necklace $CN\left(C_m; K_{t_1}, K_{t_2}, \ldots, K_{t_m}\right)$ where $n \geq 3$, $t_i \geq 3$, for $1 \leq i \leq m$, and $n = \sum_{i=1}^{m} t_i$ vertices. Then, $\gamma(G) = i(G) = m$.*

Theorem 3 ([14]) *Let G be the complete-necklace $KN\left(K_m; K_{t_1}, K_{t_2}, \ldots, K_{t_m}\right)$, where $m, t_i \geq 4$ and $1 \leq i \leq m$. Then, $\gamma(G) = \gamma_i(G) = m$.*

Theorem 4 ([14]) *For the star-necklace graph $SN\left(K_{1,m}; K_{t_1}, K_{t_2}, \ldots, K_{t_m}\right)$, where $m \geq 4$, and $t_i \geq 4$, for $1 \leq i \leq m$. γ and γ_i are equal to m.*

Theorem 5 ([15]) *In any graph G with $\delta(G) \geq 2$, the inequality $\gamma_s \leq \gamma_{st} \leq 2\,\gamma_s$ holds. Both limits are sharp.*

2 Equality of Domination, Total Domination, Inverse Domination, Secure and Secure Total Domination Numbers in Certain Necklace Graphs

In this section, we demonstrate that the domination, total domination, inverse domination, secure domination, and secure total domination numbers are all equal in (i) $PN\left(P_m; K_{t_1}, K_{t_2}, \ldots, K_{t_m}\right)$,
(ii) $CN\left(C_m; K_{t_1}, K_{t_2}, \ldots, K_{t_m}\right)$ and
(iii) $KN\left(K_m; K_{t_1}, K_{t_2}, \ldots, K_{t_m}\right)$.

Remark 1 Every vertex in a complete graph K_n constitutes a minimum dominating set of K_n.

Theorem 6 *Let G represent a P-Necklace PN, C-Necklace CN, or a K-Necklace KN. Thus the equality holds for $\gamma, \gamma', \gamma_t, \gamma_s, \gamma_{st}$ for path, cycle, and complete necklace graphs.*

Proof Let P denote the spine path of a path necklace comprising m number of vertices represented by $PN(P_m; K_{t_1}, K_{t_2}, \ldots, K_{t_m})$. From left to right, the m

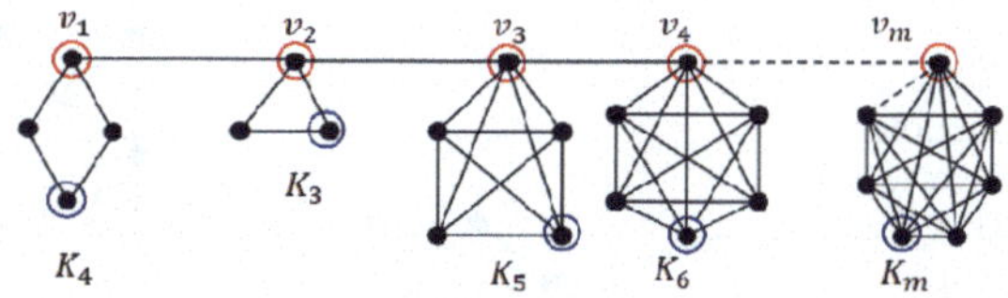

Fig. 1 The red-colored circled vertices denote the total, secure, and secure total dominating sets and blue-colored circled vertices indicate inverse dominating set of path necklace graph

Fig. 2 The red-colored circled vertices denote the total, secure, and secure total dominating sets and blue-colored circled vertices indicate inverse dominating set of C-necklace graph

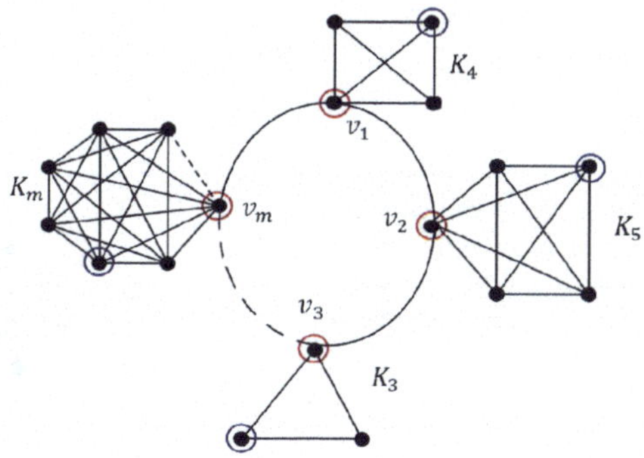

vertices of P are labeled as $v_1, v_2, \ldots, v_m$, respectively. Since $\{v_i\}$ dominates K_{t_i}, $1 \leq i \leq m$, $\{v_1, v_2, \ldots, v_m\}$ constitutes both a dominating set and a total dominating set of PN. Any set that excludes all vertices of K_{t_i} for some i, where $1 \leq i \leq m$ cannot qualify as a dominating set of PN. Therefore, $v_1, v_2, \ldots, v_m$ constitutes a minimum dominating set that generates a path. In $G \backslash \{v_1, v_2, \ldots, v_m\}$, we have $\{u_1, u_2, \ldots, u_m\}$ as a minimum dominating set. Hence, $\gamma(PN) = \gamma^{'}(PN) = \gamma_t(PN) = m$. See Fig. 1.

Let $u_i \neq v_i$ be a vertex in K_{t_i}, where $1 \leq i \leq m$. Consequently, $D = \{u_1, u_2, \ldots, u_m\}$ constitutes a minimum dominating set of PN. Every node in K_{t_i} connected u_i, where $1 \leq i \leq m$. Consequently, $(D - \{u_i\}) \cup \{v_i\}$ for any v_i in K_{t_i}, where $1 \leq i \leq m$ constitutes a dominating set of PN. Therefore, the node set $\{u_1, u_2, \ldots, u_m\}$ is a minimum secure dominating set of PN. In a similar manner, $\{v_1, v_2, \ldots, v_m\}$ constitutes a minimum secure total dominating set of PN. Thus, $\gamma(PN) = \gamma^{'}(PN) = \gamma_t(PN) = \gamma_s(PN) = \gamma_{st}(PN) = m$. A comparable argument applies to CN and KN. See Figs. 2 and 3.

Theorem 7 *Let G represent the Star-Necklace SN. Then (i) $\gamma(G) = \gamma^{'}(G) = m$ (ii) $\gamma_t(G) = \gamma_s(G) = \gamma_{st}(G) = m + 1$.*

Proof

(i) Assign the labels $v_1, v_2, \ldots, v_m$ to the rim vertices of $SN(K_{1,m}; K_{t_1}, K_{t_2}, \ldots, K_{t_m})$ in the clock-wise sense and designate v_0 as the hub vertex. Let u_i and v_i be distinct vertices in K_{t_i}, where $1 \leq i \leq m$. The sets $\{v_1, v_2, \ldots, v_{m-1}, u_m\}$ and $\{u_1, u_2, \ldots, u_{m-1}, v_m\}$constitute minimum

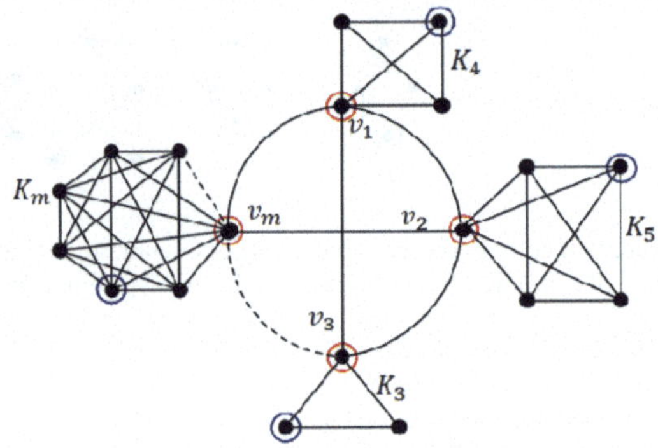

Fig. 3 The red-colored circled vertices denote the total, secure, and secure total dominating sets and inverse dominating set circled in blue color of K-necklace graph

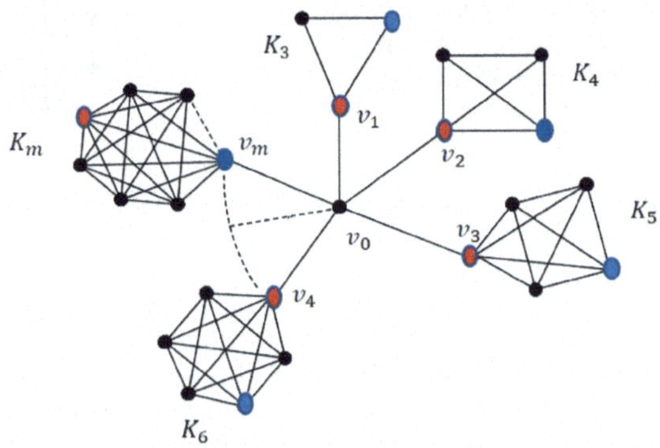

Fig. 4 Red-colored vertices denote the dominating set, and blue-colored vertices indicate inverse dominating set of star necklace graph

dominating set of G. Hence, the domination and inverse domination numbers equal to m. See Fig. 4.

(ii) We observe that $\{v_1, v_2, \ldots, v_m\}$ constitutes a minimum independent dominating set of G. Therefore the hub node must be included into any total dominating set. We ascertain that $\{v_0, v_1, v_2, \ldots, v_m\}$ constitutes a minimum total dominating set of G. Therefore$\gamma_t(G) = m+1$. For $v_0 \in V(G)$, there exists no minimum dominating set D of G containing $v \in D$ such that $(D-\{v\}) \cup \{v_0\}$ constitutes a dominating set of G. Therefore, any secure dominating set must encompass v_0 resulting in a secure domination number of $m + 1$. Further $\{v_0, v_1, v_2, \ldots, v_m\}$ gives $\gamma_t(G)$. Therefore, the domination numbers for total, secure and secure total are all equal to $m + 1$. Refer to Fig. 5.

3 Conclusion

We have computed the total, inverse, secure, and secure total domination numbers for the Path, Cycle, Complete, and Star necklace graphs. Obtaining these parameters

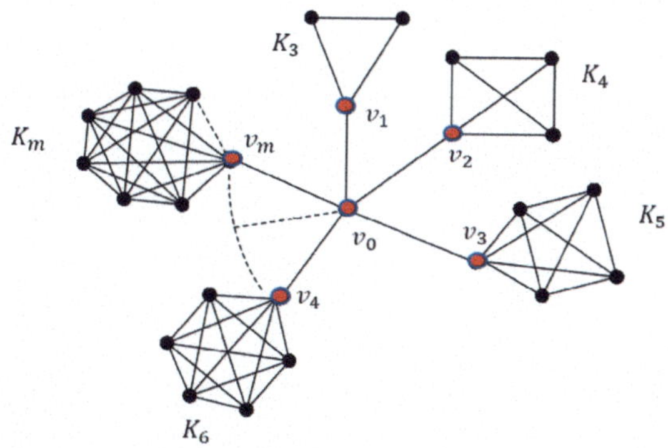

Fig. 5 Red vertices indicate the total, secure, and secure total dominating sets of the star necklace graph

for necklace graphs with cycles and complete bipartite graphs identified at vertices of paths, cycles, complete, and star graphs are yet to be studied.

References

1. Hedetniemi, S.T., Laskar, R.C.: Topics on Domination. Elsevier, Amsterdam (1991)
2. Ore, O.: Theory of Graphs. Colloquium Publications, Kolkata (1962)
3. Cockayne, E.J., Hedetniemi, S.T.: Towards a theory of domination in graphs. Networks **7**(3), 247–261 (1977)
4. Haynes, T.W., Hedetniemi, S., Slater, P.: Fundamentals of Domination in Graphs. CRC Press, Boca Raton (2013)
5. Cockayne, E.J., Dawes, R.M., Hedetniemi, S.T.: Total domination in graphs. Networks **10**(3), 211–219 (1980)
6. Kulli, V.R., Sigarkanti, S.C.: Inverse domination in graphs. Nat. Acad. Sci. Lett. **14**(12), 473–475 (1991)
7. Cockayne, E.J., Grobler, P.J.P., Gründlingh, W.R., Munganga, J., Van Vuuren, J.H.: Protection of a graph. Util. Math. **67**, 19–32 (2005)
8. Benecke, S., Cockayne, E.J., Mynhardt, C.M.: Secure total domination in graphs. Util. Math. **74**, 247–259 (2007)
9. Duginov, O.: Secure total domination in graphs: bounds and complexity. Discrete Appl. Math. **222**, 97–108 (2017)
10. Cabrera Martínez, A., Montejano, L.P., Rodríguez-Velázquez, J.A.: On the secure total domination number of graphs. Symmetry **11**(9), 1165 (2019)
11. Merouane, H.B., Chellali, M.: On secure domination in graphs. Inf. Process. Lett. **115**(10), 786–790 (2015)
12. Jha, A., Pradhan, D., Banerjee, S.: The secure domination problem in cographs. Inf. Process. Lett. **145**, 30–38 (2019)
13. Rajan, R.S., Parthiban, N., Rajalaxmi, T.M.: Embedding of recursive circulants into certain necklace graphs. Math. Comput. Sci. **9**(2), 253–263 (2015)
14. Anitha, J., Suganya, V., Jayaraman, G.: Independent domination number in necklace graph. J. Int. Pharmaceut. Res. **46**(1), 388–392 (2019)
15. Klostermeyer, W., Mynhardt, C.: Secure domination and secure total domination in graphs. Discuss. Math. Graph Theory **28**, 267–284 (2008)

Higher Order Triangular Sum Labeling of Snowflake Graph

Sharon Philomena V and J. Sweena Starlet

1 Introduction

Graph labeling is a current field of research in graph theory. A labeling of a graph is an assignment of values to the vertices and edges of graphs subject to certain conditions. Sum labeling is a fascinating and actively investigated topic in graph theory. The sum labeling issue entails assigning numeric labels to vertices and edges of a graph in such a way that specific requirements or constraints are met [1]. In 2008, Hedge and Shankaran [2] introduced the TSL (triangular sum labeling) of a graph with a vertex and edges to admit the triangular sum labeling. Centered triangular sum labeling requires finding label arrangements that satisfy both the sum constraint and the centered condition. Then we are going to show that the snowflake graph and bi-snowflake graph under the triangular and CTSL centered triangular sum labeling.

2 Prelims

Definition 1 The generalized $m, n-$ snowflake graph $G = K_{(1,m)} \odot K_{(1,n)}$, where $m \geq 3, n \geq 2$ is a graph with $in + (i+1)$ vertices and $i(n+1)$ edges where $m \geq 3$ & $n \geq 1$ where the vertices of the star graph $K_{(1,m)}$ adjacent to $u_1, u_2, \cdot, u_i$, respectively.

Sharon Philomena V (✉) · J. S. Starlet
PG. Department of Mathematics, Women's Christian College, University of Madras, Chennai, India
e-mail: sharonphilomena@wcc.edu.in

I. Rajasingh et al. (eds.), *Applied Computational Mathematics*,
Springer Proceedings in Mathematics & Statistics 480,
https://doi.org/10.1007/978-3-031-77764-6_21

Definition 2 A bi-snowflake graph is derived by joining the apex vertices of two copies of $m, n-$ snowflake graph denoted by $B(h_1, h_2, m, n)$. h_1, h_2- represents the half of the snowflake graph.

3 Main Result

In this section, we prove the newly defined m, n-snowflake graph for $m \geq 3$ is a centered triangular sum graph.

Theorem 1 *The mn-snowflake graph admits higher-order triangular sum labeling.*

Let G denote the $m, n-$*snowflake graph* $K_{1,m} o K_{1,n}$, *where* $m \geq 6$ *and* $n \geq 2$.

Let u be the central vertex and let $u_1, u_2, \ldots, u_i$ *be the pendant vertices of the star graph* $K_{1,m}$. *Let* $v_j^I, v_j^{II}, v_j^{III}, \ldots, v_j^n$ *be the vertices of the star graph* $K_{1,n}$ *adjacent to* $u_1, u_2, \ldots, u_i$, *respectively.*

Let f be injective function $f(G) : V(G) \rightarrow W$ *(where W is the set of all nonnegative integers) that induces a bijective function* $f^{+}(G) : E(G) \rightarrow \{T_1, T_2, \ldots, T_k$ *to the edges of* $\mathcal{G}$.

$$f(G) = \left\{u, u_i, v_j^m : 1 \leq i \leq m, 1 \leq j \leq n\right\}$$

$$f^{+}(G) = \left\{uu_i, u_i v_j^m : 1 \leq i \leq m, 1 \leq j \leq n\right\}$$

Since the graph obtained is a tree, clearly G has $m(n+1)+1$ *vertices and* $m(n+1)$ *edges.*

For $1 \leq i \leq m$ *and* $1 \leq j \leq n$,

Case (i): *If* T_k *are the first-order triangular numbers for* $k = 1, 2, 3, \ldots$, *we define*

$$T_k = \frac{1}{2}k(k+1)$$

Case (ii): *If* T_k *are the second-order triangular numbers for* $k = 1, 2, 3, \ldots$, *we define*

$$T_k = \frac{1}{6}k(k+1)(2k+1)$$

Case (iii): *If* T_k *are the third-order triangular numbers for* $k = 1, 2, 3, \ldots$, *we define*

$$T_k = \frac{1}{4}k^2(k+1)^2$$

Case (iv): *If* T_k *are the fourth-order triangular numbers for* $k = 1, 2, 3, \ldots$, *we define*

$$T_k = \frac{1}{30}k(k+1)(2k+1)(3k^2+3k-1)$$

Case (v): *If T_k are the fifth-order triangular numbers for $k = 1, 2, 3, \ldots,$ we define*

$$T_k = \frac{1}{12}k^2(k+1)^2(2k^2+2k-1)$$

Case (vi): *If T_k are the sixth-order triangular numbers for $k = 1, 2, 3, \ldots,$ we define*

$$T_k = \frac{1}{42}k(k+1)(2k+1)(3k^4+6k^3-3k+1)$$

Case (vii): *If T_k are the seventh-order triangular numbers for $k = 1, 2, 3, \ldots,$ we define*

$$T_k = \frac{1}{24}k^2\,(k+1)^2\,(3k^4+6k^3-k^2-4k+2)$$

Case (viii): *If T_k are the eighth-order triangular numbers for $k = 1, 2, 3, \ldots,$ we define*

$$T_k = \frac{k(k+1)(2k+1)}{90}[5k^6+15k^5+5k^4-15k^3-k^2+9k-3]$$

Case (ix): *If T_k are the ninth-order triangular numbers for $k = 1, 2, 3, \ldots,$ we define*

$$T_k = \frac{1}{20}k^2\,(k+1)^2\,(k^2+n-1)(2k^4+4k^3-k^2-3k+3)$$

Case (x): *If T_k are the tenth-order triangular numbers for $k = 1, 2, 3, \ldots,$ we define*

$$T_k = \frac{1}{66}k(k+1)(2k+1)(k^2+k-1)[3k^6+9k^5+2k^4-11k^3+3k^2+10k-5]$$

The central vertex is labeled as $f\,(u) = 0$ and $(u_i) = T_i 1 \le i \le m$.

$$\mathrm{f}\left(\mathrm{v}_{\mathrm{j}}^{\mathrm{I}}\right) = \mathrm{T}_{\mathrm{m+j}} - \mathrm{f}(\mathrm{u}_1)$$

$$\mathrm{f}\left(\mathrm{v}_{\mathrm{j}}^{\mathrm{II}}\right) = \mathrm{T}_{\mathrm{m+n+j}} - \mathrm{f}(\mathrm{u}_2)$$

$$\mathrm{f}\left(\mathrm{v}_{\mathrm{j}}^{\mathrm{III}}\right) = \mathrm{T}_{\mathrm{m+2n+j}} - \mathrm{f}(\mathrm{u}_3)$$

$$\mathrm{f}\left(\mathrm{v}_{\mathrm{j}}^{\mathrm{IV}}\right) = \mathrm{T}_{\mathrm{m+3n+j}} - \mathrm{f}(\mathrm{u}_4)$$

$$\vdots$$

$$f\left(v_j^m\right) = T_{m+(n-1)n+j} - f(u_m)$$

Thus, the edge labels are the first n triangular numbers and the vertices u, u_i, v_j^m obtained are all distinct and in nondecreasing order.

Hence, the snowflake graph G admits triangular sum labeling.

Proof Let G denote the m, n-snowflake graph $K_{(1,m)} \odot K_{(1,n)}$ for $m \geq 3$ and $n \geq 2$.

Let $f(V)$ and $f^*(E)$ denote the number of vertices and edges, respectively.

Let u be the middle vertex.

Let $u_1, u_2, \ldots, u_i$ be the vertices adjacent to the central vertex u and are labeled in a clockwise direction.

Let $v_j^I, v_j^{II}, v_j^{III}, v_j^{IV}, v_j^V, v_j^{VI}, \ldots, v_j^m$, respectively (Fig. 1).

The vertex set is given by

$$f(V) = \left\{u, u_i, v_j^I, v_j^{II}, v_j^{III}, v_j^{IV}, v_j^V, v_j^{VI}, \ldots, v_j^m : 1 \leq i \leq m, 1 \leq j \leq n\right\}$$

The edge set is given by

$$f^*(E) = \left\{uu_i, u_i v_j^I, u_i v_j^{II}, u_i v_j^{III}, u_i v_j^{IV}, u_i v_j^V, u_i v_j^{VI}, \ldots, u_i v_j^m \right.$$
$$\left. : 1 \leq i \leq m, 1 \leq j \leq n\right\}$$

Clearly, G has $V = in + (i + 1)$ and $E = i(n + 1)$. For $m \leq i \leq n, n \geq 2$, and $1 \leq j \leq n$, define

$$C_n = \frac{1}{2}(3n^2 - 3n + 2)$$

The central vertex is labeled as $f(u) = 0$ and

$$f(u_i) = C_i, \tag{1}$$

The end vertices of the first star label:

Fig. 1 $m, n-$ snowflake graph

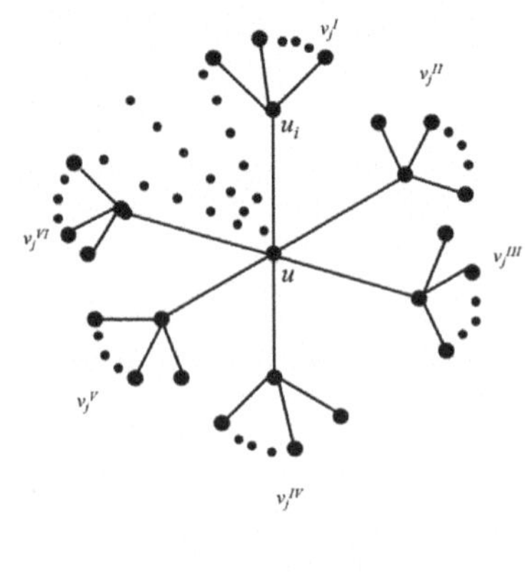

$$f\left(v_j^I\right) = C_{(m+j)} - f(u_1), \tag{2}$$

The end vertices of the second star label:

$$f\left(v_j^{II}\right) = C_{(m+n+j)} - f(u_2), \tag{3}$$

The end vertices of the third star label:

$$f\left(v_j^{III}\right) = C_{(m+2n+j)} - f(u_3), \tag{4}$$

The end vertices of the fourth star label:

$$f\left(v_j^{IV}\right) = C_{(m+3n+j)} - f(u_4), \tag{5}$$

The end vertices of the fifth star label:

$$f\left(v_j^{V}\right) = C_{(m+4n+j)} - f(u_5), \tag{6}$$

The end vertices of the mth star label:

$$f\left(v_j^{m}\right) = C_{(m+kn+j)} - f(u_m), \tag{7}$$

Thus, the edge labels and the vertices $u, u_i, v_j^I, v_j^{II}, v_j^{III}, v_j^{IV}, v_j^{V}, v_j^{VI}, \ldots, v_j^m$ all are distinct.

Hence, the snowflake graph G admits the CTSL.

An illustration of 6, 5-snowflake graph is shown in Fig. 2.

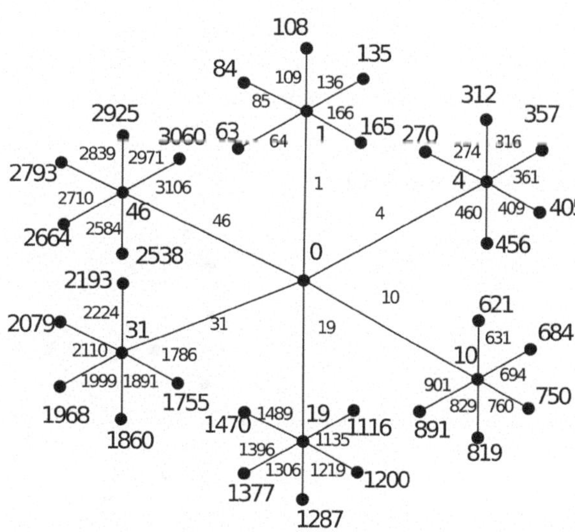

Fig. 2 6, 5-snowflake graph

Theorem 2 *The bi-snowflake graph admits the CTSL.*

Proof Let G denote the m, n-snowflake graph for $m \geq 3$ and $n \geq 2$.

Then the bi-snowflake graph is denoted as $B(h_1, h_2, m, n)$.

Let $f(V)$ and $f^*(E)$ denote the number of vertices and edges, respectively (Fig. 3).

The vertex set is given by

$$f(V) = \left\{u, u_i, v, v_i, u_j^I, u_j^{II}, \ldots, u_j^m, v_j^I, v_j^{II}, \ldots, v_j^m : 1 \leq i \leq k, 1 \leq j \leq n\right\}$$

The edges set is given by

$$f^*(E) = \left\{uv, uu_i, vv_i, u_iu_j^m, v_iv_j^m : 1 \leq i \leq k, 1 \leq j \leq n, 1 \leq m \leq n\right\}$$

Clearly, G has $V = i(n+1) - 1$ and $E = in(i+1)$ For $m \leq i \leq n, n \geq 2$, and $1 \leq j \leq n$, define

$$C_n = \frac{1}{2}(3n^2 - 3n + 2)$$

The vertex is labeled as $u = 0$ and $v = 1$:

$$f(u_i) = C_i, \tag{8}$$

$$f(v_i) = C_{i+1}, \tag{9}$$

The end vertices of the first star in h_1 label:

$$f\left(u_j^I\right) = C_{(m+j+1)} - f(u_1), \tag{10}$$

The end vertices of the second star in h_1 label:

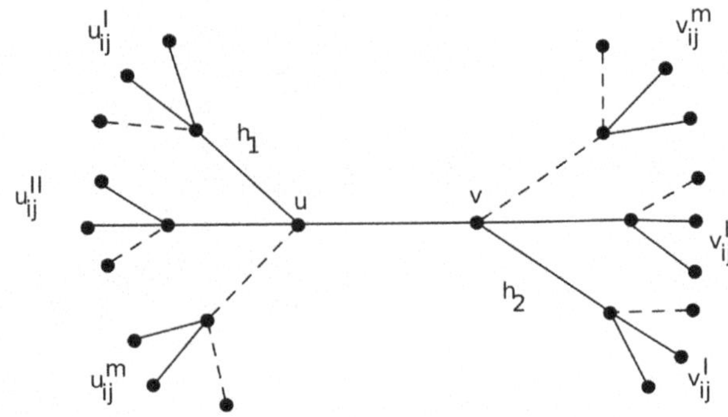

Fig. 3 Bi-snowflake graph $B(h_1, h_2, m, n)$

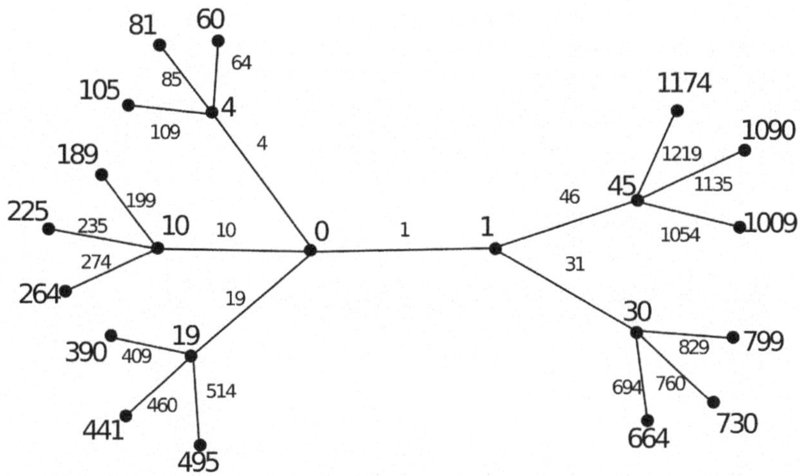

Fig. 4 Bi-snowflake graph $B(3, 2, 5, 3)$

$$f\left(u_j^{II}\right) = C_{(m+j+n+1)} - f(u_2), \tag{11}$$

The end vertices of the u^i star in h_1 label:

$$f\left(u_j^{i}\right) = C_{(m+j+kn+1)} - f(u_i), \tag{12}$$

The end vertices of the first star in h_2 label:

$$f\left(v_j^{I}\right) = C_{(m+j+(k+1)n+2)} - f(v_1), \tag{13}$$

The end vertices of the second star in h_2 label:

$$f\left(v_j^{II}\right) = C_{(m+j+(k+2)n+1)} - f(v_2), \tag{14}$$

The end vertices of the v^i star in h_2 label:

$$f\left(v_j^{i}\right) = C_{(m+j+sn+1)} - f(v_i), \tag{15}$$

Thus, the edge and vertex $u, v, u_i, v_i, u_j^I, u_j^{II}, u_j^{III}, \ldots, u_j^m, v_j^I, v_j^{II}, v_j^{III}, \ldots, v_j^m$ all are distinct.

Hence, the bi-snowflake graph G admits the CTSL.

An illustration of bi-snowflake graph $B(3, 2, 5, 3)$ is shown in Fig. 4.

Theorem 3 *The bi-snowflake graph admits the TSL.*

Proof Let G denote the m, n-snowflake graph for $m \geq 3$ and $n \geq 2$.

Then the bi-snowflake graph is denoted as $B(h_1, h_2, m, n)$.

The vertex set is given by

$$f(V) = \left\{u, u_i, v, v_i, u_j^I, u_j^{II}, \ldots, u_j^m, v_j^I, v_j^{II}, \ldots, v_j^m : 1 \leq i \leq k, 1 \leq j \leq n\right\}$$

The set of edges is given by

$$f^*(E) = \left\{uv, uu_i, vv_i, u_i u_j^m, v_i v_j^m : 1 \leq i \leq k, 1 \leq j \leq n, 1 \leq m \leq n\right\}$$

Clearly, G has $V = i(n+1) - 1$ and $E = in(i+1)$. For $m \leq i \leq n, n \geq 2$, and $1 \leq j \leq n$, define

$$T_n = \frac{1}{2}n(n+1)$$

The vertex is labeled as $f(u) = 0$ and $f(v) = 1$:

$$f(u_i) = T_i, \tag{16}$$

$$f(v_i) = T_{i+1}, \tag{17}$$

The end vertices of the first star in h_1 label:

$$f\left(u_j^I\right) = T_{(m+j+1)} - f(u_1), \tag{18}$$

The end vertices of the second star in h_1 label:

$$f\left(u_j^{II}\right) = T_{(m+j+n+1)} - f(u_2), \tag{19}$$

The end vertices of the u^i star in h_1 label:

$$f\left(u_j^i\right) = T_{(m+j+kn+1)} - f(u_i), \tag{20}$$

The end vertices of the first star in h_2 label:

$$f\left(v_j^I\right) = T_{(m+j+(k+1)n+2)} - f(v_1), \tag{21}$$

The end vertices of the second star in h_2 label:

$$f\left(v_j^{II}\right) = T_{(m+j+(k+2)n+1)} - f(v_2), \tag{22}$$

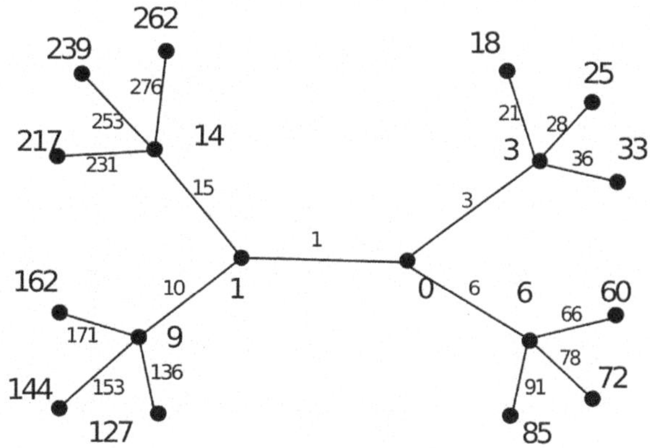

Fig. 5 Bi-snowflake graph $B(h_2, h_2, 4, 3)$

The end vertices of the v^i star in h_2 label:

$$f\left(v_j^i\right) = T_{(m+j+sn+1)} - f(v_i), \tag{23}$$

where $1 \le k \le s \le n$ Thus, the edge labels and the vertices $u, v, u_i, v_i, u_j^I, u_j^{II}, u_j^{III}, \ldots, u_j^m, v_j^I, v_j^{II}, v_j^{III}, \ldots, v_j^m$ are all distinct.

Hence, the bi-snowflake graph G admits the TSL.

An illustration of bi-snowflake graph $B(h_1, h_2, m, n)$ is shown in Fig. 3 (Fig. 5).

4 Conclusion

In this paper, we looked into the CTSL of the snowflake and the bi-snowflake graph. This study adds the new findings and the ideas of graph labeling. The CTSL can be proven for many graphs. Also, more CTSL can be studied.

References

1. Harary. F.: Graph Theory. Adision – Wesley, Reading Mass (1969)
2. Hegde, S.M., Shankaran, P.: On Triangular Sum Labeling of Graphs in Labeling of Discrete Structures and Applications, pp. 109–115. Narosa Publishing House, New Delhi (2008)

Paired Transmission in Networks

Jessy Sujana G, Jane Olive Sharon, and J. Anitha

1 Introduction

In theoretical computer science, *transmission* is considered as one of the basic descriptors of fixed interconnection networks because it provides the average distance between any two nodes of the network [5, 6]. With Wiener's discovery of a close correlation between the boiling points of certain alkanes and the sum of the distances between vertices in the graphs representing their molecular structures, it became apparent that graph parameters and topological indices can potentially be used to predict properties of chemical compounds.

In the classical study of distances in graph theory, the main focus has been on the study of the two main graph parameters concerned with distance, namely, the diameter and the radius. Methods developed to determine the diameter of a graph became the foundation for most research in other distance-related graph parameters [9, 10].

J. Sujana G (✉)
Department of Computer Science, SRM Institute of Science and Technology, Vadapalani, Chennai, India
e-mail: jessysug@srmist.edu.in

J. O. Sharon
Department of Computing Technologies, SRM Institute of Science and Technology, Chennai, India

J. Anitha
Department of Mathematics, Rajalakshmi Institute of Technology, Chennai, India

I. Rajasingh et al. (eds.), *Applied Computational Mathematics*,
Springer Proceedings in Mathematics & Statistics 480,
https://doi.org/10.1007/978-3-031-77764-6_22

1.1 Transmission in Networks

The *transmission* of a vertex u, also called *farness* or *vertex Wiener value* of u in the literature, is defined as the sum of the lengths of all shortest paths between the chosen vertex u and all other vertices in G [7, 8]. In other words, $T(u) = \sum_{v \in V(G)} d(u, v)$.

Using transmission, we can define the Wiener index $W(G)$ of a graph G. This graph index is defined for a connected graph G as the sum of the lengths of shortest paths between all unordered pairs of vertices in G. It is the oldest topological index related to molecular branching, and based on its success, many other topological indices correlating to the distance matrix of chemical graphs have been developed subsequent to Wiener's work [1, 13]. The Wiener index was at first used for predicting the boiling points of paraffins, but later a strong correlation between the Wiener index and other chemical and physical properties of a compound was found, such as critical points in general, the density, surface tension, viscosity of compound's liquid phase, and the van der Waals surface area of the molecule [2–4]. In theoretical computer science, the Wiener index is known as the *transmission index* and is considered as one of the basic descriptors of fixed interconnection networks because it provides the average distance between any two nodes of the network [14].

The transmission $T(u)$ of a vertex is a concept closely related to the Wiener index but localized to the selected vertex. The reciprocal of $T(u)$ is defined as closeness of vertex u. This is a useful measure that establishes how fast the flow of information would be through a give node to other nodes [11, 12].

In location theory, sets of vertices with the minimum (or maximum) distance in a graph play a special role because they form target sets for locations of facilities. The framework can be made more general by considering the sum of distances to a specified multiset of vertices referred to as profiles. In real life, a school and a hospital complex that are located close to each other are preferred to be with no communication link between them. See Fig. 1a, b.

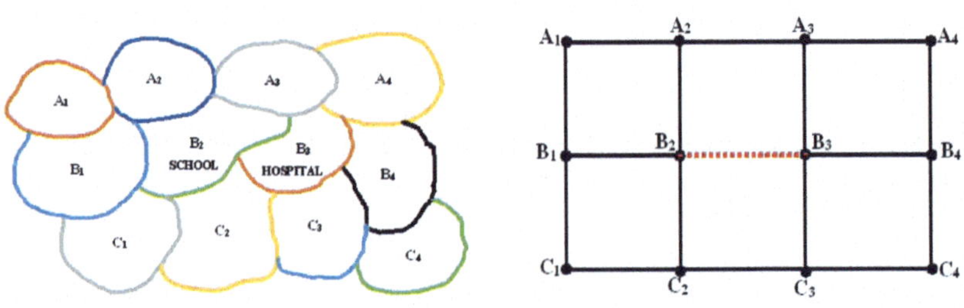

(a) (b)

Fig. 1 (**a**) Map of a region. (**b**) Mathematical model of the network for paired facility locations

In this paper, we consider exactly two vertices u and v in a profile of a network G and impose the condition that u and v are not adjacent in G. We call it a *paired transmission* in the network G. The same has been considered for a grid network.

2 Grid Network

Let P_r denote a path of length r. Then the Cartesian product $P_m \times P_n$ is called a grid network. This grid is denoted as $M_{(m \times n)}$ with m rows and n columns. A vertex v in the ith row and j^{th} column is denoted as $v \equiv (i, j)$, $1 \leq i \leq, 1 \leq j \leq n$. An edge in a row is called a *horizontal edge* and an edge in a column is called a *vertical edge*.

Definition 1 ([15]) Let M be grid network of order $m \times n$, $m \leq n$. For an edge, $e = (a, b)$ in M, the paired transmission T_e is defined as the sum of the transmissions from a and b in $M - e$.

Theorem 1 *Let M be grid network of order $m \times n$, $m \geq n$ and let $e = (a, b)$ be an edge of M, where $a \equiv (p, q)$ and $b \equiv (p, q+1)$. Then, $T_e = [(p-1)p + (m-p)(m-p+1)]n + [(q-1)q + (n-q-1)(n-q)]m + 2q(n-q) + mn$.*

Proof Suppose $e = (a, b)$ is a horizontal edge of M and $a \equiv (p, q), b \equiv (p, q+1)$, then $T_e = [(p-1)p + (m-p)(m-p+1)]n + [(q-1)q + (n-q-1)(n-q)]m + 2q(n-q) + mn$. Consider vertices u and v in M which are not in the same row. Then there exists a shortest path between u and v which does not pass through e. See Fig. 2a. If u and v are in the same row, which does not contain e, then the unique shortest path between u and v does not contain e. See Fig. 2b. Suppose u and v are in the same row which contains e. If the y coordinates of both u and v are less than or equal to q or if the y coordinates of both u and v are greater than or equal to $q+1$, then no shortest path between u and v passes through e. See Fig. 2c. We now consider the case when the y coordinate of u is less than or equal to q and the y coordinate of v is greater than or equal to $q+1$. In this case, every shortest paths between u and v passes through e. Every shortest path between such pairs of points in M contribute an addition of length 2 in $M - e$. There are q vertices in row p to the left of vertex a and $(n-q)$ vertices in row p to the right of vertex b. In this case, any shortest path between u and v not passing through e contributes an additional length of 2. See Fig. 2d. We now proceed to compute T_e. Let $S_1, S_2, \ldots, S_{m-1}$ be the horizontal convex cuts as shown in Fig. 3a. Their contribution to $T(a)$ and $T(b)$ is

$$= 2[(n + 2n + \cdots + (p-1)n) + (n + 2n + \cdots + (m-p)n)]$$

$$= 2n\Big[\frac{(p-1)p}{2} + \frac{(m-p)(m-p+1)}{2}\Big]$$

$$= n[(p-1)p + (m-p)(m-p+1)]$$

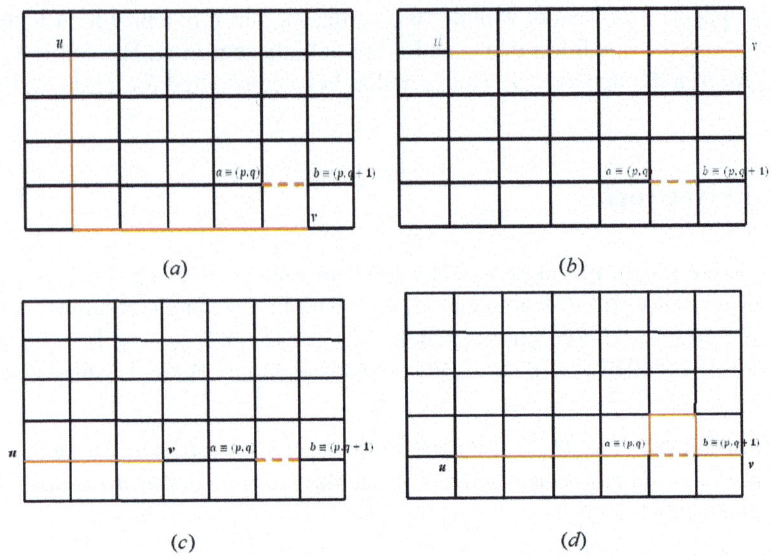

Fig. 2 (**a**) u and v are not in the same row, (**b**) u and v are in the same row, (**c**) y coordinates of u and v are less than q, and (**d**) y coordinates of u is less than q and the y coordinates of v is greater than q+1

Further, let $S'_1, S'_2, \ldots, S'_{n-1}$ be the vertical convex cuts as shown in Fig. 3b. Their contribution to $T(a)$ and $T(b)$ is

$$
\begin{aligned}
&= (m + 2m + \cdots + (q-1)m) + (m + 2m + \cdots + (n-q)m) \\
&\quad + (m + 2m + \cdots + qm) + (m + 2m + \cdots + (n-q-1)m) + 2(q)(n-q) \\
&= 2m\frac{(q-1)q}{2} + qm + (n-q)m + 2m\frac{(n-q-1)(n-q)}{2} + 2q(n-q) \\
&= (q-1)qm + mn + (n-q-1)(n-q)m + 2q(n-q)
\end{aligned}
$$

Thus, $T_e = [(p-1)p + (m-p)(m-p+1)]n + [(q-1)q + (n-q-1)(n-q)]m + 2q(n-q) + mn$.

Remark 1 If edge $e = (a, b)$ is a vertical edge of $M_{m\times n}$, then interchanging the roles of rows and columns in $M_{m\times n}$, T_e can be easily derived.

Open Problem Suppose e_i, $i = 1, 2, \ldots, k$ is a collection of independent edges in $M_{m\times n}$ and the profile of $M \setminus \{e_1, e_2, \ldots, e_k\}$ is defined to be profile $(M \setminus \{e_1, e_2, \ldots, e_k\}) = \sum_{i=1}^{k} T_{e_i}$ where T_{e_i} is the sum of the transmissions at the end vertices of e_i in $\{e_1, e_2, \ldots, e_k\}$, $1 \le i \le k$. The problem is to derive an expression for the profile of $(M \setminus \{e_1, e_2, \ldots, e_k\})$.

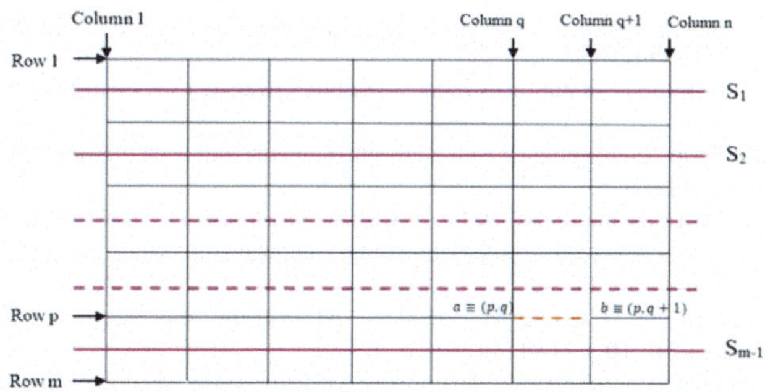

(*a*)

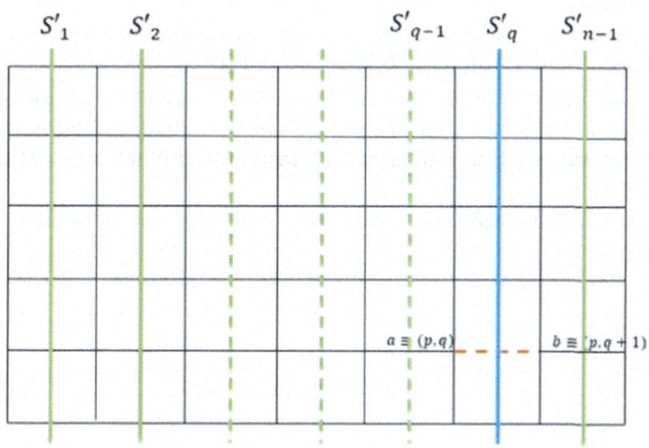

(*b*)

Fig. 3 (**a**) Horizontal edge cuts $S_1, S_2, \ldots, S_{(m-1)}$. (**b**) Vertical edge cuts $S'_1, S'_2, \ldots, S'_{(n-1)}$

3 Conclusion

Paired transmissions are to be studied in networks such as honeycomb networks, Sierpinski-like network, fractals, hypercubes, and butterfly networks.

References

1. Ahmadi-Javid, A., Seyedi, P., Syam, S.S.: A survey of healthcare facility location. Comput. Oper. Res. **79**, 223–263 (2017)
2. Al-Yakoob, S., Stevanović, D.: On interval transmission irregular graphs. J. Appl. Math. Comput. **68**, 1–24 (2022)

3. Babujee, J.B., Senbagamalar, J.: Wiener index of graphs using degree sequence. Appl. Math. Sci. **6**(88), 4387–4395 (2012)
4. Bavelas, A.: Communication patterns in task-oriented groups. J. Acoust. Soc. Am. **22**(6), 725–730 (1950)
5. Bienstock, D., Györi, E.: Average distance in graphs with removed elements. J. Graph Theory **12**(3), 375–390 (1988)
6. Borgatti, S.P.: Centrality and network flow. Soc. Networks **27**(1), 55–71 (2005)
7. Brandes, S.U.: A faster algorithm for betweenness centrality J. Math. Sociol. **25**(2), 163–177 (2001)
8. Bruno, G., Cerulli, R., Gentili, M.: Models. Algorithms and Applications for Location Problems. Optim. Lett. **10**, 871–873 (2018)
9. Dobrynin, A.A.: Infinite family of 2-connected transmission irregular graphs. Appl. Math. Comput. **340**, 1–4 (2019)
10. Dobrynin, A.A., Gutman, I.: The wiener index for trees and graphs of hexagonal systems. Diskretnyi Analiz i Issledovanie Operatsii **5**(2), 34–60 (1998)
11. Dobrynin, A.A., Sharafdini, R.: Stepwise transmission irregular graphs. Appl. Math. Comput. **371**, 124949 (2020)
12. Fontaine, F., Minner, S.: A dynamic discrete network design problem for maintenance planning in traffic networks. Ann. Oper. Res. **253**, 757–772 (2017)
13. Freeman, L.C., et al., Centrality in Social Networks: Conceptual Clarification, Social Network: Critical Concepts in Sociology, vol. 1, pp. 238–263. Routledge, Milton Park (2002)
14. Krnc, M., Škrekovski, R.: Centralization of transmission in networks. Discrete Math. **338**(12), 2412–2420 (2015)
15. Sharon P, J.O.: Transmission and closeness centrality in networks. Ph.D. thesis, Anna University (2020)

Part II
Fuzzy Mathematics

An Application of Generalized Fuzzy Soft Sets in Analyzing Student's Performance Level During Online Mode of Education

P. Miya Fathima Benazir and G. Gomathi

1 Introduction

In the modern new-fangled situations, hassle solving and decision-making are the complex areas that entail a vital endeavor to resolve commercial enterprise and existence complexities. Problem-solving obviously requires prioritizing and decision-making crucial for management. There are tactics and strategies to enable quick decision-making and the first-class of choices. A wide range of choices has rendered resolving administrative problems unanswered. The enigmatic nature of opportunities thrown open does impact delay or indecision. Most mathematical tools, specially statistical data or probability functions, come handy reconciling such issues. But then, those intricate concerns that never get conciliated require a higher-order interpretation, thus paving a way for modern strategies such as fuzzy rough set [9], and soft set theories. Uncertainties, arising due to partial information about the problem or inadequate receipt of information from a couple of supply, or facts that are not always completely dependable, pose hurdles to decision-making. Fuzzy set theory, rough set theory, vague set theory, theory of probability are some mathematical tools that can handle uncertainty arising due to vagueness. Fuzzy set concept, rough set theory, vague set concept, and concept of probability are a few mathematical gear that can handle uncertainty bobbing up because of vagueness.

Fuzzy set theory [13] has been a subtle disguise of the probability theory and is quite aptly suitable for modeling any actual lifestyle phenomenon. Zadeh [12] conceived the concept of fuzzy sets and later the concept of soft computing to deal with impreciseness and uncertainty concerned in all decision-making issues. In the 1980s and 1990s, a vast pool of researchers deduced the idea of fuzzy sets

P. M. F. Benazir · G. Gomathi (✉)
Department of Mathematics, Saveetha School of Engineering, SIMATS, Chennai, Tamil Nadu, India

I. Rajasingh et al. (eds.), *Applied Computational Mathematics*,
Springer Proceedings in Mathematics & Statistics 480,
https://doi.org/10.1007/978-3-031-77764-6_23

and soft computing to reconcile many problems in economics and sociological, scientific, and environmental concerns [2, 3]. In 1999, Molodtsov [8] initiated the unconventional idea of soft set theory as a competitor to those theories. Maji, Biswas, and Roy [4] studied the theory of soft set units initiated with the aid of Moldstov in 2003 and put forward numerous results. In 2010, Majumdar and Samanta [6] introduced the notion of generalized fuzzy soft sets and successfully carried out their notion in a decision problem. P.K. Maji et al. [5] also discuss the theory of soft sets to solve a decision-making problem using rough mathematics, and this paved way for any related researches. The much required comparison of fuzzy sets and rough sets has been reviewed and presented by Y.Y. Yao [11] and gave the researchers an insight on the extension of classical set algebra. D. Dubois et al. [1] clarified the difference between fuzzy sets and rough sets and presented few works that helped to get a clear understanding about each in detail. W.Z. Wu et al. [10] devised a general framework for studying fuzzy rough sets that incorporate both constructive and axiomatic methods. The link between fuzzy relations and fuzzy rough approximation operators is investigated. The presence of specific types of fuzzy relations yielding the same operators is guaranteed by axioms of fuzzy approximation operators. As the current pandemic situation has turned the lifestyle of people upside down, the new global online era has its impact majorly on education system. Hence, we have analyzed and studied the advantages and disadvantages of online education and how it started to evolve the mindset of both students and teachers. The performance of students differs in online mode comparable with regular physical mode of education, and even the experts predict that this new normal will surely disturb the future and career of students both in positive and negative ways. Henceforth, we feel that it is very important to research the performance level of students in online mode, and thus we have framed our paper accordingly. As mentioned earlier, we have used the concept of generalized fuzzy soft set and designed an algorithm that helps us to decide the performance level of students based on specific attributes or criteria and given input data. The present study attempts to apply the concept of generalized fuzzy soft sets in online education system as a tool to resolve complex decision taking situations, and the technique upholds efficiency for sure and thus more area of relevant problems can be solved. As the technological development reached all the fields of engineering and science, we have used the large number of data and predict the performance level of students by implementing our technique in large input data with some predefined value of attributes.

2 Preliminaries

The following definitions and preliminaries are required in the sequel of our work and hence presented below in brief.

Definition 1 (Soft Set [2]) Let U be an initial universal set and E be a set of parameters. Let P(U) denotes the power set of U and $A \subseteq E$. A pair (F, A) is called a soft set over U, where F is a mapping given by $F : A \to P(U)$.

Definition 2 (Fuzzy Soft Set [6]) A pair (F, A) is called a fuzzy soft set over U where $F : A \to \widetilde{P}(U)$ is a mapping from A to $\widetilde{P}(U)$, where $\widetilde{P}(U)$ is the set of all fuzzy subsets of U.

Definition 3 (Generalized Fuzzy Soft Sets [1]) Let $U = \{x_1, x_2 \ldots x_n\}$ be the universal set of elements and $E = \{e_1, e_2 \ldots e_m\}$ be the universal set of parameters. The pair (U, E) will be called a soft universe. Let $F : E \to I^U$ and μ be a fuzzy subset of E, i.e., $\mu : E \to I = [0, 1]$ where I^U is the collection of all fuzzy subsets of U. Let F_μ be the mapping $F_\mu : E \to I^U \times I$ be a function defined as follows: $F_\mu(e) = (F(e), \mu(e))$, where $F(e) \in I^U$. Then F_μ is called generalized fuzzy soft sets over the soft universe (U, E). Here for each parameter, e_i $F_\mu(e_i) = (F(e_i), \mu(e_i))$ indicates not only the degree of belongingness of the elements of U in $F(e_i)$ but also the degree of possibility of such belongingness which is represented by $\mu(e_i)$.

Definition 4 ([1]) Let F_μ be a generalized fuzzy soft set over (U, E). Then F_μ^c is said to be complement of F_μ and is defined as $F_\mu^c(e) = (F^c(e), \mu^c(e))$.

Definition 5 Let F_μ and G_δ be two generalized fuzzy soft sets over (U, E). Now F_μ is said to be a generalized fuzzy soft subset of G_δ if (i) μ is a fuzzy subset of δ (ii) $F(e)$ is also a fuzzy subset of $G(e)$, $\forall e \in E$ In this case, we write $F_\mu \widetilde{\subseteq} G_\delta$.

3 The Importance of Online Mode of Education

Online learning becomes one of the imminent trends in the education section nowadays all around the globe; we see that even higher learning institutions prefer online education mode. The most important aspects of online learning are its inherent flexibility and the student and teacher need to be extremely self-motivated to catch this new normal situation and to utilize it in a positive way. Online learning experience provides a high-quality learning opportunity, and it helps improve the student's outcomes and skills, and the location, time, and quality are no longer considered to be the major factors in attaining any degree courses or higher education. This seems to improve the value of education but just like every new upgrade, this mode also has its disadvantages.

Humans always have the ability to adapt to any change and this adaptive nature should get inherited in our new generation students as this becomes vital in this pandemic era. Of course, our students are always eager to learn new technology; it becomes mandatory for a teacher and parent to open up new learning tools and techniques to the student. It is the major responsibility of a student nowadays to get adapted to this new mode of education and develops them by utilizing the available open-source Internet. Students should be ready to drive along with the technology as it gets updated each moment. And we see that the performance of a student is not just calculated by the knowledge they showcase in the class, but they should be ready to accomplish more, as there are many attributes involved in determining their performance. As the whole world is stepping into this evolution phase together, as a research person, we feel responsible to analyze this subject and present our work to motivate others to proceed with an insight.

Few of the recent research papers have helped us in understanding the related concepts and techniques. Pinaki Majumdar et al. [7] used generalized fuzzy soft sets to grade the students in a class, and this paved way to solve some student ranking problems that were not able to be solved by standard methods. Manash Jyoti Borah et al. [22] attempted their work on application of generalized fuzzy soft sets in the teaching evaluation process, and analyzing both these papers motivated us to research more on the concept of generalized fuzzy soft sets and thus we came up with our paper work. In the recent days, many researches have made decision-making and ranking problems in the education field, but many of that kind are of statistical techniques. Here, we made an attempt by generalized fuzzy soft set and analyzed online mode of education system.

4 An Application of Generalized Fuzzy Soft Set in Decision-Making

In this section, we discuss the design of an algorithm using the generalized fuzzy soft set theory, which can be used in ranking the objects of the universe U. Also we have illustrated this algorithm in ranking the students according to their performance level.

Let $U = \{S_1, S_2, S_3, \ldots, S_r\}$ be the set of object and $Q = \{Q_1, Q_2, Q_3, \ldots, Q_k\}$ be the set of parameters associated with these objects. Let $\mu : Q \to I$, be the fuzzy set on Q where $\mu(Q_k)$ is the fuzzy weight associated with each parameter Q_k and the compliment of μ is the fuzzy $\mu^C : Q \to I$ defined by $\mu^c(Q_k) = 1 - \mu(Q_k)$. Let us consider the mapping, $F : Q \to I^U$, where I^U is the collection of all fuzzy soft set F_s as $F_S : Q \to I^U \times I$ and its compliment is $F^S = 1 - F_s$. Here for each parameter (Q_i), $F_S(Q_i)$ indicates not only the degree of the elements of U in F_S but also the degree of preference of such elements which is represented by $\mu(Q_i)$.

Then, this leads us to define a fuzzy set $\eta : U \to I$ defined by

$$\eta(S_r) = \frac{\sum_{i=1}^{k} F(Q_i)(S_r) * \mu(Q_i)}{\sum_{i=1}^{k} \mu(Q_i)}$$

Here $\eta(S_r)$ is called the membership value. Then,we define $\eta_1 : U \to I$ by

$$\eta_1(S_r) = \frac{\sum_{i=1}^{k} F^C(Q_i)(S_r) * \mu^C(Q_i)}{\sum_{i=1}^{k} \mu^C(Q_i)}$$

Here $\eta_1(S_r)$ is called as the non-membership value.

4.1 Algorithm for Ranking the Objects Using Generalized Fuzzy Soft Sets

Step 1 Obtain the fuzzy sets μ,F,$F(s)$.
Step 2 Compute the value of η and η_1.
Step 3 Find $\eta(S_r)$, $\eta_1(S_r)$for each S_r belong to U.
Step 4 Using these values, rank the elements of U.

4.2 Illustration

Let $U = \{S_1, S_2, \ldots, S_{65}\}$ be the set of 65 students in an Institution and $Q = \{Q_1, Q_2, \ldots, Q_{12}\}$ is the set of parameters indicating the questions to evaluate the Student's performance level during online mode of education. The Questions are pertaining to Doubt clarification(Q_1), Attitude toward online exams (Q_2), Hard work in an academics(Q_3), Taking notes in class(Q_4), Subject understanding(Q_5), Attendance perfection(Q_6), Interaction in class(Q_7), Assessments and Assignments(Q_8), Student's staff relationship(Q_9), Proper usage of technical gadgets and the Internet connections(Q_{10}), Regular study updates to teachers (Q_{11}), and Basic knowledge on subjects(Q_{12}). Let $\mu : Q \rightarrow I$, be a fuzzy set on Q defined by $\mu(Q_1) = 0.7$,$\mu(Q_2) = 0.5$,$\mu(Q_3) = 0.9$,$\mu(Q_4) = 0.6$,$\mu(Q_5) = 0.9$,$\mu(Q_6) = 0.8$,$\mu(Q_7) = 0.8$,$\mu(Q_8) = 0.7$,$\mu(Q_9) = 0.5$,$\mu(Q_{10}) = 0.7$,$\mu(Q_{11}) = 0.7$,$\mu(Q_{12}) = 0.8$. Consider F_S as the generalized fuzzy soft set representing the student's performance level during online mode of education. Values of μ can be arrived at from the expert knowledge by considering the weightage of the questions.

The values of the generalized fuzzy soft set F_S for each of the parameters are calculated and tabulated below. Generalized fuzzy soft set values are the obtained student's performance during the online mode of education (Table 1).

Compute the $\otimes$ operation table by multiplying each value of generalized fuzzy soft set by the corresponding value of $\mu(Q)$ (Table 2). This table shows the values of the satisfaction level of the students according to the importance of the question.

Obtain F_S^C and compute the $\otimes$ operation table for F_S^C (Table 3).

The calculated membership value η and non-membership value η_1 of the generalized fuzzy soft set and the decision score table "$\eta - \eta_1$"are evaluated (Table 4). The largest value of η denotes the good performance by the students, and the highest value of η_1 indicates the low-average performance by the students. Hence, "$\eta - \eta_1$" indicates the good accuracy score of student's performance during the online mode of education. Highest value describes the student's good performance level during

Table 1 Operation table

	Q_1	Q_2	Q_3	Q_4	Q_5	Q_6	Q_7	Q_8	Q_9	Q_{10}	Q_{11}	Q_{12}
$\mu(Q_i)$	0.7	0.5	0.9	0.6	0.9	0.8	0.8	0.7	0.5	0.7	0.7	0.8
S_1	0.7	0.8	0.8	0.8	0.8	0.6	0.7	0.7	0.6	0.6	0.8	0.7
S_5	0.3	0.5	0.4	0.8	0.8	0.7	0.8	0.5	0.2	0.5	0.8	0.6
S_{12}	0.9	1	0.9	1	1	0.8	0.8	0.8	0.8	0.9	1	0.9
S_{18}	0.9	1	1	1	1	0.9	1	0.8	0.8	0.7	1	0.9
S_{20}	0.9	0.9	0.8	0.6	0.8	0.8	0.9	0.8	0.8	0.7	0.7	0.8
S_{24}	1	1	0.7	0.6	1	1	1	1	0.6	1	1	0.8
S_{26}	0.8	0.8	0.6	0.9	0.9	0.8	0.7	0.7	0.8	0.6	0.7	0.9
S_{31}	0.8	0.6	0.6	0.4	0.7	0.5	0.6	0.7	0.7	0.4	0.8	0.8
S_{34}	0.9	0.9	0.9	0.8	0.8	0.8	1	0.8	0.8	1	0.9	0.9
S_{46}	0.8	0.8	0.7	0.7	0.8	0.8	0.8	0.7	0.6	0.7	0.7	0.7
S_{51}	0.8	0.8	0.8	0.8	0.9	0.8	0.8	0.8	0.8	0.6	0.8	0.8
S_{58}	0.8	0.8	0.8	0.7	0.8	0.7	0.7	0.7	0.7	0.6	0.7	0.8
S_{59}	0.8	0.8	0.7	0.8	0.8	0.8	0.8	0.8	0.8	0.4	0.8	0.7
S_{62}	0.8	0.8	0.8	0.8	0.9	0.8	0.8	0.8	0.8	0.6	0.8	0.7
S_{65}	0.8	0.9	0.9	0.8	0.9	0.9	0.8	0.8	0.8	0.6	0.8	0.9

Table 2 $\otimes$ Operation table

	Q_1	Q_2	Q_3	Q_4	Q_5	Q_6	Q_7	Q_8	Q_9	Q_{10}	Q_{11}	Q_{12}
$\mu(Q_i)$	0.7	0.5	0.9	0.6	0.9	0.8	0.8	0.7	0.5	0.7	0.7	0.8
S_1	0.49	0.4	0.72	0.48	0.72	0.48	0.56	0.49	0.3	0.42	0.56	0.56
S_5	0.21	0.25	0.36	0.48	0.72	0.56	0.64	0.35	0.1	0.35	0.56	0.48
S_{12}	0.63	0.5	0.81	0.6	0.9	0.64	0.64	0.56	0.4	0.63	0.7	0.72
S_{18}	0.63	0.5	0.9	0.6	0.9	0.72	0.8	0.56	0.4	0.49	0.7	0.72
S_{20}	0.63	0.45	0.72	0.36	0.72	0.64	0.72	0.56	0.4	0.49	0.49	0.64
S_{24}	0.7	0.5	0.63	0.36	0.9	0.8	0.8	0.7	0.3	0.7	0.7	0.64
S_{26}	0.56	0.4	0.54	0.54	0.81	0.64	0.56	0.49	0.4	0.42	0.49	0.72
S_{31}	0.56	0.3	0.54	0.24	0.63	0.4	0.48	0.49	0.35	0.28	0.56	0.64
S_{34}	0.63	0.45	0.81	0.48	0.72	0.64	0.8	0.56	0.4	0.7	0.63	0.72
S_{46}	0.56	0.4	0.63	0.42	0.72	0.64	0.64	0.49	0.3	0.49	0.49	0.56
S_{51}	0.56	0.4	0.72	0.48	0.81	0.64	0.64	0.56	0.4	0.42	0.56	0.64
S_{58}	0.56	0.4	0.72	0.42	0.72	0.56	0.56	0.49	0.35	0.42	0.49	0.64
S_{59}	0.56	0.4	0.63	0.48	0.72	0.64	0.64	0.56	0.4	0.28	0.56	0.56
S_{62}	0.56	0.4	0.72	0.48	0.81	0.64	0.64	0.56	0.4	0.42	0.56	0.56
S_{65}	0.56	0.45	0.81	0.48	0.81	0.72	0.64	0.56	0.4	0.42	0.56	0.72

the online mode of education. The highest score is 0.83 and it corresponds to the Student S_{18}. Thus, S_{18} has performed well during online mode among the entire students under consideration.

Table 3 Operation table

	Q_1	Q_2	Q_3	Q_4	Q_5	Q_6	Q_7	Q_8	Q_9	Q_{10}	Q_{11}	Q_{12}
S_1	0.09	0.1	0.02	0.08	0.02	0.08	0.06	0.09	0.2	0.12	0.06	0.06
S_5	0.21	0.25	0.06	0.08	0.02	0.06	0.04	0.15	0.4	0.15	0.06	0.08
S_{12}	0.03	0	0.01	0	0	0.04	0.04	0.06	0.1	0.03	0	0.02
S_{18}	0.03	0	0	0	0	0.02	0	0.06	0.1	0.09	0	0.02
S_{20}	0.03	0.05	0.02	0.16	0.02	0.04	0.02	0.06	0.1	0.09	0.09	0.04
S_{24}	0	0	0.03	0.16	0	0	0	0	0.2	0	0	0.04
S_{26}	0.06	0.1	0.04	0.04	0.01	0.04	0.06	0.09	0.1	0.12	0.09	0.02
S_{31}	0.06	0.2	0.04	0.24	0.03	0.1	0.08	0.09	0.15	0.18	0.06	0.04
S_{34}	0.03	0.05	0.01	0.08	0.02	0.04	0	0.06	0.1	0	0.03	0.02
S_{46}	0.06	0.1	0.03	0.12	0.02	0.04	0.04	0.09	0.2	0.09	0.09	0.06
S_{51}	0.06	0.1	0.02	0.08	0.01	0.04	0.04	0.06	0.1	0.12	0.06	0.04
S_{58}	0.06	0.1	0.02	0.12	0.02	0.06	0.06	0.09	0.15	0.12	0.09	0.04
S_{59}	0.06	0.1	0.03	0.08	0.02	0.04	0.04	0.06	0.1	0.18	0.06	0.06
S_{62}	0.06	0.1	0.02	0.08	0.01	0.04	0.04	0.06	0.1	0.12	0.06	0.06
S_{65}	0.06	0.5	0.01	0.08	0.01	0.02	0.04	0.06	0.1	0.12	0.06	0.02

Table 4 Operation table

	η	η_1	$\eta - \eta_1$
S_1	0.72	0.29	0.43
S_5	0.59	0.46	0.13
S_{12}	0.9	0.1	0.8
S_{18}	0.92	0.09	0.83
S_{20}	0.79	0.21	0.58
S_{24}	0.9	0.13	0.77
S_{26}	0.76	0.23	0.53
S_{31}	0.64	0.37	0.27
S_{34}	0.88	0.13	0.75
S_{46}	0.74	0.28	0.46
S_{51}	0.79	0.21	0.58
S_{58}	0.74	0.27	0.47
S_{59}	0.75	0.24	0.51
S_{62}	0.78	0.22	0.56
S_{65}	0.83	0.19	0.64

5 Conclusion

The devised algorithm using generalized fuzzy soft set is to deal with the complexities in decision-making. This algorithm reduces the complexity in obtaining the student's performance level during the online mode of education. The application of this algorithm is explained with live examples satisfying the need and its validity has been testified for students in an Institution. The level of satisfaction depicts that the implementation of the algorithm holds good to challenge complexities of decision-making.

Funding
No funding was received to assist with the preparation of this manuscript.

Data Availability
Not applicable.

Conflict of Interest
The authors declare that they have no conflict of interests regarding the publication of this paper.

Acknowledgments The authors thank the management, SIMATS, for their constant support toward the successful completion of this work.

References

1. Dubois, D., Prade, H.: Rough fuzzy set and fuzzy rough sets. Int. J. Gen. Syst. **17**, 191–209 (1990)
2. Feng, F.: Soft rough sets applied to multicriteria group decision making. Ann. Fuzzy Math. Inform. **2**, 69–80 (2011)
3. Feng, F., Liu, X., Violeta, F.L., Young, J.B.: Soft sets and soft rough sets. Inform. Sci. **181**, 1125–1137 (2011)
4. Maji, P.K., Roy, A.R., Biswas, R.: Fuzzy soft sets. J. Fuzzy Math. **9**, 589–602 (2001)
5. Maji, P.K., Roy, A.R., Biswas, R.: An application of soft sets in a decision making problem. Comput. Math. Appl. **44**, 1077–1083 (2002)
6. Majumdar, P., Samanta, S.K: Generalised fuzzy soft sets. Comput. Math. Appl. **59**, 1425–1432 (2010)
7. Majumdar, P., Samantha, S.K.: A generalised fuzzy soft set based student ranking system. Int. J. Adv. Soft Comput. Appl. **3**(3), 37–45 (2011).
8. Molodstov, D.A.: Soft set theory-first result. Comput. Math. Appl. **37**, 19–31 (1999)
9. Nanda, S., Manjumdar, S.: Fuzzy rough sets. Fuzzy Sets Syst. **45**, 157–160 (1992)
10. Wu, W.Z., Mi, J.S., Zhang, W.X.: Generalized fuzzy rough sets. Inf. Sci. **151**, 263–282 (2003)
11. Yao, Y.Y.: A comparative study of fuzzy sets and rough sets. Inf. Sci. **109**, 227–242 (1998)
12. Zadeh L.A.: Fuzzy sets. Inform. Control **8**, 338–353 (1965)
13. Zadeh L.A.: Fuzzy logic. Neural Networks Soft Commun. ACM **37**(3), 77–84 (1994)

Novel Similarity Measures for Linguistic Intuitionistic Fuzzy TOPSIS Method for Decision Support System

P. John Robinson and A. Leonishiya

1 Introduction

Whenever attributes involving differences of opinion are involved in a decision system, one of the DSS's working principles is to rank the best option among those that are available. This is the Technique of Order Preference by Similarity to Ideal Solution (TOPSIS) technique. Based on ranking techniques that gauge proximity to either the positive or negative ideal answer, the TOPSIS method is where the decision-making problem will focus its methodology [3, 4, 6, 7, 10]. In recent days, intuitionistic fuzzy data [1, 2] has gained the attention of researchers to a large extent. Similarity measures are extremely important in fuzzy decision-making situations [5, 8, 9, 11–15]. In this work, we have suggested several similarity measures for Linguistic Intuitionistic Fuzzy Sets (LIFSs), and we employ them in the Linguistic Intuitionistic Fuzzy TOPSIS method for both attribute weight determination and alternative ranking. Utilizing the suggested similarity metrics, various calculations are carried out, and contrasts with current decision algorithms are established. According to the study, our novel similarity measures are less susceptible to modifications that are permitted in the weight vectors that are obtained from the various suggested similarity measures for LIFNs.

P. J. Robinson (✉) · A. Leonishiya
Department of Mathematics, Bishop Heber College, Affiliated Bharathidasan University, Tiruchirappalli, India
e-mail: johnrobinson.ma@bhc.edu.in; leonishiya.ma@bhc.edu.in

I. Rajasingh et al. (eds.), *Applied Computational Mathematics*,
Springer Proceedings in Mathematics & Statistics 480,
https://doi.org/10.1007/978-3-031-77764-6_24

2 Linguistic Intuitionistic Fuzzy Sets

Definition 2.1 Let $\tilde{\sigma}_1 = \langle l_{\theta(\sigma_1)}, (\alpha(\alpha_1), \gamma(\sigma_1))\rangle$ and $\tilde{\sigma}_2 = \langle l_{\theta(\sigma_2)}, (\alpha(\sigma_2), \gamma(\alpha_2))\rangle$ be two LIFNs and $\lambda \geq 0$. Then the operations of LIFNs are defined as follows:

$$\tilde{\alpha}_1 + \tilde{\sigma}_2 =< l_{\theta(\sigma_1)+\theta(\sigma_2)}, (\alpha(\sigma_1) + \alpha(\sigma_2) - \alpha(\sigma_1)\alpha(\sigma_2), \gamma(\sigma_1)\gamma(\sigma_2))>,$$

$$\tilde{\sigma}_1 \otimes \tilde{\sigma}_2 =< l_{\theta(\sigma_1)\times\theta(\sigma_2)}, (\alpha(\sigma_1)\alpha(\sigma_2), \gamma(\sigma_1) + \gamma(\sigma_2) - \gamma(\sigma_1)\gamma(\sigma_2))>,$$

$\lambda\tilde{\sigma}_1 = \langle l_{\lambda\times\theta(\sigma_1)}, (1-(1-\alpha(\sigma_1))^{\lambda}, (\gamma(\sigma_1))^{\lambda})\rangle$ and $\tilde{\sigma}_1^{\lambda} = \Big\langle l_{\theta(\sigma_1)^{\lambda}}, (\alpha(\sigma_1))^{\lambda}, 1-(1-(\gamma(\sigma_1))^{\lambda})\Big\rangle$.

3 Similarity Measures for Linguistic Intuitionistic Fuzzy Sets

Definition 3.1 Let $M = \{\langle x, S_{\theta_M}(x), \mu_M(x), \gamma_M(x)\rangle / x \in X\}$ and $N = \{\langle x, S_{\theta_N}(x), \mu_N(x), \gamma_N(x)\rangle / x \in X\}$be any two LIFNs, where $S_{\theta_M}(x)$ denotes the linguistic degree, $\mu_M(x)$ denotes the membership degree, and $\gamma_M(x)$ denotes the non-membership degree. Letting $\delta : \Phi(X)^2 \rightarrow [0, 1]$, then the degree of similarity between $M \in \Phi(X)$ and $N \in \Phi(X)$ is defined as $\delta(M, N)$, which satisfies the following conditions:

(a) $0 \leq \delta(M, N) \leq 1$.
(b) $M = N \iff \delta(M, N) = 1$.
(c) $\delta(M, N) = \delta(N, M)$.

We are proposing the following new similarity measures for LIFSs:

(a) $S_{l_{HK}}(M, N) = 1 - \frac{\sum_{i=1}^{n} |S_{\theta_M}(t_i) - S_{\theta_N}(t_i)| + |\mu_M(t_i) - \mu_N(t_i)| + |\gamma_M(t_i) - \gamma_N(t_i)|}{n(m+2)}$.

(b) $S_{l_{LX}}(M, N) = 1 - \frac{\sum_{i=1}^{n} \left(\begin{array}{l} |S_{\theta_M}(t_i) - S_{\theta_N}(t_i)| + |(\mu_M(t_i) - (\mu_N(t_i)| \\ + |(\gamma_M(t_i) - \gamma_N(t_i))| + |(\eta_M(t_i) - \eta_N(t_i)| \end{array} \right)}{n(m+2)}$.

(c) $S_{l_O}(M, N) = 1 - \sqrt{\frac{\sum_{i=1}^{n} (S_{\theta_M}(t_i) - S_{\theta_N}(t_i))^2 + (\mu_M(t_i) - \mu_N(t_i))^2 + (\gamma_M(t_i) - \gamma_N(t_i))^2}{n(m^2+2)}}$.

(d) $S_{l_{DC}}(M, N) = 1 - \sqrt[p]{\frac{\sum_{i=1}^{n} |\Phi_{l_M}(t_i) - \Phi_{l_N}(t_i)|^p + |S_{\theta_M}(t_i) - S_{\theta_N}(t_i)|^p}{2n}}$,
where $\Phi_{l_M}(t_i) = \frac{\mu_M(t_i) + 1 - \gamma_M(t_i)}{2}$, $\Phi_{l_N}(t_i) = \frac{\mu_N(t_i) + 1 - \gamma_N(t_i)}{2}$

(e) $S^p{}_{l_L}(M, N)$
$= 1 - \sqrt[p]{\sum_{i=1}^{n} \frac{1}{(n^p + m^p)} \left\{ \begin{array}{l} |S_{\theta_M}(t_i) - S_{\theta_N}(t_i)|^p + |\mu_M(t_i) - \mu_N(t_i)|^p \\ + |\gamma_M(t_i) - \gamma_N(t_i)|^p + |\eta_M(t_i) - \eta_N(t_i)|^p \end{array} \right\}}$.

Definition 3.2 Let $M = \{\langle x, S_{\theta_M}(x), \mu_M(x), \gamma_M(x)\rangle / x \in X\}$ and $N = \{\langle x, S_{\theta_N}(x), \mu_N(x), \gamma_N(x)\rangle / x \in X\}$be any two LIFNs, where $S_{\theta_M}(x)$ denotes the linguistic degree, $\mu_M(x)$ denotes the membership degree, and $\gamma_M(x)$ denotes the non-membership degree. Letting $\delta' : \Phi(X)^2 \to [0, 1]$, then the degree of similarity between $M \in \Phi(X)$ and $N \in \Phi(X)$ is defined as $\delta'(M, N)$, which satisfies the following conditions:

(a) $0 \leq \delta'(M, N) \leq 1$.
(b) if $M = N$ *then* $\delta'(M, N) = 1$.
(c) $\delta'(M, N) = \delta'(N, M)$.

We are proposing the following similarity measures:

(a) $S_{l_{HK1}}(M, N) = 1 - \frac{\sum_{i=1}^{n} \left|\left(S_{\theta_M}(t_i) - S_{\theta_N}(t_i)\right) + \left(\mu_M(t_i) - \mu_N(t_i)\right) + \left(\gamma_M(t_i) - \gamma_N(t_i)\right)\right|}{n(m+2)}$.
(b) $S_{l_m}(M, N) = \frac{1}{3}\left(A'_\theta(M, N) + A'_{md}(M, N) + A'_{nd}(M, N)\right)$

$$\text{where } A'_\theta(M, N) = 1 - \sqrt{\sum_{i=1}^{n} \frac{\left|S_{\theta_M}(t_i) - S_{\theta_N}(t_i)\right|^p}{m^p n}},$$

$$A'_{md}(M, N) = 1 - \sqrt{\sum_{i=1}^{n} \frac{\left|\mu_M(t_i) - \mu_N(t_i)\right|^p}{n}},$$

$$A'_{nd}(M, N) = 1 - \sqrt{\sum_{i=1}^{n} \frac{\left|\gamma_M(t_i) - \gamma_N(t_i)\right|^p}{n}}.$$

Cosine Similarity Measures

Definition 3.3 Let $M = \left\{\left\langle x, S_{\theta_M}(x), \mu_M(x), \gamma_M(x)\right\rangle / x \in X\right\}$ and $N = \left\{\left\langle x, S_{\theta_N}(x), \mu_N(x), \gamma_N(x)\right\rangle / x \in X\right\}$ be any two LIFNs, where $S_{\theta_M}(x)$ denotes the linguistic degree, $0 \leq S_{\theta_M}(x) \leq 1$, $\mu_M(x)$ denotes the membership degree, and $\gamma_M(x)$ denotes the non-membership degree.

Letting $\delta' : C_l(X)^2 \to [0, 1]$, then the degree of cosine similarity between $M \in \Phi(X)$ and $N \subset \Phi(X)$ is defined as $C_l(M, N)$, which satisfies the following conditions:

(a) $0 \leq C_l(M, N) \leq 1$.
(b) if $M = N$ *then* $C_l(M, N) = 1$.
(c) $C_l(M, N) = C_l(N, M)$.

We are proposing the following cosine similarity measures for LIFSs:

(a) $C_{l_Y}(M, N) = \frac{1}{n}\sum_{i=1}^{n} \frac{S_{\theta_M}(t_i)S_{\theta_N}(t_i) + \mu_M(t_i)\mu_N(t_i) + \gamma_M(t_i)\gamma_N(t_i)}{\sqrt{S^2_{\theta_M}(t_i) + \mu^2_{\theta_M}(t_i) + \gamma^2_{\theta_M}(t_i)}\sqrt{S^2_{\theta_N}(t_i) + \mu^2_{\theta_N}(t_i) + \gamma^2_{\theta_N}(t_i)}}$.
(b) $C_{l_{SY}}(M, N)$
$= \frac{1}{n}\sum_{i=1}^{n} \frac{S_{\theta_M}(t_i)S_{\theta_N}(t_i) + \mu_M(t_i)\mu_N(t_i) + \gamma_M(t_i)\gamma_N(t_i) + \eta_M(t_i)\eta_N(t_i)}{\sqrt{S^2_{\theta_M}(t_i) + \mu^2_{\theta_M}(t_i) + \gamma^2_{\theta_M}(t_i) + \eta^2_{\theta_M}(t_i)}\sqrt{S^2_{\theta_N}(t_i) + \mu^2_{\theta_N}(t_i) + \gamma^2_{\theta_N}(t_i) + \eta^2_{\theta_M}(t_i)}}$.

Now we propose the following cosine similarity measures for linguistic vague sets: Let $M = \{\langle x, S_{\theta_M}(x), \mu_M(x), 1-\gamma_A(x)\rangle / x \in X\}$ and $N = \{\langle x, S_{\theta_N}(x), \mu_N(x), 1-\gamma_N(x)\rangle / x \in X\}$be any two LIFVNs.

(a) $S_{l_{vag}}(A, B) = \frac{1}{n}\sum_{i=1}^{n}\left\{\frac{S_{\theta_M}(t_i)S_{\theta_N}(t_i)+\mu_M(t_i)\mu_N(t_i)+(1-\gamma_M(t_i))(1-\gamma_N(t_i))}{\sqrt{S^2_{\theta_M}(t_i)+\mu^2_M(t_i)+(1-\gamma_M(t_i))^2}\sqrt{S^2_{\theta_N}(t_i)+\mu^2_N(t_i)+(1-\gamma_N(t_i))^2}}\right\}.$

Definition 3.4 Let $M = \{\langle x, S_{\theta_M}(x), \mu_M(x), \gamma_M(x)\rangle / x \in X\}$ and $N = \{\langle x, S_{\theta_N}(x), \mu_N(x), \gamma_N(x)\rangle / x \in X\}$be any two LIFNs, where $S_{\theta_M}(x)$ denotes the linguistic degree, $0 \le S_{\theta_M}(x) \le 1$,$\mu_M(x)$ denotes the membership degree, and $\gamma_M(x)$ denotes the non-membership degree. Letting $C_l : \Phi(X)^2 \to [0, 1]$, then the degree of cosine similarity between $M \in \Phi(X)$ and $N \in \Phi(X)$ is defined as $C_l(M, N)$, which satisfies the following conditions:

(a) $0 \le C_l(M, N) \le 1$.
(b) $M = N \iff C_l(M, N) = 1$.
(c) $C_l(M, N) = C_l(N, M)$.

We are proposing the following cosine similarity measures for LIFSs:

(a) $S_{l_{YC}}(M, N) = \frac{1}{n}\sum_{i=1}^{n} cos\left[\frac{\pi}{6}\left(\begin{matrix}|S_{\theta_M}(t_i) - S_{\theta_N}(t_i)| + |\mu_M(t_i) - \mu_N(t_i)| + \\ |\gamma_M(t_i) - \gamma_N(t_i)| + |\eta_M(t_i) - \eta_N(t_i)|\end{matrix}\right)\right].$

(b) $S_{l_{GV}}(M, N) = 1 - log_2\left\{1 + \frac{1}{4n}\sum_{i=1}^{n}\left(\begin{matrix}\left|\sqrt{S_{\theta_M}(t_i)} - \sqrt{S_{\theta_N}(t_i)}\right| \\ + \left|\sqrt{\mu_M(t_i)} - \sqrt{\mu_N(t_i)}\right| + \\ \left|\sqrt{\gamma_M(t_i)} - \sqrt{\gamma_N(t_i)}\right| \\ + \left|\sqrt{\eta_M(t_i)} - \sqrt{\eta_N(t_i)}\right|\end{matrix}\right)\right\},$

where $0 \le S_{\theta_M}(x) \le 1, 0 \le \mu_M(x), \gamma_M(x), \eta_M(x) \le 1$.

3.1 Properties of the Proposed Novel Similarity Measures

In this section, we have proved the properties of the proposed similarity measures.

Theorem 3.1 *Let X be the Universe of discourse. M and N are LIFSs. Then the functions* $S_{l_{HK}}(M, N)$, $S_{l_{LX}}(M, N)$, $S_{l_O}(M, N)$, $S_{l_{DC}}(M, N)$, $S^{p}{}_{l_L}(M, N)$: $\Phi(X)^2 \to [0, 1]$ *similarity measures.*

Proof

(a) Clearly, $0 \le S_{l_{HK}}(M, N) \le 1$, from the definition of $S_{l_{HK}}(M, N)$.
(b) $M = N \Leftrightarrow S_{l_{HK}}(M, N) = 1$.

$$S_{l_{HK}}(M, N) = 1 \Leftrightarrow 1$$
$$-\frac{\sum_{i=1}^{n}\left|S_{\theta_M}(t_i) - S_{\theta_N}(t_i)\right| + |\mu_M(t_i) - \mu_N(t_i)| + |\gamma_M(t_i) - \gamma_N(t_i)|}{n(m+2)} = 1$$

$$\Leftrightarrow \frac{\sum_{i=1}^{n} \left|S_{\theta_M}(t_i) - S_{\theta_N}(t_i)\right| + |\mu_M(t_i) - \mu_N(t_i)| + |\gamma_M(t_i) - \gamma_N(t_i)|}{n(m+2)} = 0$$

$$\Leftrightarrow \sum_{i=1}^{n} \left|S_{\theta_M}(t_i) - S_{\theta_N}(t_i)\right| + |\mu_M(t_i) - \mu_N(t_i)| + |\gamma_M(t_i) - \gamma_N(t_i)| = 0$$

$$\Leftrightarrow \left|S_{\theta_M}(t_i) - S_{\theta_N}(t_i)\right| = 0,\ |\mu_M(t_i) - \mu_N(t_i)| = 0 and\ |\gamma_M(t_i) - \gamma_N(t_i)| = 0,\ \forall i$$

$$\Leftrightarrow S_{\theta_M}(t_i) = S_{\theta_N}(t_i),\ \mu_M(t_i) = \mu_N(t_i) and \gamma_M(t_i) = \gamma_N(t_i),\ \forall i$$

$$\Leftrightarrow M = N$$

(c) $S_{l_{HK}}(M, N) = 1 - \frac{\sum_{i=1}^{n} \left|S_{\theta_M}(t_i) - S_{\theta_N}(t_i)\right| + |\mu_M(t_i) - \mu_N(t_i)| + |\gamma_M(t_i) - \gamma_N(t_i)|}{n(m+2)}$

$$= 1 - \frac{\sum_{i=1}^{n} \left|S_{\theta_N}(t_i) - S_{\theta_M}(t_i)\right| + |\mu_N(t_i) - \mu_M(t_i)| + |\gamma_N(t_i) - \gamma_M(t_i)|}{n(m+2)}$$

$$\therefore S_{l_{HK}}(M, N) = S_{l_{HK}}(N, M).$$

Hence, $S_{l_{HK}}(M, N)$ is a similarity measure. Similarly, we can prove that all other functions proposed above are similarity measures.

Theorem 3.2 *Let X be the Universe of discourse. M and N are LIFSs. Then the functions* $S_{l_{HK}}(M, N)$, $S_{l_m}(M, N)$: $\Phi(X)^2 \to [0, 1]$ *similarity measures.*

Proof

(a) Clearly, $0 \le S_{l_{HK1}}(M, N) \le 1$, from the definition of $S_{l_{HK1}}(M, N)$.
(b) $M = N \Rightarrow S_{l_{HK1}}(M, N) = 1$.

$$M = N \Rightarrow S_{\theta_M}(t_i) = S_{\theta_N}(t_i),\ \mu_M(t_i) = \mu_N(t_i) and \gamma_M(t_i) = \gamma_N(t_i),\ \forall i$$

$$\Rightarrow \left|S_{\theta_M}(t_i) - S_{\theta_N}(t_i)\right| = 0,\ |\mu_M(t_i) - \mu_N(t_i)| = 0 and\ |\gamma_M(t_i) - \gamma_N(t_i)| = 0,\ \forall i$$

$$\Rightarrow \sum_{i=1}^{n} \left|S_{\theta_M}(t_i) - S_{\theta_N}(t_i)\right| + |\mu_M(t_i) - \mu_N(t_i)| + |\gamma_M(t_i) - \gamma_N(t_i)| = 0$$

$$\Rightarrow \frac{\sum_{i=1}^{n} \left|(S_{\theta_M}(t_i) - S_{\theta_N}(t_i)) + (\mu_M(t_i) - \mu_N(t_i)) + (\gamma_M(t_i) - \gamma_N(t_i))\right|}{n(m+2)} = 0$$

$$\Rightarrow 1 - \frac{\sum_{i=1}^{n} \left|(S_{\theta_M}(t_i) - S_{\theta_N}(t_i)) + (\mu_M(t_i) - \mu_N(t_i)) + (\gamma_M(t_i) - \gamma_N(t_i))\right|}{n(m+2)} = 1$$

$M = N \Rightarrow S_{l_{HK1}}(M, N) = 1$.

(c) $S_{l_{HK1}}(M, N) = 1 - \frac{\sum_{i=1}^{n} \left|(S_{\theta_M}(t_i) - S_{\theta_N}(t_i)) + (\mu_M(t_i) - \mu_N(t_i)) + (\gamma_M(t_i) - \gamma_N(t_i))\right|}{n(m+2)}$

$$= 1 - \frac{\sum_{i=1}^{n} \left|(S_{\theta_N}(t_i) - S_{\theta_M}(t_i)) + (\mu_N(t_i) - \mu_M(t_i)) + (\gamma_N(t_i) - \gamma_M(t_i))\right|}{n(m+2)}$$

$$\therefore S_{l_{HK1}}(M, N) = S_{l_{HK1}}(N, M).$$

Hence, $S_{l_{HK1}}(M, N)$ is a similarity measure. Similarly, we can prove that all other functions proposed above are similarity measures.

Theorem 3.3 *Let X be the Universe of discourse. M and N are LIFSs. Then the functions,* $C_{l_Y}(M, N)$, $C_{l_{SY}}(M, N)$, $S_{l_{vag}}(M, N)$*:* $\Phi(X)^2 \rightarrow [0, 1]$ *similarity measures.*

Proof

(a) Clearly, $0 \leq C_{l_Y}(M, N) \leq 1$, from the definition of $C_{l_Y}(M, N)$.
(b) $M = N \Rightarrow C_{l_Y}(M, N) = 1$.

$$M = N \Rightarrow S_{\theta_M}(t_i) = S_{\theta_N}(t_i), \mu_M(t_i) = \mu_N(t_i) and \gamma_M(t_i) = \gamma_N(t_i), \forall i$$

$$\Rightarrow C_{l_Y}(M, N)$$

$$= \frac{1}{n}\sum_{i=1}^{n} \frac{S_{\theta_M}(t_i)S_{\theta_N}(t_i) + \mu_M(t_i)\mu_N(t_i) + \gamma_M(t_i)\gamma_N(t_i)}{\sqrt{S_{\theta_M}^2(t_i) + \mu_{\theta_M}^2(t_i) + \gamma_{\theta_M}^2(t_i)}\sqrt{S_{\theta_N}^2(t_i) + \mu_{\theta_N}^2(t_i) + \gamma_{\theta_N}^2(t_i)}}$$

$$\Rightarrow \frac{1}{n}\sum_{i=1}^{n} \frac{(S_{\theta_M}(t_i))^2 + (\mu_M(t_i))^2 + (\gamma_M(t_i))^2}{\sqrt{S_{\theta_M}^2(t_i) + \mu_{\theta_M}^2(t_i) + \gamma_{\theta_M}^2(t_i)}.\sqrt{S_{\theta_M}^2(t_i) + \mu_{\theta_M}^2(t_i) + \gamma_{\theta_M}^2(t_i)}} = 1$$

$$\Rightarrow C_{l_Y}(M, N) = 1, \therefore M = N \Rightarrow C_{l_Y}(M, N) = 1.$$

(c) $C_{l_Y}(M, N) = \frac{1}{n}\sum_{i=1}^{n} \frac{S_{\theta_M}(t_i)S_{\theta_N}(t_i) + \mu_M(t_i)\mu_N(t_i) + \gamma_M(t_i)\gamma_N(t_i)}{\sqrt{S_{\theta_M}^2(t_i) + \mu_{\theta_M}^2(t_i) + \gamma_{\theta_M}^2(t_i)}\sqrt{S_{\theta_N}^2(t_i) + \mu_{\theta_N}^2(t_i) + \gamma_{\theta_N}^2(t_i)}}$

$$= \frac{1}{n}\sum_{i=1}^{n} \frac{S_{\theta_N}(t_i)S_{\theta_M}(t_i) + \mu_N(t_i)\mu_M(t_i) + \gamma_N(t_i)\gamma_M(t_i)}{\sqrt{S_{\theta_N}^2(t_i) + \mu_{\theta_N}^2(t_i) + \gamma_{\theta_N}^2(t_i)}\sqrt{S_{\theta_M}^2(t_i) + \mu_{\theta_M}^2(t_i) + \gamma_{\theta_M}^2(t_i)}}$$

$$\therefore C_{l_Y}(M, N) = C_{l_Y}(N, M).$$

Hence, $C_{l_Y}(M, N)$ is a similarity measure. Similarly, we can prove that all other functions proposed above are similarity measures.

Theorem 3.4 *Let X be the Universe of discourse. M and N are LIFSs. Then the functions,* $S_{lYC}(M,N)$, $S_{lGV}(M,N)$: $\Phi(X)^2 \rightarrow [0,1]$ *similarity measures.*

Proof

(a) Clearly, $0 \le S_{lYC}(M,N) \le 1$, from the definition of $S_{lYC}(M,N)$.
(b) $M = N \Leftrightarrow S_{lYC}(M,N) = 1$.

$$S_{lYC}(M,N) = \frac{1}{n}\sum_{i=1}^{n}\cos\left[\frac{\pi}{6}\left(\begin{matrix}\left|S_{\theta_M}(t_i) - S_{\theta_N}(t_i)\right| + |\mu_M(t_i) - \mu_N(t_i)| + \\ |\gamma_M(t_i) - \gamma_N(t_i)| + |\eta_M(t_i) - \eta_N(t_i)|\end{matrix}\right)\right]$$

$$S_{lYC}(M,N)$$

$$= 1 \Leftrightarrow \frac{1}{n}\sum_{i=1}^{n}\cos\left[\frac{\pi}{6}\left(\begin{matrix}\left|S_{\theta_M}(t_i) - S_{\theta_N}(t_i)\right| + |\mu_M(t_i) - \mu_N(t_i)| + \\ |\gamma_M(t_i) - \gamma_N(t_i)| + |\eta_M(t_i) - \eta_N(t_i)|\end{matrix}\right)\right] = 1$$

$$\Leftrightarrow \sum_{i=1}^{n}\cos\left[\frac{\pi}{6}\left(\begin{matrix}\left|S_{\theta_M}(t_i) - S_{\theta_N}(t_i)\right| + |\mu_M(t_i) - \mu_N(t_i)| + \\ |\gamma_M(t_i) - \gamma_N(t_i)| + |\eta_M(t_i) - \eta_N(t_i)|\end{matrix}\right)\right] = n$$

$$\Leftrightarrow \left[\frac{\pi}{6}\left(\begin{matrix}\left|S_{\theta_M}(t_i) - S_{\theta_N}(t_i)\right| + |\mu_M(t_i) - \mu_N(t_i)| + \\ |\gamma_M(t_i) - \gamma_N(t_i)| + |\eta_M(t_i) - \eta_N(t_i)|\end{matrix}\right)\right] = 0$$

$$\Leftrightarrow \left|S_{\theta_M}(t_i) - S_{\theta_N}(t_i)\right| = 0, |\mu_M(t_i) - \mu_N(t_i)| = 0, |\gamma_M(t_i) - \gamma_N(t_i)| = 0, \text{ and } |\eta_M(t_i) - \eta_N(t_i)| = 0 \forall i$$

$$\Leftrightarrow S_{\theta_M}(t_i) = S_{\theta_N}(t_i), \mu_M(t_i) = \mu_N(t_i), \gamma_M(t_i) = \gamma_N(t_i) and \eta_M(t_i) = \eta_N(t_i) \forall i$$

$$\Leftrightarrow M = N$$

(c) $S_{lYC}(M,N) = \frac{1}{n}\sum_{i=1}^{n} cos\left[\frac{\pi}{6}\left(\begin{matrix}\left|S_{\theta_M}(t_i) - S_{\theta_N}(t_i)\right| + |\mu_M(t_i) - \mu_N(t_i)| + \\ |\gamma_M(t_i) - \gamma_N(t_i)| + |\eta_M(t_i) - \eta_N(t_i)|\end{matrix}\right)\right]$

$$= \frac{1}{n}\sum_{i=1}^{n} cos\left[\frac{\pi}{6}\left(\begin{matrix}\left|S_{\theta_N}(t_i) - S_{\theta_M}(t_i)\right| + |\mu_N(t_i) - \mu_M(t_i)| + \\ |\gamma_N(t_i) - \gamma_M(t_i)| + |\eta_N(t_i) - \eta_M(t_i)|\end{matrix}\right)\right]$$

$$\therefore S_{lYC}(M,N) = S_{lYC}(N,M).$$

Hence, $S_{lYC}(M,N)$ is a similarity measure. Similarly, we can prove that all other functions proposed above are similarity measures.

4 The Linguistic Intuitionistic Fuzzy TOPSIS Method

Let $E = \{E_1, E_2, \ldots E_p\}$ be a set of experts, $L = (L_1, L_2, \ldots L_m)$ be a discrete set of alternatives, $C = (C_1, C_2, \ldots C_n)$ be the set of attributes, and $W = (\omega_1, \omega_2, \ldots \omega_n)^T$ be the weighting vector of the attributes, $\sum_{j=1}^{n} \omega_j = 1, \omega_j \geq 0$. Let $\lambda = (\lambda_1, \lambda_2, \ldots \lambda_p)$ be the expert's weighting vector, $\sum_{k=1}^{p} \lambda_k = 1$. Suppose that $\tilde{R}^k = \left[\tilde{t}_{ij}^k\right]_{m\times n}$ is the decision matrix, where $\tilde{t}_{ij}^k = \langle a^k{}_{ij,}(\alpha_{ijk}, \gamma_{ijk})\rangle$ takes the form of the linguistic intuitionistic number, given by the decision-maker E_k, for alternative L_i with respect to attribute C_j. Rank the alternatives by using the steps below:

Step 1: Make the integrated matrix

Integrate the matrix $\tilde{R}^k = \left[\tilde{t}_{ij}^k\right]_{mXn}$ given by the decision-maker E_k into the integrated matrix

$\tilde{R}^k = \left[\tilde{t}_{ij}^k\right]_{mXn} : \tilde{t}_{ij} = \sum_{k=1}^{p} \lambda_k \tilde{t}_{ij}^k$, where $\tilde{t}_{ij} = \langle a_{ij,}(\alpha_{ij}\gamma_{ij})\rangle$.

Step 2: Evaluate the attribute weights

For the attribute C_j, the deviation values of alternative L_i to all the other alternatives can be defined as $D_{ij}(\omega_j) = \sum_{l=1}^{m} (1 - \delta(\tilde{t}_{ij}, \tilde{t}_{lj}))\omega_j$, where $D_j(\omega_j) = \sum_{i=1}^{m} D_{ij}(\omega_j) = \sum_{i=1}^{m} \sum_{l=1}^{m} (1 - \delta(\tilde{t}_{ij}, \tilde{t}_{lj}))\omega_j$ indicates the total deviation values of all alternatives to the other alternatives for the attribute C_j. $D(\omega_j) = \sum_{j=1}^{n} D_j(\omega_j) = \sum_{j=1}^{n} \sum_{i=1}^{m} \sum_{l=1}^{m} (1 - \delta(\tilde{t}_{ij}, \tilde{t}_{lj}))\omega_j$ represents the deviation of all attributes to all alternatives. The optimum model is built as follows:

$$\begin{cases} \max D(\omega_j) = \sum_{j=1}^{n} \sum_{i=1}^{m} \sum_{l=1}^{m} (1 - \delta(\tilde{t}_{ij}, \tilde{t}_{lj}))\omega_j \\ s.t \sum_{j=1}^{n} \omega_j^2 = 1, \omega_j \geq 0, j = 1, 2, \ldots n. \end{cases}$$

We can get $\omega_j = \frac{\sum_{i=1}^{m} \sum_{l=1}^{m} (1-\delta(\tilde{t}_{ij}, \tilde{t}_{lj}))}{\sqrt{\sum_{j=1}^{n} \sum_{i=1}^{m} \sum_{l=1}^{m} (1-\delta(\tilde{t}_{ij}, \tilde{t}_{lj}))^2}}$. Furthermore, the normalized attribute weights are as follows:

$\omega_j = \frac{\sum_{i=1}^{m} \sum_{l=1}^{m} (1-\delta(\tilde{t}_{ij}, \tilde{t}_{lj}))}{\sum_{j=1}^{n} \sum_{i=1}^{m} \sum_{l=1}^{m} (1-\delta(\tilde{t}_{ij}, \tilde{t}_{lj}))}$.

Step 3: To rank the alternatives, use the TOPSIS method

The fundamental principle of TOPSIS is that the chosen alternative ought to be closest to the positive ideal solution and most far away from the negative optimal solution.

(1) Construct the weighted matrix: $\tilde{P} = (\tilde{p}_{ij})_{m\times n}$

$$= \begin{bmatrix} \omega_1\tilde{t}_{11} & \omega_2\tilde{t}_{12} & \ldots & \omega_n\tilde{t}_{1n} \\ \omega_1\tilde{t}_{21} & \omega_2\tilde{t}_{22} & \ldots & \omega_n\tilde{t}_{2n} \\ \ldots & \ldots & \ldots & \ldots \\ \omega_1\tilde{t}_{m1} & \omega_2\tilde{t}_{m2} & \ldots & \omega_n\tilde{t}_{mn} \end{bmatrix},$$

where $\tilde{p}_{ij} = \langle b_{ij,}(\alpha_{ij}, \gamma_{ij})\rangle$.

(2) Decide the positive and negative ideal solution: $\begin{cases} \tilde{p}_j^+ = \langle l_h, (1,0)\rangle \\ \tilde{p}_j^- = \langle l_1, (0,1)\rangle \end{cases}$ $j = 1, 2, \ldots, n$.

(3) Compute the distance between the alternative and the positive/negative optimal solution as follows: $\delta_i{}^+ = \sum_{j=1}^{n} (1 - \delta(\tilde{p}_{ij}, \tilde{p}_j^+))$ and $\delta_i{}^- = \sum_{j=1}^{n} (1 - \delta(\tilde{p}_{ij}, \tilde{p}_j^-)), i = 1, 2, \ldots, m$.

(4) Compute the relative closeness coefficient as follows: $O_i = \frac{\delta_i{}^+}{\delta_i{}^+ + \delta_i{}^-} (i = 1, 2, \ldots, m)$

(5) To rank the alternatives, apply the relative closeness coefficient. The smaller O_i, the better is the alternative.

5 Numerical Illustration

Assume there are four industries (alternatives) $\{L_1, L_2, L_3, L_4\}$ to be weighed against certain criteria. Evaluate industries in terms of their technological innovation capability, evaluating "factors" such as resource ability for digitalization (C_1), organizational innovation (C_2), innovation centers (C_3), and innovative products (C_4). Consider a group of experts whose weights are given as $\lambda = (0.4, 0.32, 0.28)$. The experts assessment of the four industries are listed in Tables 1, 2, and 3.

Make the integrated matrix as mentioned in the algorithm:

$$\tilde{R} = \begin{bmatrix} \langle l_{4.68}, (0.169, 0.670)\rangle & \langle l_{2.6}, (0.313, 0.658)\rangle & \langle l_{4.08}, (0.388, 0.581)\rangle & \langle l_{4.24}, (0.270, 0.591)\rangle \\ \langle l_{4.32}, (0.374, 0.591)\rangle & \langle l_{4.36}, (0.342, 0.558)\rangle & \langle l_{3.04}, (0.133, 0.730)\rangle & \langle l_{3.4}, (0.388, 0.586)\rangle \\ \langle l_{3.6}, (0.200, 0.666)\rangle & \langle l_{4.28}, (0.229, 0.670)\rangle & \langle l_{2.52}, (0.285, 0.692)\rangle & \langle l_{4.08}, (0.233, 0.700)\rangle \\ \langle l_{4.84}, (0.365, 0.533)\rangle & \langle l_{2.92}, (0.246, 0.663)\rangle & \langle l_{2.96}, (0.262, 0.600)\rangle & \langle l_{3.88}, (0.300, 0.570)\rangle \end{bmatrix}$$

Table 1 Decision matrix I

Industries	Digitalization C_1	Organizational innovation C_2	Innovation centers C_3	Innovative products C_4
L_1	$\langle l_5, (0.2, 0.7)\rangle$	$\langle l_2, (0.4, 0.6)\rangle$	$\langle l_5, (0.5, 0.5)\rangle$	$\langle l_3, (0.2, 0.6)\rangle$
L_2	$\langle l_4, (0.4, 0.6)\rangle$	$\langle l_5, (0.4, 0.5)\rangle$	$\langle l_3, (0.1, 0.8)\rangle$	$\langle l_4, (0.5, 0.5)\rangle$
L_3	$\langle l_3, (0.2, 0.7)\rangle$	$\langle l_4, (0.2, 0.7)\rangle$	$\langle l_4, (0.3, 0.7)\rangle$	$\langle l_5, (0.2, 0.7)\rangle$
L_4	$\langle l_6, (0.5, 0.4)\rangle$	$\langle l_2, (0.2, 0.8)\rangle$	$\langle l_3, (0.2, 0.6)\rangle$	$\langle l_3, (0.3, 0.6)\rangle$

Table 2 Decision matrix II

Industries	Digitalization C_1	Organizational innovation C_2	Innovation centers C_3	Innovative products C_4
L_1	$\langle l_4, (0.1, 0.7)\rangle$	$\langle l_3, (0.2, 0.7)\rangle$	$\langle l_3, (0.2, 0.8)\rangle$	$\langle l_6, (0.4, 0.5)\rangle$
L_2	$\langle l_5, (0.4, 0.5)\rangle$	$\langle l_3, (0.3, 0.6)\rangle$	$\langle l_4, (0.2, 0.6)\rangle$	$\langle l_3, (0.2, 0.7)\rangle$
L_3	$\langle l_3, (0.2, 0.7)\rangle$	$\langle l_4, (0.2, 0.7)\rangle$	$\langle l_2, (0.4, 0.6)\rangle$	$\langle l_3, (0.3, 0.7)\rangle$
L_4	$\langle l_5, (0.3, 0.6)\rangle$	$\langle l_4, (0.4, 0.5)\rangle$	$\langle l_2, (0.3, 0.6)\rangle$	$\langle l_4, (0.2, 0.6)\rangle$

Table 3 Decision matrix III

Industries	Digitalization C_1	Organizational innovation C_2	Innovation centers C_3	Innovative products C_4
L_1	$\langle l_5, (0.2, 0.6)\rangle$	$\langle l_3, (0.3, 0.7)\rangle$	$\langle l_4, (0.4, 0.5)\rangle$	$\langle l_4, (0.2, 0.7)\rangle$
L_2	$\langle l_4, (0.3, 0.7)\rangle$	$\langle l_5, (0.3, 0.6)\rangle$	$\langle l_2, (0.1, 0.8)\rangle$	$\langle l_3, (0.4, 0.6)\rangle$
L_3	$\langle l_4, (0.2, 0.7)\rangle$	$\langle l_5, (0.3, 0.6)\rangle$	$\langle l_1, (0.1, 0.8)\rangle$	$\langle l_4, (0.2, 0.7)\rangle$
L_4	$\langle l_3, (0.2, 0.7)\rangle$	$\langle l_3, (0.1, 0.7)\rangle$	$\langle l_4, (0.3, 0.6)\rangle$	$\langle l_5, (0.4, 0.5)\rangle$

Calculate the attribute weights using the similarity measures $(1 - S_{l_{HK}}(A, B))$ proposed in the paper and the model mentioned in the algorithm. Hence, $\omega = (0.2383, 0.2714, 0.3294, 0.1609)^T$. To rank the alternatives, use the TOPSIS method: Compute the similarity measure $(1 - S_{l_{HK}}(A, B))$ between the alternative and the positive/negative ideal solution as follows-make the weighted matrix:

$$\tilde{P} = \begin{bmatrix} \langle l_{1.115}, (0.0433, 0.9091)\rangle & \langle l_{0.7055}, (0.0969, 0.8927)\rangle & \langle l_{1.3439}, (0.1495, 0.8363)\rangle & \langle l_{0.6823}, (0.0495, 0.9188)\rangle \\ \langle l_{1.0296}, (0.1055, 0.8822)\rangle & \langle l_{1.1831}, (0.1073, 0.8535)\rangle & \langle l_{1.0013}, (0.0460, 0.9014)\rangle & \langle l_{0.5471}, (0.0761, 0.9176)\rangle \\ \langle l_{0.8580}, (0.0518, 0.9078)\rangle & \langle l_{1.1614}, (0.0683, 0.8972)\rangle & \langle l_{0.8300}, (0.1047, 0.8857)\rangle & \langle l_{0.6566}, (0.0419, 0.9442)\rangle \\ \langle l_{1.1536}, (0.1025, 0.8606)\rangle & \langle l_{0.7924}, (0.0737, 0.8945)\rangle & \langle l_{0.9750}, (0.0951, 0.8451)\rangle & \langle l_{0.6244}, (0.0558, 0.9135)\rangle \end{bmatrix}$$

Compute the similarity $(1 - S_{l_{HK}}(A, B))$ between the alternative and the positive/negative ideal solution:

$$d_1{}^+ = 27.3706; \quad d_2{}^+ = 27.4586; \quad d_3{}^+ = 27.8622; \quad d_4{}^+ = 27.6413;$$

$$d_1{}^- = 1.8537; \quad d_2{}^- = 1.4471; \quad d_3{}^- = 1.4485; \quad d_4{}^- = 1.5752.$$

The relative closeness coefficient is calculated as follows:

$$O_1 = 0.9442; \quad O_2 = 0.9560; \quad O_3 = 0.9565; \quad O_4 = 0.9526.$$

Hence, we conclude that the ranking of the best alternative is $O_3 > O_2 > O_4 > O_1$. Based on the order of ranking, L_1 is observed to be the best alternative. Then calculating the attribute weights from using the proposed similarity measures and the ranking of the alternatives are recorded in Table 4.

The comparison study reveals that there is no change in the choice of the best alternative.

Results The format used for the data in this study is Linguistic Intuitionistic Fuzzy Numbers. We introduce several similarity measures and develop corresponding theorems for each measure. The weights in the TOPSIS technique are mostly determined by these suggested similarity measurements. A decision support system (DSS) called TOPSIS is used in decision systems with attribute-based differences to determine which option is the best one. The approach works by trying to figure out which option, out of all the options, is the best.

Table 4 Weight vector obtained from different similarity measures

S. no.	Similarity measures	Weights using similarity measures	Ranking of alternatives
1	$S_{l_{HK}}(M, N)$	$\omega_1 = 0.2383; \omega_2 = 0.2714; \omega_3 = 0.3294; \omega_4 = 0.1609$	$L_1 \succ L_4 \succ L_2 \succ L_3$
2	$S_{l_{LX}}(M, N)$	$\omega_1 = 0.2405; \omega_2 = 0.2734; \omega_3 = 0.3183; \omega_4 = 0.1678$	$L_1 \succ L_4 \succ L_2 \succ L_3$
3	$S_{l_O}(M, N)$	$\omega_1 = 0.2287; \omega_2 = 0.2632; \omega_3 = 0.3559; \omega_4 = 0.1502$	$L_1 \succ L_4 \succ L_2 \succ L_3$
4	$S_{l_{DC}}(M, N)$	$\omega_1 = 0.2468; \omega_2 = 0.2593; \omega_3 = 0.2922; \omega_4 = 0.2017$	$L_1 = L_4 \succ L_2 \succ L_3$
5	$S^{p}{}_{l_L}(M, N)$	$\omega_1 = 0.2306; \omega_2 = 0.2627; \omega_3 = 0.3539; \omega_4 = 0.1528$	$L_1 \succ L_4 \succ L_2 \succ L_3$
6	$S_{l_{HK1}}(M, N)$	$\omega_1 = 0.2294; \omega_2 = 0.2686; \omega_3 = 0.3635; \omega_4 = 0.1385$	$L_1 \succ L_4 \succ L_2 \succ L_3$
7	$S_{l_m}(M, N)$	$\omega_1 = 0.2614; \omega_2 = 0.2769; \omega_3 = 0.2429; \omega_4 = 0.2188$	$L_2 > L_4 > L_1 > L_3$
8	$C_{l_Y}(M, N)$	$\omega_1 = 0.1849; \omega_2 = 0.5264; \omega_3 = 0.2053; \omega_4 = 0.0834$	$L_1 \succ L_4 \succ L_3 \succ L_2$
9	$C_{l_{SY}}(M, N)$	$\omega_1 = 0.1948; \omega_2 = 0.5104; \omega_3 = 0.1955; \omega_4 = 0.0993$	$L_1 \succ L_3 \succ L_4 \succ L_2$
10	$S_{l_{vag}}(M, N)$	$\omega_1 = 0.1719; \omega_2 = 0.2567; \omega_3 = 0.1834; \omega_4 = 0.3880$	$L_2 > L_1 > L_4 > L_3$
11	$S_{l_{HK}}(M, N)$	$\omega_1 = 0.2336; \omega_2 = 0.4856; \omega_3 = 0.1652; \omega_4 = 0.1156$	$L_2 > L_3 > L_4 > L_1$
12	$S_{l_{GV}}(M, N)$	$\omega_1 = 0.2583; \omega_2 = 0.3542; \omega_3 = 0.1973; \omega_4 = 0.1902$	$L_2 > L_4 > L_1 > L_3$

Conclusions Given that linguistic intuitionistic approaches to many real-world issues, because fuzzy sets are inherently complex, the suggested methodology will undoubtedly lessen the decision-maker's partiality. To compute the attribute weights involved in the decision issue, several variants of similarity measures for LIFNs are proposed. The closeness coefficient in the TOPSIS decision-making process is also determined using each of the suggested distance measurements. To ensure the stability of the suggested similarity measures, a few fundamental theorems are established. A comparative analysis is conducted between the suggested similarity measure and the current approach for determining weights using similarity measurements, using LIFN decision data for numerical illustration. The article concludes with a comparison of the many computational methods that have been suggested, highlighting their consistency.

References

1. Atanassov, K.: Intuitionistic fuzzy sets. Fuzzy Sets Syst. **20**, 87–96 (1986)
2. Atanassov, K.: More on intuitionistic fuzzy sets. Fuzzy Sets Syst. **33**, 37–46 (1989)

3. Cables, E., García-Cascales, M.S., Lamata, M.T.: The LTOPSIS: an alternative to TOPSIS decision-making approach for linguistic variables. Expert Syst. Appl. **39**(2), 2119–2126 (2012). https://doi.org/10.1016/j.eswa.2011.07.119
4. Du, Y., Zuo, J.: An extended TOPSIS method for the multiple attribute group decision making problems based on intuitionistic linguistic numbers. Sci. Res. Essays **6**(19), 4125–4132 (2011). https://doi.org/10.5897/SRE11.849
5. Gora, P.: Intuitionistic fuzzy modulus similarity measure. Int. J. Decis. Support Syst. Technol. **15**(1), 1–22 (2023). https://doi.org/10.4018/IJDSST.315757
6. Lei, F., Wei, G., Gao, H., Wu, J., Wei, C. TOPSIS method for developing supplier selection with probabilistic linguistic information. Int. J. Fuzzy Syst. **22**, 749–759 (2020). https://doi.org/10.1007/s40815-019-00797-6
7. Leonishiya, A., Robinson, J.: Varieties of linguistic intuitionistic fuzzy distance measures for linguistic intuitionistic fuzzy TOPSIS method. Indian J. Sci. Technol. **16**(33), 2653–2662 (2023). https://doi.org/10.17485/IJST/v16i33.640-icrsms
8. Liu, D., Chen, X., Peng, D.: The intuitionistic fuzzy linguistic cosine similarity measure and its application in pattern recognition. Complexity (2018). https://doi.org/10.1155/2018/9073597
9. Liu, D., Liu, Y., Chen, X.: The new similarity measure and distance measure of a hesitant fuzzy linguistic term set based on a linguistic scale function. Symmetry **10**(9), 367 (2018). https://doi.org/10.1002/int.22162
10. Pei, Z. A note on the TOPSIS method in MADM problems with linguistic evaluations. Appl. Soft Comput. **36**, 24–35 (2015). https://doi.org/10.1016/j.asoc.2015.06.042
11. Song, Y., Wang, X., Lei, L., Xue, A.: A novel similarity measure on intuitionistic fuzzy sets with its applications. Appl. Intell. **42**, 252–261 (2015). https://doi.org/10.1007/s10489-014-0596-z
12. Song, Y., Wang, X., Quan, W., Huang, W. A new approach to construct similarity measure for intuitionistic fuzzy sets. Soft Comput. **23**(6), 1985–1998 (2019). https://doi.org/10.1007/s00500-017-2912-0
13. Xu, Z.: Some similarity measures of intuitionistic fuzzy sets and their applications to multiple attribute decision making. Fuzzy Optim. Decis. Making **6**, 109–121 (2007). https://doi.org/10.1007/s10700-007-9004-z
14. Ye, J.: Cosine similarity measures for intuitionistic fuzzy sets and their applications. Math. Comput. Modell. **53**(1–2), 91–97 (2011). https://doi.org/10.1016/j.mcm.2010.07.022
15. Zhang, Q., Hu, J., Feng, J., Liu, A., Li, Y. New similarity measures of Pythagorean fuzzy sets and their applications. IEEE Access **7**, 138192–138202 (2019). https://doi.org/10.1109/ACCESS.2019.2942766

Part III
Statistics

The Analysis of the Value of π by Simulation Approach and Non-parametric Statistical Testing of Hypothesis

Sanjay Kulkarni and Sandeep Kulkarni

1 Introduction

The technique of simulation has been used by the designers and analysis in physical sciences, and it promises to become an important tool for tackling the complicated problems of managerial decision-making. Scale models of machines have been used to simulate the plant layouts, and models of aircraft have been tested in wind tunnels to determine their aerodynamic characteristics. Simulation, which can appropriately be called management laboratory, determines the effect of several alternate policies without disturbing the real system. Techniques such as linear programming, dynamic programming, queuing theory, and network models are not enough to tackle all the important managerial problems requiring data analysis. Each technique has its own limitations. When the characteristics such as uncertainty, complexity, dynamic interaction between the decision and subsequent event, and the need to develop detailed procedures and finely divided time intervals combine in one situation, it becomes too complex to be solved by any of the techniques of mathematical programming and probabilistic models. It must be analyzed by some other kind of quantitative technique which may give accurate and reliable results.

The aim of the present paper is to study the value of π obtained by the Monte Carlo simulation method and compare the results with the experimental value. Then the Monte Carlo simulation distribution is tested by applying nonparametric testing of hypothesis such as Friedman's test and chi-square test discussed in detail.

S. Kulkarni (✉)
Department of First Year Engineering, Hope Foundation's Finolex Academy of Management and Technology, Ratnagiri, Maharashtra, India
e-mail: sanjay.kulkarni@famt.ac.in

S. Kulkarni
Tata Consultancy Services, Mumbai, India

I. Rajasingh et al. (eds.), *Applied Computational Mathematics*,
Springer Proceedings in Mathematics & Statistics 480,
https://doi.org/10.1007/978-3-031-77764-6_25

In computer science and its applications, when a new algorithm or an improved version is proposed, it is quite important to compare it with existing algorithms on several datasets, to prove that the control algorithm performs better than the benchmark ones, i.e., the results of the control algorithm are statistically superior to the benchmark algorithms with respect to the considered datasets. Let x_1 represent the control algorithm and $x_2.x_3.x_4 \cdots .x_k$ represent the $(k-1)$ benchmark algorithms. The challenge lies in the test to better judge whether the control algorithm x_1 has a significant advantage over other benchmark algorithms on some experimental datasets. Due to the diversity of datasets and various random factors in training and testing processes, it is rare that the control algorithm performs better on all experimental datasets. Therefore, it is difficult to draw a meaningful conclusion without adopting a statistical hypothesis test to favor the decision process [1]. Statistical hypothesis tests can be divided into two categories parametric tests and nonparametric tests [2–4].

In practical applications, it is barely possible to verify that the results obtained by the algorithms on different datasets fall into these assumptions. Therefore, nonparametric statistical tests are commonly considered [5]. When selecting a nonparametric statistical test, it is also necessary to distinguish the pairwise comparison and the multiple comparison. In pairwise comparison, nonparametric statistical tests include the Wilcoxon signed rank test [6]. For multiple comparison, nonparametric statistical tests include Friedman's test, multiple sign test [7], and contrast estimation based on medians [8]. However, nonparametric statistical tests have been widely adopted in the published papers [9–12]. Among the previous nonparametric statistical tests, Friedman's test is more effective and is widely used by many scholars [13, 14]. Some similar work has been reported on the analysis of the nonparametric statistical tests for multiple comparison. O Gorman [15] compared the F-test, Friedman's test, and several aligned Friedman's tests with Monte Carlo simulation.

2 The Value of π Experimentally

Draw the coordinate axes OX and OY with center O drawn on an arc PR of unit radius as shown in Fig. 1 and complete the square $OPQR$. The equation of the circle is $x^2 + y^2 = 1$. From the random number table, select any two random numbers, say 0.2068 and 0.7295, and let $x = 0.2068$ and $y = 0.7295$. Plot the point $P_1(0.2068, 0.7295)$. Obviously, $x^2 + y^2 \leq 1$, P_1 will lie inside or on the arc of the circle if $x^2 + y^2 > 1$, and P_1 will lie outside the arc but within the square.

In this manner, hundreds or thousands of pairs of random numbers are selected, and it is ascertained whether the points representing them lie in/on the arc or beyond the arc but inside the square. Suppose N is the total number of points considered, out of which n lie in/on the arc. Then,

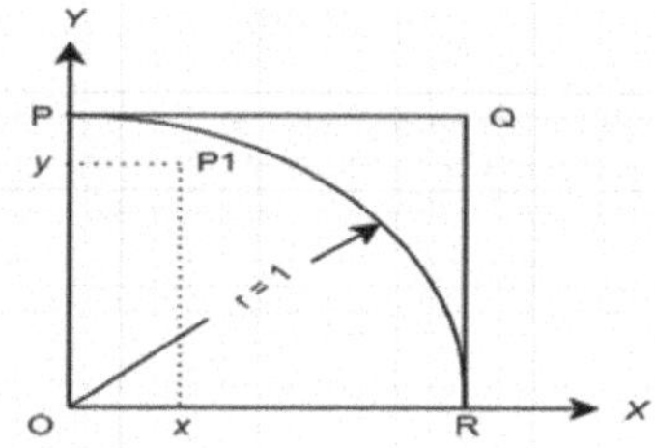

Fig. 1 To find experimental value of Pi

$$\frac{n}{N} = \frac{\text{Area enclosed by the arc}}{\text{Area of the Square}}$$

$$\frac{n}{N} = \frac{\frac{\pi}{4}}{1}$$

$$\pi = \frac{4n}{N}$$

The above equation gives the experimental value of π. Obviously, the larger the sample size N, the closer will be the obtained value to the true value of π.

3 The Monte Carlo Simulation Approach

The technique employs random numbers and is impracticable and problems that involve probability and wherein physical experimentation is impracticable and formation of mathematical model is impossible. It is a method of simulation by sampling technique. The steps involved in carrying out the Monte Carlo simulation are as follows:

1. Select the measure of effectiveness of the problem.
2. Identify the variables that affect the measure of effectiveness significantly.
3. Determine the cumulative probability distribution of each variable selected in step (b).
4. Get a set of random numbers.
5. Consider each random number as a decimal value of the cumulative probability distribution.
6. Record the value or values if several variables are being simulated/generated in step 5. Substitute in the formula chosen for the measure of effectiveness and find its simulated value.

Repeat steps 5 and 6 until the sample is large enough to the satisfaction of the decision-maker.

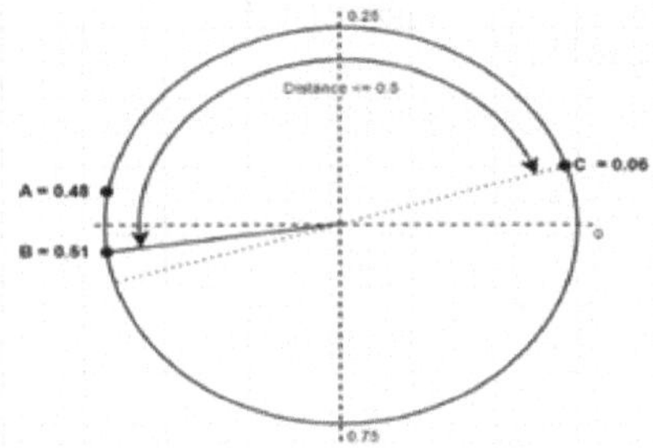

Fig. 2 To find three points randomly on the circumference of a circle by using Monte Carlo Method

Three points are chosen at random on the circumference of a circle by using Monte Carlo methods the probability that they lie on the same semicircle. Draw a circle of circumference unity, i.e., of radius $\frac{1}{2\pi}$ as shown in Fig. 2. Consider a triplet of three random numbers 0.48, 0.51, and 0.06 from the random number table. These are plotted as points A, B, and C in Fig. 2, the distance of a point from O along the circumference anticlockwise being the value of random number. Since for this triplet, the difference between the maxima (0.51) and minima (0.06) is less than 0.50, the triplet lies obviously on a semicircle.

The general rule for ascertaining whether a triplet of random numbers lies on a semicircle or not is as follows:

1. Find the difference between the maxima and minima. If this difference is $\leq$0.50, the triplet lies on a semicircle.
2. If this difference is >0.50, add unity to those random numbers in the triplet which is <0.50. Now find the difference between the new maxima and the minima. If this difference is $\leq$0.50, the triplet lies on a semicircle; otherwise, the triplet will not lie on a semicircle.

Table 1 gives the details the above rule it is ascertained whether each of the following 32 triplets lies or does not lie on the semicircle.

Thus, out of the 32 triplets, 25 lie on a semicircle, yielding the required probability $\frac{25}{32} = 0.78125$. However, if many triplets are considered, the probability obtained will be very close to the theoretical/experimental value of π. The experimental value of π is in this case is or for the present case is $N = 32$ and $n = 25$ is given by $\pi = \frac{4n}{N} = \frac{4\times 25}{32} = 3.125$. If we increase the sample size N, we get the value of π very close to the experimental value.

4 Friedman's Test/Two-Way Analysis of Variance by Ranks

Friedman's test is used to compare the distributions of the two or more quantitative variables. We have assigned the ranks for the above selected 32 random numbers row wise for each Triplet −1, Triplet −2, and Triplet −3 as shown in the following Table 2.

Table 1 The triplet of random numbers lies on a semicircle

Sr. no	Triplet −1	Triplet −2	Triplet −3	Difference of maxima and minima	Difference of new maxima and minima	The Number of triplets lies on semicircle
1	0.21	0.11	0.71	0.71 − 0.11 = 0.60	1.11 − 0.21 = 0.90	
2	0.65	0.41	0.35	0.65 − 0.35 = 0.30		0.3
3	0.17	0.91	0.07	0.91 − 0.07 = 0.84	1.07 − 0.17 = 0.90	
4	0.34	0.12	0.43	0.43 − 0.12 = 0.31		0.31
5	0.38	0.49	0.13	0.49 − 0.13 = 0.36		0.36
6	0.05	0.96	0.76	0.96 − 0.05 = 0.91	1.05 − 0.76 = 0.29	0.26
7	0.85	0.69	0.57	0.85 − 0.57 = 0.28		0.28
8	0.63	0.41	0.03	0.63 − 0.03 = 0.60	1.03 − 0.41 = 0.62	
9	0.91	0.58	0.62	0.91 − 0.58 = 0.33		0.33
10	0.75	0.89	0.23	0.89 − 0.23 = 0.66	1.23 − 0.75 = 0.48	0.48
11	0.21	0.36	0.59	0.59 − 0.21 = 0.38		0.38
12	0.39	0.19	0.21	0.39 − 0.19 = 0.20		0.2
13	0.74	0.86	0.9	0.90 − 0.74 = 0.16		0.16
14	0.64	0.18	0.67	0.67 − 0.18 = 0.49		0.49
15	0.2	0.72	0.34	0.72 − 0.20 = 0.52	1.20 − 0.34 = 0.86	
16	0.54	0.3	0.22	0.54 − 0.22 = 0.32		0.32
17	0.48	0.74	0.76	0.76 − 0.48 = 0.28		0.28
18	0.02	0.07	0.64	0.64 − 0.02 = 0.62	1.02 − 0.07 = 0.95	
19	0.95	0.23	0.91	0.95 − 0.23 = 0.72	1.23 − 0.91 = 0.32	0.32
20	0.48	0.55	0.91	0.91 − 0.48 = 0.43		0.43
21	0.48	0.51	0.06	0.51 − 0.06 = 0.45		0.45
22	0.56	0.06	0.91	0.91 − 0.06 = 0.85	1.06 − 0.56 = 0.50	0.5
23	0.51	0.13	0.65	0.65 − 0.13 = 0.52	1.13 − 0.51 = 0.62	
24	0.65	0.6	0.51	0.65 − 0.51 = 0.14		0.14
25	0.64	0.89	0.74	0.89 − 0.64 = 0.25		0.25
26	0.99	0.53	0.58	0.99 − 0.53 = 0.46		0.46
27	0.83	0.44	0.64	0.83 − 0.44 = 0.39		0.39
28	0.36	0.6	0.82	0.82 − 0.36 − 0.46		0.46
29	0.34	0.46	0.31	0.46 − 0.31 = 0.15		0.15
30	0.82	0.63	0.54	0.82 − 0.54 = 0.28		0.28
31	0.94	0.32	0.68	0.94 − 0.32 = 0.62	1.32 − 0.68 = 0.64	
32	0.58	0.92	0.69	0.92 − 0.58 = 0.34		0.34
Total						25

Setting Testing of Hypothesis

(a) The Null Hypothesis: All triplets have identical effects.
The Alternative Hypothesis: Triplets have different effects.

(b) Test Statistic:
N = The total number of triplets = 32

Table 2 The row wise rank assigned for triplet of random numbers

Sr.	Triplet			Rank of		
no	−1	−2	−3	Triplet −1	Triplet −2	Triplet −3
1	0.21	0.11	0.71	2	1	3
2	0.65	0.41	0.35	3	2	1
3	0.17	0.91	0.07	2	3	1
4	0.34	0.12	0.43	2	1	3
5	0.38	0.49	0.13	2	3	1
6	0.05	0.96	0.76	1	3	2
7	0.85	0.69	0.57	3	2	1
8	0.63	0.41	0.03	3	2	1
9	0.91	0.58	0.62	3	1	2
10	0.75	0.89	0.23	2	3	1
11	0.21	0.36	0.59	1	2	3
12	0.39	0.19	0.21	3	1	2
13	0.74	0.86	0.9	1	2	3
14	0.64	0.18	0.67	2	1	3
15	0.2	0.72	0.34	1	3	2
16	0.54	0.3	0.22	3	2	1
17	0.48	0.74	0.76	1	2	3
18	0.02	0.07	0.64	1	2	3
19	0.95	0.23	0.91	3	1	2
20	0.48	0.55	0.91	1	2	3
21	0.48	0.51	0.06	2	1	3
22	0.56	0.06	0.91	2	3	1
23	0.51	0.13	0.65	2	3	1
24	0.65	0.6	0.51	1	2	3
25	0.64	0.89	0.74	3	1	2
26	0.99	0.53	0.58	1	3	2
27	0.83	0.44	0.64	1	3	2
28	0.36	0.6	0.82	3	2	1
29	0.34	0.46	0.31	2	1	3
30	0.82	0.63	0.54	1	2	3
31	0.94	0.32	0.68	1	3	2
32	0.58	0.92	0.69	3	1	2
Total of ranks for each of three columns				62	64	66

$k =$ The number of triplets considered $= 3$
$R =$ Total of ranks for each of three columns 62, 64, & 66
Friedman calculated value is given by the following formula:
$FM = \left(\frac{N}{N \times k \times (k+1)}\right)\left(\sum R^2 - 3 \times N \times (k+1)\right)$

$FM = \left(\frac{32}{32\times3\times(3+1)}\right)\left((62^2+64^2+66^2)-(3\times32\times(3+1))\right)$
$FM_{\text{Calculated Value}} = 992.66$

(c) Level of Significance: Setting LOS (level of significance) of 5%, the degrees of freedom $= n - 1 = 3 - 1 = 2$
(d) Critical Value: The chi-square table at 5% LOS for the degrees of freedom 2 is $FM_{\text{table}} = 5.99$.
(e) Decision: Since the computed value of Friedman's test is $FM_{\text{Calculated}} = 992.66$, which is greater than the table value $FM_{\text{table}} = 5.99$, the null hypothesis is rejected, i.e., the triplets have different effects.

5 Chi-Square (χ^2) Test to Test the Goodness of Fit

χ^2- test is very commonly known as χ^2-test of goodness of fit because it enables us to ascertain how well the theoretical distributions fit the observed values or frequencies. This test does not deal with the parameter; therefore, it is also called nonparametric test. In the following Table 3:

N = The total number of triplets = 32
O = Observed minimum difference between maxima and minima
E = Expected minimum difference between maxima and minima
$E = \sum \frac{O}{N}$

Setting Testing of Hypothesis

(a) The Null Hypothesis: Triplets of random number will follow the goodness of fit.
The Alternative Hypothesis: Triplets of random number will not follow the goodness of fit.
(b) Test Statistic: N = The total number of triplets = 32.
O = Observed minimum difference between maxima and minima. E = Expected minimum difference between maxima and minima. $E = \frac{\sum O}{N}$
$E = \frac{\sum O}{N} = \frac{13.88}{32} = 0.434$ Chi-square (χ^2) Test calculated value is given by the following formula:

$$\chi^2 = \sum \frac{(O-E)^2}{E} = \frac{1.5562}{0.434}$$

$$\chi^2_{\text{(Calculated Value)}} = 3.5857$$

(c) Level of Significance: Setting level of significance (LOS) of 5% the degrees of freedom $= n - 1 = 32 - 1 = 31$.
(d) Critical Value: The chi-square table at 5% LOS for the degrees of freedom 31 is $\chi^2_{\text{table}} = 43.77$.

Table 3 The list of observed and expected difference between maxima and minima of triplet of random numbers

Sr. no	Triplet −1	Triplet −2	Triplet −3	Difference of maxima and minima	Difference of new maxima and new minima	O	$O-E$	$(O-E)^2$
1	0.21	0.11	0.71	$0.71-0.11=0.60$	$1.11-0.21=0.90$	0.9	0.47	0.2209
2	0.65	0.41	0.35	$0.65-0.35=0.30$		0.3	−0.13	0.0169
3	0.17	0.91	0.07	$0.91-0.07=0.84$	$1.07-0.17=0.90$	0.94	0.51	0.2601
4	0.34	0.12	0.43	$0.43-0.12=0.31$		0.31	−0.12	0.0144
5	0.38	0.49	0.13	$0.49-0.13=0.36$		0.36	−0.07	0.0049
6	0.05	0.96	0.76	$0.96-0.05=0.91$	$1.05-0.76=0.29$	0.29	−0.14	0.0196
7	0.85	0.69	0.57	$0.85-0.57=0.28$		0.28	−0.15	0.0225
8	0.63	0.41	0.03	$0.63-0.03=0.60$	$1.03-0.41=0.62$	0.62	0.19	0.0361
9	0.91	0.58	0.62	$0.91-0.58=0.33$		0.33	−0.1	0.01
10	0.75	0.89	0.23	$0.89-0.23=0.66$	$1.23-0.75=0.48$	0.48	0.05	0.0025
11	0.21	0.36	0.59	$0.59-0.21=0.38$		0.38	−0.05	0.0025
12	0.39	0.19	0.21	$0.39-0.19=0.20$		0.2	−0.23	0.0529
13	0.74	0.86	0.9	$0.90-0.74=0.16$		0.16	−0.27	0.0729
14	0.64	0.18	0.67	$0.67-0.18=0.49$		0.49	0.06	0.0036
15	0.2	0.72	0.34	$0.72-0.20=0.52$	$1.20-0.34=0.86$	0.86	0.43	0.1849
16	0.54	0.3	0.22	$0.54-0.22=0.32$		0.32	−0.11	0.0121
17	0.48	0.74	0.76	$0.76-0.48=0.28$		0.28	−0.15	0.0225
18	0.02	0.07	0.64	$0.64-0.02=0.62$	$1.02-0.07=0.95$	0.95	0.52	0.2704
19	0.95	0.23	0.91	$0.95-0.23=0.72$	$1.23-0.91=0.32$	0.32	−0.11	0.0121
20	0.48	0.55	0.91	$0.91-0.48=0.43$		0.43	0	0
21	0.48	0.51	0.06	$0.51-0.06=0.45$		0.45	0.02	0.0004
22	0.56	0.06	0.91	$0.91-0.06=0.85$	$1.06-0.56=0.50$	0.5	0.07	0.0049
23	0.51	0.13	0.65	$0.65-0.13=0.52$	$1.13-0.51=0.62$	0.62	0.19	0.0361

24	0.65	0.6	0.51	0.65 − 0.51 = 0.14		0.14	−0.29	0.0841
25	0.64	0.89	0.74	0.89 − 0.64 = 0.25		0.25	−0.18	0.0324
26	0.99	0.53	0.58	0.99 − 0.53 = 0.46		0.46	0.03	0.0009
27	0.83	0.44	0.64	0.83 − 0.44 = 0.39		0.39	−0.04	0.0016
28	0.36	0.6	0.82	0.82 − 0.36 = 0.46		0.46	0.03	0.0009
29	0.34	0.46	0.31	0.46 − 0.31 = 0.15		0.15	−0.28	0.0784
30	0.82	0.63	0.54	0.82 − 0.54 = 0.28		0.28	−0.15	0.0225
31	0.94	0.32	0.68	0.94 − 0.32 = 0.62	1.32 − 0.68 = 0.64	0.64	0.21	0.0441
32	0.58	0.92	0.69	0.92 − 0.58 = 0.34		0.34	−0.09	0.0081
Total						13.88		1.5562

(e) Decision: Since the computed value of the chi-square (χ^2) test is $\chi^2_{(\text{Calculated Value})} = 3.5857$ less than the table value $\chi^2_{\text{table}} = 43.77$, the null hypothesis is accepted, i.e., triplets of random number will follow the goodness of fit.

6 Conclusions

The present paper is to study the value of π obtained by the Monte Carlo simulation method and compare the results with the experimental value of π. Then the Monte Carlo simulation distribution is tested by applying nonparametric testing of hypothesis such as Friedman's test and chi-square test discussed in detail. This type of data analysis will be helpful for students and researchers, and data analysis in computer engineering and data science makes a suitable decision whether the hypothesis is accepted or rejected. From this study, the following points are observed. If the sample size is equal to 20, the results are identical to that of [16]:

- The larger the sample size N closer will be the obtained value to the true value of π.
- The triplets of random numbers have different effects.
- Triplets of random number will follow the goodness of fit.

Acknowledgments The authors are thankful to the Management of Hope Foundation's Finolex Academy of Management and Technology, Ratnagiri, for providing financial support and excellent computational facilities. The authors are also thankful to unknown referees for their comments and suggestions for the substantial improvement of the quality of the paper.

References

1. Liu, Z., Blasch, F., Jhon, V.: Statistical comparison of image fusion algorithms recommendations. Inf. Fusion 2017. Smith, T.F., Waterman, M.S.: Identification of common molecular subsequences. Inf. Fusion **36**, 251–260 (2017)
2. Couch, S., Kazan, Z., Shi, K., Bray, A., Groce, A.: Differentially private nonparametric hypothesis testing. Proc. ACM Conf. Comput. Commun. Secur. 737–750 (2019)
3. García, S., Fernández, A., Luengo, J., Herrera, F.: Advanced non- parametric tests for multiple comparisons in the design of experi- ments in computational intelligence and data mining: experimental analysis of power. Inf. Sci. **180**(10), 2044–2064 (2010)
4. Pawar, S.D., Shirke, D.T.: Nonparametric tests for multivariate multi-sample locations based on data depth. J. Stat. Comput. Simul. **89**(9), 1574–1591 (2019)
5. Derrac, J., García, S., Hui, S., Suganthan, P.N., Herrera, F.: Analyzing convergence performance of evolutionary algorithms: a statistical approach. Inf. Sci. **289**, 41–58 (2014)
6. Demšar, J.: Statistical comparisons of classifiers over multiple data sets. J. Mach. Learn. Res. **7**, 1–30 (2006)
7. Steel, R.G.D.: A multiple comparison sign test: treatments versus control. J. Am. Stat. Assoc. **54**, 767–775 (1959)

8. Doksum, K.: Robust procedures for some linear models with one observation per cell. Ann. Math. Stat. **38**, 878–883 (1967)
9. Chandra, T.B. Verma, K., Singh, B.K., Jain, D., Netam, S.S.: Coronavirus disease (COVID-19) detection in chest X-Ray images using majority voting-based classifier ensemble. Expert Syst. Appl. **65**, 1–13 (2021)
10. De Gregorio, M., Giordano, M.: An experimental evaluation of weightless neural networks for multi-class classification. Appl. Soft Comput. J. **72**, 338–354 (2018)
11. Pulgar, F.J., Charte, F., Rivera, A.J., Del Jesus, M.J.: Choosing the proper autoencoder for feature fusion based on data complexity and classifiers: analysis, tips and guidelines. Inf. Fusion **54**, 44–60 (2020)
12. Shi, S., Ding, S., Zhang, Z., Jia, W.: Energy-based structural least squares MBSVM for classification. Appl. Intell. **50**(3), 681–697 (2020)
13. Liu, J.: Fuzzy support vector machine for imbalanced data with borderline noise. Fuzzy Sets Syst. **413**, 64–73 (2020)
14. Petrovic, M., Miljkovic, Z., Jokic, A.: A novel methodology for optimal single mobile robot scheduling using whale optimization algorithm. Appl. Soft Comput. J. **81**(2), (2019)
15. O'Gorman, T.W.: A comparison of the F-test, Friedman's test, and several aligned rank tests for the analysis of randomized complete blocks. J. Agric. Biol. Environ. Stat. **6**, 367–378 (2001)
16. Kulkarni, S.B., Kulkarni, S.S.: Analyzing π through Monte Carlo Simulations: a comparative study with experimental approaches. J. Comput. Mech. Manage. **3**(1), 22–29 (2024)

Optimizing Quality of Service in Internet and Web Services Through a Two-Stage Retrial Tandem Queuing Model with Working Vacation Techniques

R. Keerthika and S. P. Niranjan

1 Introduction

Queueing theory is a powerful tool for modeling and analyzing various real-world systems involving the flow of entities through a series of service points. In many practical applications, such as call centers, telecommunication networks, manufacturing processes, and service centers, it is common to encounter complex dynamics that go beyond the classical queueing models. Specifically, the combination of unreliable single service, retrial behavior, and working vacations introduces a level of intricacy that demands systematic analysis and understanding. The motivation behind this research stems from the significant practical implications of unreliable batch service, retrials, and server vacations. When a task fails due to transient issues such as network errors, resource unavailability, or temporary system glitches, it is moved to the retrial queue. The system then periodically reprocesses or retries the tasks in the retrial queue, hoping that the conditions causing the failure have been resolved. During a working vacation, individuals may choose to continue working remotely, handle essential tasks, or stay connected to work-related responsibilities, even while physically away from their usual workplace. This could involve checking emails, attending virtual meetings, or completing specific work assignments. The fundamental objectives of this study are twofold: (i) to develop a comprehensive understanding of unreliable batch service tandem queues with retrial and vacation dynamics and (ii) to develop mathematical models and analytical methodologies for evaluating the performance of these systems, including critical metrics such as queue lengths, wait times, and system reliability.

R. Keerthika (✉) · S. P. Niranjan
Department of Mathematics, Vel Tech Rangarajan Dr. Sagunthala R & D Institute of Science and Technology, Chennai, Tamil Nadu, India

I. Rajasingh et al. (eds.), *Applied Computational Mathematics*,
Springer Proceedings in Mathematics & Statistics 480,
https://doi.org/10.1007/978-3-031-77764-6_26

Our model is motivated by the Internet and web services: In the context of the Internet and web services, TQVR systems can model users' interactions with web servers, content delivery networks, or cloud-based services. Interruptions in service availability (vacations) and retries due to slow response times can be important considerations for system design and optimization. Retrial queues with working vacations.

Recent advancements in queuing theory have significantly contributed to optimizing Quality of Service (QoS) in communication systems. Agrawal et al. [1] introduced an M/M/1 model with working vacations and server breakdowns, while Chaudhry [4] analyzed transient and steady-state behaviors of intermittently available servers. Aguir et al. [2] and Gomez-Corral & Garcia (2014) examined retrial phenomena, focusing on user behavior and queue lengths in M/M/c systems. Baumann & Sandmann [3] studied multi-server tandem queues with finite buffers, relevant to 4G/5G networks, and Gubbi et al. [5] addressed IoT traffic challenges. Together, these works enhance modeling and QoS optimization in modern networks. Recent studies have expanded the application of queuing theory to address complex network scenarios. Gun & Makowski [6] developed matrix geometric solutions for tandem queuing systems with phase-type servers, incorporating blocking and failures. Jeganathan et al. [7] applied matrix-analytic methods for steady-state analysis of Markovian queuing systems, focusing on scrap items. Manikandan & Nair [8] explored M/M/1 queuing-inventory models with working vacations and vacation interruptions, highlighting the impact on inventory and service performance. Tran-Gia & Mandjes [9] modeled customer retrial behaviors in cellular networks, providing insights into traffic dynamics and retrial impacts. These contributions enhance the understanding and optimization of queuing systems in modern communication networks. Yang and Templeton [10] surveyed retrial queues, analyzing their impact on system performance, stability, and congestion, providing a foundational framework for modeling retrial behaviors in queuing systems. However, recent applications to network and web sciences show that there are queuing situations in which the number of clients in the orbit has no effect on the retrial rate. There is the retrial tandem queuing model with working vacation techniques treated in this article that differs from the above models in the service discipline. Figure 1 represents the diagram of cloud performance in the tandem queuing systems.

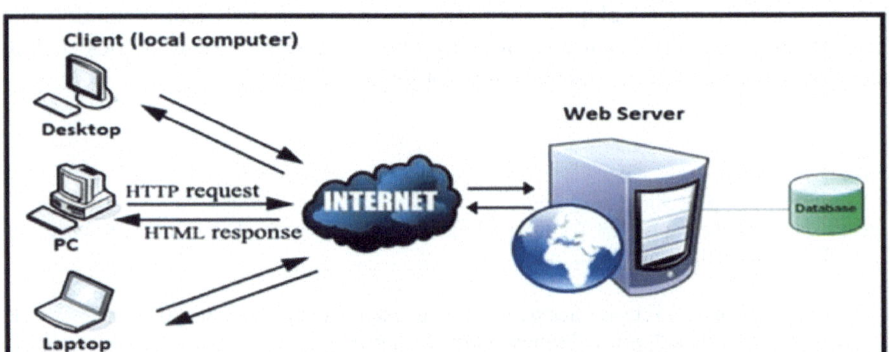

Fig. 1 Internet and web server performance in the tandem queuing systems

2 System and Notations

There are two stage tandem queues, labeled stage 1 and stage 2. Customers arrive at stage 1 and immediately join the queue at stage 2 to receive service. The arrival process of entities is assumed to be Markovian with an arrival rate χ, meaning that the probability of arrivals in any small time interval depends only on the current system state. The two servers serve parameters μ_1 for stage 1 and μ_2 for stage 2, where $\mu_1 \neq \mu_2$. It is considered that service times at both stages of the tandem queue follow exponential distributions. The queues at each stage are assumed to be independent of each other. The state of the first stage does not directly influence the state of the second stage. The queuing system operates under the assumption of a first-come-first-serve discipline, where entities are served in the order of their arrival. Retrial times for entities in retrial status are considered to have an exponential distribution. Working vacation durations have an exponential distribution with parameter γ at Node 1 and $\gamma\mu_1$ for low service rate on working vacation at Node 1. If a client shows up when the server is away on vacation, the server will return to service that customer. However, consumers are only provided during vacation at lower rates compared to regular service. To be precise, Node 2 serves this customer at the rate $\theta\mu_2, 0 < \theta < 1$. If a customer is not served immediately, they may either retry (retrial) or go on a working vacation before returning for service. The system operates under certain assumptions regarding vacation rules, retrial rates, arrival rates, and service rates. The diagram showing the system is given in Fig. 2.

Notations

χ—Arrival rate at stage 1
μ_1—Service rate at stage 1
θ_1—Retrial rate at stage 1
γ—Working vacation rate at stage 1

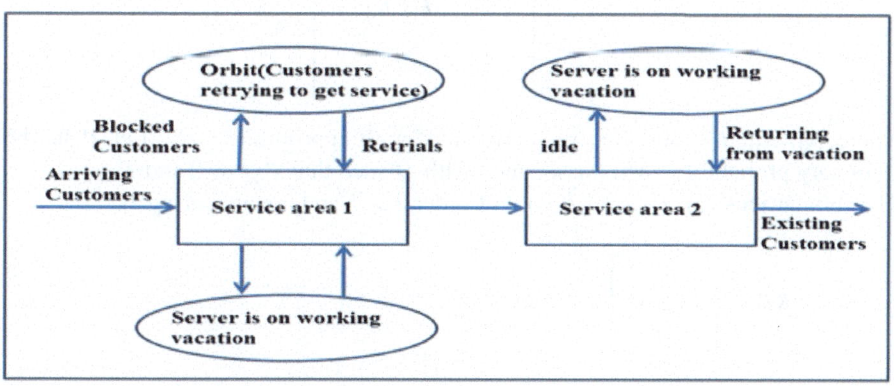

Fig. 2 Tandem retrial queue with working vacation

$\gamma\mu_1$—Low service rate on working vacation at stage 1
μ_v—Service rate at working vacation at stage 1
μ_b—Service rate during the busy period at stage 1
μ_2—Service rate at stage 2
θ_2—Retrial rate at stage 2

3 Analysis of Steady-State Probabilities

Denote $N_1(t)$ to be the quantity of clients at Server 1 and $N_2(t)$ to be the quantity of clients at Server 2. Letting $S_1(t)$ and $S_2(t)$ be the state of the server at time t, then

$$S_1(t) = \begin{cases} 0, \text{ if the server is in a working vacation period at time } t \\ 1, \text{ if the server is a regularly busy at time } t \\ 2, \text{ if the server is in a retrial orbit at } t \end{cases}$$

$$S_2(t) = \begin{cases} 0, \text{ if the server is in a working vacation period at time } t \\ 1, \text{ otherwise} \end{cases}$$

$\{N_1(t), N_2(t), S_1(t), S_2(t)\}$ is a level- independent QBD process with state space $\aleph = \bigcup_{i=0}^{\infty} c(k)$
where $c(k) = \{(i, j, k, l, m)/0 \le i \le N_1, 0 \le j \le N_2, 0 \le k, l \le N_1^r, 0 \le m \le N_2^w\}$
This Markov chain's infinitesimal producing matrix Q is provided by

$$Q = \begin{bmatrix} B_0 & K_0 & & & \\ K_2 & K_1 & K_0 & & \\ & K_2 & K_1 & K_0 & \\ & & K_2 & K_1 & \ddots \\ & & & \ddots & \ddots \\ & & & & \ddots \end{bmatrix} \tag{1}$$

where B_0, K_0, K_1, and K_2 are submatrices with appropriate sizes. Let P be the stationary probability vector associated with Q such that $PQ = 0$ and $Pe = 1$
The submatrices K_0, K_1, and K_2 are of order 3×3 and are appearing as

$$K_0 = \begin{bmatrix} \chi & 0 & 0 \\ 0 & \chi & 0 \\ 0 & 0 & \chi \end{bmatrix},$$

$$K_1 = \begin{bmatrix} -(\chi + \mu_1 + \gamma) & \gamma & 0 \\ 0 & -(\chi + \mu_v + \theta_1) & 0 \\ \gamma\mu_1 & \theta_1 & -(\chi + \mu_b + \theta_2) \end{bmatrix},$$

$$K_2 = \begin{bmatrix} \mu_1 & 0 & 0 \\ \mu_2 & \mu_v + \mu_b & 0 \\ 0 & 0 & \mu_2 \end{bmatrix}$$

Also, the definition of the boundary matrices is

$$B_0 = \begin{bmatrix} -(\chi + \gamma) & \gamma & 0 \\ 0 & -\chi & 0 \\ \mu\theta_2 & \theta_1 & -(\chi + \theta_1) \end{bmatrix} \tag{2}$$

4 Busy Period Analysis

B represents a typical busy period, which is the duration of time that the server operates at a rate of μ_1 constantly. A typical busy period can begin after the end of a working vacation or idle period. Among the components of a busy cycle are a working vacation V, a period of inactivity following the conclusion of this V (if any), and a usual hectic period that follows B be aware that this is merely a working vacation γ in a busy cycle C and $E(\gamma) = \frac{1}{\xi}$. By applying the alternate renewal process's limiting theorem, we have $P(L = 0) = \frac{E(\gamma)}{E(C)} = \Omega_0 \frac{1}{1-r}$ (3)
We therefore predict a busy cycle. $E(C) = \frac{1-r}{\xi\Omega_0}$ (4)
Likewise, applying the alternative renewal theorem, we obtain
$P(L = 1) = \frac{E(\gamma)}{E(C)} = \sum_{k=1}^{\infty} \Omega_{k,1}$ (5)
By changing the value of $E(C)$ in the above expression, we may determine the planned busy time B:

$$E(B) = E(C)\sum_{k=1}^{\infty} \Omega_{k,1} = \frac{1}{\xi(1-\rho)}\left[\frac{\chi - r\mu_v}{\mu} + \frac{\chi + \xi - r\mu_v}{\theta}\right] \tag{6}$$

We conclude the expected idle period in a busy cycle, $\Omega_{0,1} = \frac{E(I)}{E(C)}$.
Substituting the E(C) value into the above expression, we can conclude the expected idle period in a busy cycle:

$$E(I) = \frac{1-r}{\xi\theta}(\chi + \xi - r\mu_v) \tag{7}$$

5 Performance Measures

With the help of the abovementioned steady-state probability, we can determine various system performance metrics. Given that Server 1 expects a certain number of consumers in the queue,
$E(L_1) = \sum_{i=0}^{\infty} i\Omega_i e + \Omega_1 \left[l^2(1-l)^{-2} + 2l(1-l)^{-1}\right] e$ (8)
given that Server 2 expects a certain number of consumers in the queue:

$$E(L_2) = \sum_{i=0}^{\infty} i\Omega_i e + \Omega_1 \left[l^2(1-l)^{-2} + 2l(1-l)^{-1}\right] e \quad (9)$$

The average wait time during regular busy server use in the system is given by

$$\begin{aligned} P_{R1} &= \Pr\{J = 1\} = \sum_{i=1}^{\infty} \Omega_{i1} \\ P_{R2} &= pr\{J = 1) = \sum_{i=1}^{\infty} \Omega_{i1} \end{aligned} \quad (10)$$

Chance that the server is on vacation while working $P_{wv} = pr\{J = 0\} = \sum_{i=1}^{\infty} \Omega_{i0}$ (11)

6 Model of Costs (Specific Cases)

The following is the development of a steady-state anticipated cost function:
Total Expense (TC) $= C_1 * MNS + C_2 * P_{busy} + C_3 * WV + C_4 P_{idle}$ (12), where C_1 the cost per unit during the system's waiting period for clients, for service; C_2 when the server is busy, the cost per unit time; C_3 the price per unit during the server's working vacation; and C_4 the price per unit of idle server time. We have performed numerical computations. $C_1 = 50$ units, $C_2 = 100$ units, $C_3 = 150$ units, and $C_4 = 200$ units. The codes created in MATLAB software have been used to calculate the stationary system size probability. Five decimal places have been added to all computations. Even though we are solving an infinite capacity model, we have limited the system capacity to N for computational reasons.

7 Numerical Results

The findings are displayed in the following Table 1 for different parameter settings. In Fig. 3, we fix ($\mu_1 = 3, \mu_2 = 5, \theta_1 = 0.5, \theta_2 = 0.2, \gamma = 0.5$)

Table 1 Steady-state system size probabilities and performance measure values for various values of the arrival rate χ ($\mu_1 = 3, \mu_2 = 5, \theta_1 = 0.5, \theta_2 = 0.2, \gamma = 0.5$)

	$\chi = 1$		$\chi = 2$		$\chi = 3$		$\chi = 4$	
	$l = 1$	$l = 2$	$l = 1$	$l = 2$	$l = 1$	$l = 2$	$l = 1$	$l = 2$
Ω_0	0.21512	0.21132	0.33099	0.31531	0.38679	0.35549	0.40447	0.35844
Ω_1	0.76358	0.75597	0.59719	0.59719	0.47439	0.44757	0.38074	0.35007
Ω_2	0.00175	0.00268	0.01052	0.05902	0.02698	0.03827	0.48667	0.06603
Ω_3	0.01938	0.02977	0.05902	0.01052	0.102155	0.14492	0.14028	0.19038
Ω_4	0.00004	0.00012	0.00033	0.00033	0.001882	0.00267	0.05856	0.00795
Ω_5	0.00016	0.00002	0.00060	0.00060	0.007126	0.01019	0.0168	0.02298
Value of R	0.08009	0.08009	0.07830	0.07830	0.16411	0.16411	0.24688	0.24688
Sum of probabilities	1.00000	1.00000	1.00000	1.00000	1.00000	1.00000	1.00000	1.00000
MNS	0.249833	0.26998	0.48021	0.54105	0.70429	0.81005	0.93816	1.09624
P_{busy}	0.22642	0.23403	0.40281	0.41163	0.55243	0.61926	0.64993	0.70125
P_{idle}	0.76357	0.75596	0.59718	0.57836	0.47439	0.44756	0.38074	0.35007
Total cost	253.5954	150.234	305.105	199.234	387.248	235.678	489.610	280.609

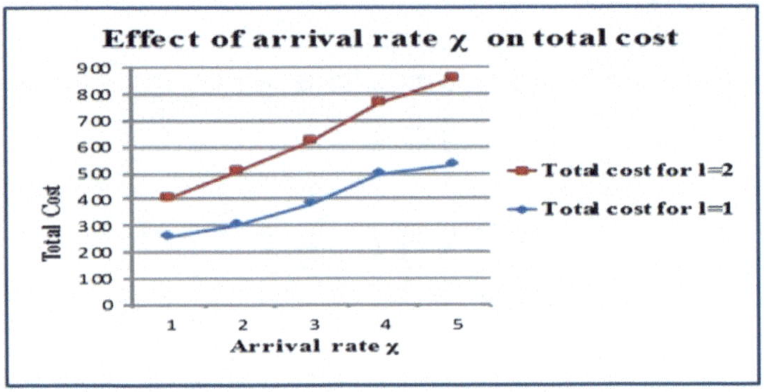

Fig. 3 Arrival rate versus total cost

and display the total cost by varying the arrival rate. Figure 3 shows that the total cost increases as the arrival rate increases for both $l = 1$ and $l = 2$. The total cost $l = 2$ is always greater than the total cost $l = 1$.

8 Conclusion

In this study, we investigated the effectiveness of a two-stage retrial tandem queuing model with working vacation techniques in optimizing quality of service (QoS) for Internet and web services. The novelty of the research lies in the combination of a two-stage retrial tandem queuing model, working vacation techniques, and the specific optimization strategies employed, all tailored for the challenges presented by the Internet and web services. The novelty and significance of this paper lie in its innovative approach to addressing the challenges of optimizing QoS in Internet and web services through a realistic queuing model and the incorporation of working vacation techniques, ultimately contributing to a more reliable and responsive digital infrastructure. In this paper, we have obtained the steady-state probability vector, busy period analysis, and some performance measures by applying the matrix analytic method. Furthermore, we have performed numerical analysis by assuming particular values for the parameter.

References

1. Agrawal, P.K., Jain, A., Jain, M.: M/M/1 Queuing model with working vacation and two type of server Breakdown. J. Phys. Conf. Ser. **1849**, 1742–6596 (2020)
2. Aguir, M.S., Karaesman, S., Akshin, O.Z.: The impact of retrials on callcenter performance. OR Spectrum. **26**, 353–376 (2004)

3. Baumann, H., Sandmann, W.: Multi – server tandem queue with Markovian arrival process, phase –type service times, and finite buffers. Eur. J. Oper. Res. **256**(1), 187–195 (2017)
4. Chaudhry, M.L.: Transient/steady state solution of a single channel queue with arrival and intermittently available server. Math. Oper. Forschung Stat. **5**, 307–315 (1974)
5. Gubbi, J., Buyya, R., Marusic, S., Palaniswami, M.: Internet of things (IoT): a vision, architectural elements, and future directions. Future Gener. Comput. Syst. **29**(7), 1645–60 (2013)
6. Gun, L., Makowski, A.M.: Matrix geometric solution for two node tandem queuing systems with phase-type servers subject to blocking and failures. Commun. Stat. Stoch. Models **5**(2), 431–457 (1989)
7. Jeganathan, K., Koffer, A., Sugapriya, C., Nagarajan, D.: A matrix–analytic method for the steady state analysis of a Markovian queuing system with scrap items. Decis. Anal. J. **7**, 407–427 (2023)
8. Manikandan, R., Nair, S.: An M/M/1 queuing – inventory system with working vacations, vacation interruptions and Lost sales. Autom. Remote Control **81**(4), 746–759 (2020)
9. Tran-Gia, P., Mandjes, M.: Modeling of customer retrial phenomenon in cellularmobile networks. IEEE J. Sel. Areas Commun. **15**, 1406–1414 (1997)
10. Yang, T., Templeton, J.G.C.: A survey on retrial queues. Queueing Syst. **2**, 201–233 (1987)

Part IV
Physics

An Analysis of MHD Thermally and Chemically Reactant Williamson Nanofluid Over an Exponentially Stretched Sheet with Joule's Heating

Pradeep G. Janthe, Jagadish V. Tawade, and Suresh Biradar

1 Introduction

Particles smaller than 100 nanometers, known as nanofluids, are added to base fluids such as water, oil, biofluids, ethylene, and lubricants. More researchers are becoming interested in nanofluids than in other fluids, despite the fact that they are vital to business, medicine, and several additional promising scientific and technological domains. Nanofluids, however, continue to play a crucial role in the medical industry, processing tiny explosives that are used to destroy cancerous tumors and screening cancerous tumors with gold nanoparticles, among other applications.

Sakiadis [1] studied the behaviour of boundary layers for two-dimensional and axisymmetric flow on continuous solid surfaces. Bhatti and Rashidi [2] described the combined results of heat radiation and thermal diffusion on a porous stretched sheet filled with the Williamson nanofluid. Tamilzharasan et al. [3] studied a Williamson nanofluid flow by magneto-mixed convection across a double-stratified porous media with activation energy present. Sonu Kumar et al. [4] studied the non-Newtonian Williamson fluid's time-dependent axisymmetric flow when nanoparticles are present in the porous media, resulting from a radially stretching surface. Rafiqul Islam et al.'s [5] investigation presents using MHD non Newtonian free convective unstable fluid flow computational models over an infinitely tall porous plate with mass and heat transfer properties.

Hussain et al.'s [6] goal was to investigate the effect on heat transfer in CFF via stretching the cylinder of an uneven heat source/sink. Vishwambhar et al. [7] studied

P. G. Janthe · J. V. Tawade (✉)
Department of Mathematics, Vishwakarma University, Pune, India
e-mail: jagadish.tawade@vupune.ac.in

S. Biradar
Department of Mathematics, Sharanbasva University, Kalaburagi, India

I. Rajasingh et al. (eds.), *Applied Computational Mathematics*,
Springer Proceedings in Mathematics & Statistics 480,
https://doi.org/10.1007/978-3-031-77764-6_27

mass and heat transfer of porous-stretched surface filled with MHD Williamson nanofluid flow under thermal radiation and chemical reactions. Walelign et al. [8] examined a chemical procedure used to analyze mass transfer and heat production in a radiant Williamson nanofluid flow with unstable MHD across the stretched sheet.

After a rigorous literature survey, authors such as Cheela Sunitha et al. [9], Nazar and Bidin [10], Ishak [11], and Reddy and Shankar [12] have discussed only on without chemical reactions. These authors have neglected the combined effects of magnetic field and Joule's heating as both are very important physical parameters in the heat transport processes.

After conducting a rigorous literature survey, it is observed that no authors have attempted to explore the combined effects of magnetic fields and Joule's heating, despite both being crucial physical parameters in heat transport processes for exponentially porous stretching sheet passed by a chemically reactive Williamson nanofluid. The goal of the current study is to fill this gap by providing a reliable numerical scheme using MATLAB's bvp4c. The flow problem's governing equations are written in the mathematical formulation section. Using similarity transformations, the proposed equations are generated and transformed from partial differential equations (PDEs) into ordinary differential equations (ODEs). The numerical techniques are used to solve these odes, represented in the section numerical solution. The consequences of several physical characteristics, including thermal radiation, viscous dissipation, Brownian motion, chemical reaction, porous parameter, etc., are shown with the aid of graphs and discussed in detail in the section results and discussion. The investigation of important findings is also mentioned in the concussion section.

2 Mathematical Formulation

The governing equation of the flow problems are as follows:

$$\frac{\partial u}{\partial x}+\frac{\partial v}{\partial y}=0. \tag{1}$$

$$u\frac{\partial u}{\partial x}+v\frac{\partial u}{\partial y}=\sqrt{2}\nu\gamma\frac{\partial u}{\partial y}\frac{\partial^2 u}{\partial y^2}+u_\infty\frac{du_\infty}{dx}+v\frac{\partial^2 u}{\partial y^2}+\sigma\frac{B^2(x)}{\rho_f}(u_\infty-u) \tag{2}$$

$$u\frac{\partial T}{\partial x}+v\frac{\partial T}{\partial y}=\alpha\frac{\partial^2 T}{\partial y^2}-\frac{1}{(\rho Cp)_f}\frac{\partial q_r}{\partial y}+\tau\left[D_B\left(\frac{\partial C}{\partial y}\frac{\partial T}{\partial y}\right)+\frac{D_T}{T_\infty}\left(\frac{\partial T}{\partial y}\right)^2\right]$$
$$+\frac{\mu}{(\rho Cp)_f}\left[\left(\frac{\partial u}{\partial y}\right)^2+\sqrt{2}\gamma\left(\frac{\partial u}{\partial y}\right)^3\right]+\frac{\sigma B_0^2}{(\rho Cp)_f}(u-u_\infty)^2 \tag{3}$$

$$u\frac{\partial C}{\partial x}+v\frac{\partial C}{\partial y}=D_B\left(\frac{\partial^2 C}{\partial y^2}\right)+\frac{D_T}{T_\infty}\left(\frac{\partial^2 T}{\partial y^2}\right)-K_r(C-C_\infty) \tag{4}$$

According to the boundary conditions,

$$
\begin{aligned}
u = u_w(x) = ae^{\frac{x}{l}}; v = 0; T = T_w(x); C = C_w(x); at y = 0 \\
u \to u_\infty(x); v = 0; T \to T_\infty(x); T \to T_\infty(x); at y = \infty
\end{aligned}
\tag{5}
$$

On the sheet's surface, the designated temperature and concentration are taken to be $T_w(x) = T_\infty + T_0 e^{\frac{x}{2l}}$ and $C_w(x) = C_\infty + C_0 e^{\frac{x}{2l}}$, where T_0 is the temperature and $C_0$is the concentration; now stream function =(x,y)is characterized as $u = \frac{\partial \psi}{\partial y} v = -\frac{\partial \psi}{\partial x}$, where Eq. (1) is exactly satisfied. Here, x and y are the velocity components denoted by letters u and v, respectively: $\nu = \frac{\mu}{\rho}$,

where μ is the coefficient of viscosity, ν is the kinematic viscosity, ρ is the fluid density, $\alpha = \frac{k}{(\rho Cp)_f}$ is the thermal diffusivity, $\tau = \frac{\rho Cp)_P}{(\rho Cp)_f}$ is the ratio of the nanofluid's heat capacity to that of the base fluid, K is the thermal conductivity, Cp is the specific heat at constant pressure, q_r is the radioactive heat flux, T and C are the temperature and concentration of nanoparticles, T_w is the wall surface temperature of the nanoparticle, C_w is the nanoparticle concentration at the wall, T_∞ is the nanoparticle fluid temperature away from the sheet, C_∞ is the concentration of nanoparticles away from the sheet, γ is the time constant, D_T is the thermophoresis diffusion, D_B is the Brownian motion, and K_r is the chemical reaction parameter.

According to Roselands' assumption, the heat flow from radioactivity is

$$
q_r = -\frac{4\sigma^*}{3k^*}\frac{\partial T^4}{\partial y},
\tag{6}
$$

where σ^* is the Stefan-Boltzmann constant and k^* represents the mean absorption coefficient. It is also assumed that T^4 may be represented as a temperature function that is linear, due to sufficiently significant temperature differential within the flow. Therefore, by increasing T^4 in the Taylor series without taking into the account of higher-order terms, we get

$T^4 \cong 4T_\infty^3 T - 3T_\infty^4$

The succeeding similarity transformations are now initiated:

$$
\Psi = \sqrt{2l\nu a}e^{\frac{x}{2l}}, \theta(\eta) = \frac{T - T_\infty}{T_w - T_\infty}, \phi(\eta) = \frac{C - C_\infty}{C_w - C_\infty}
\tag{7}
$$

Now, Eq. (2) to (5) become

$$
f''' + \lambda f'' f''' + ff'' - 2f'^2 + 2\epsilon^2 + M(\epsilon - f') = 0
\tag{8}
$$

$$
\left(1 + \frac{4R}{3}\right)\theta'' + Pr(f\theta' + Ec(f''^2 + Wef''^3 + Mf'^2) + Nb\phi'\theta' + Nt\theta'^2) = 0
\tag{9}
$$

$$
\phi'' + Lef\phi' - Lef'\phi + \frac{Nt}{Nb}\theta'' - \gamma Le\phi = 0
\tag{10}
$$

The boundary conditions are

$$\begin{aligned} f(\eta) = 0, f'(\eta) = 1, \theta(\eta) = 1, \phi(\eta) = 1; \text{ at } \eta = 0, \\ f'(\eta) \to \epsilon, \theta(\eta) \to 0, \phi(\eta) \to 0 at \eta = \infty \end{aligned} \tag{11}$$

where $\epsilon = \frac{b}{a}$ represents velocity ratio parameter, $\lambda = \gamma\sqrt{\frac{a^3 e^{\frac{3x}{l}}}{vl}}$ is the Williamson parameter, $M = \frac{2l\sigma B_0^2}{a\rho_f}$ is the magnetic parameter, $R = \frac{4\sigma^* T_\infty^3}{3k^* k}$ is the radiation parameter, $Pr = \frac{\nu}{\alpha}$ is the Prandtl number, $Nb = \frac{\tau D_T(T_W - T_\infty)}{T_\infty \nu}$ is the thermophoresis parameter, $Ec = \frac{u_w^2}{(C_P)_f(T_W - T_\infty)}$ is the Eckert number, $Le = \frac{\nu}{D_B}$ is the Lewis number, $W_e = x\gamma\sqrt{\frac{2a^3}{\nu}}$ is the Weissenberg number, and $\gamma = \frac{2lK_r}{uW}$ is the chemical reaction parameter. $C_f = \frac{\tau_w}{\rho u_w^2}$ is the nondimensional skin friction coefficient, where $\tau_w = \mu\frac{\partial u}{\partial y}\left(1 + \frac{\gamma}{\sqrt{2}}\left(\frac{\partial u}{\partial y}\right)^2\right)_{y=0}$, the Nusselt number $Nu_x = \frac{xq_w}{k(T_w - T_\infty)}$, and the Sherwood number $Sh_x = \frac{xq_m}{D_B(C_w - C_\infty)}$, where q_w is heat flux and q_m is the mass flux at the surface given by

$$q_w = \left(-\left(k + \frac{16\sigma^* T_\infty^3}{3k^*}\right)\left(\frac{\partial T}{\partial y}\right)\right)_{y=0}, q_m = -\left(D_B\left(\frac{\partial u}{\partial y}\right)\right)_{y=0} \tag{12}$$

Substituting q_w and q_m in the preceding equation, we get

$$\sqrt{2}Re_x^{\frac{1}{2}}C_f = \left(f''(0) + \frac{\lambda}{2}(f''(0))^2\right) \text{ and } \sqrt{2}Re_x^{\frac{1}{2}}Sh_x = -\phi'(0), \tag{13}$$

where $Re_x = \frac{u_w x}{\nu}$ is the local Reynolds number.

3 Numerical Methods

The numerical solutions of Eqs. (8) through (10) under the given boundary conditions (11) are generated using the bvp4c MATLAB solver. We may use this numerical method to assist in solving a given set of ODEs. The system of Eqs. (7) through (10) is transformed to first ODEs (14) through (23) as added below in order to apply the bvp4c coding:

$$f = f \tag{14}$$

$$f' = f1 \tag{15}$$

$$f'' = f2 \tag{16}$$

$$\theta = f3 \tag{17}$$

$$\theta' = f4 \tag{18}$$

$$\phi = f5 \tag{19}$$

$$\phi' = f6 \tag{20}$$

$$f''' = -\frac{1}{(1+\lambda)} * [f1 * f2 - 2 * f1 * f1 + 2 * \epsilon^2 + M * (\epsilon - f1)] \tag{21}$$

$$\begin{aligned} \theta'' = \frac{(-3)}{(3+4R)} &* [Pr * (f * f4 + Ec * (f2 * f3 + We * f2 * f2 * f2 \\ &+ M * f1 * f1) + Nb * f6 * f4 + Nt * f4 * f4] \end{aligned} \tag{22}$$

$$\phi'' = \left[Le * f * f6 - Le * f1 * f5 + \frac{Nt}{Nb} * \theta'' - \gamma * Le * f5 \right] \tag{23}$$

with boundary conditions

$$\begin{aligned} &\text{for } \eta = 0;\ f = 0,\ f1 = 1,\ f3 = 1,\ f5 = 1. \\ &\text{for } \eta = \infty;\ f1 = \epsilon,\ f3 = 0,\ f5 = 0. \end{aligned} \tag{24}$$

Seven prerequisites must be met in order to obtain the answer because there are seven ODEs. However, there are just five starting conditions in Eq. (24). Thus, for y(3) and y(7), suitable first estimations are expected.

4 Results and Discussion

Graphs representing separate profiles are displayed for varying values of the governing factors. We include relevant values in the overall evaluation, as displayed in the tables and graphs, with the exception of rebuilt values:

$\lambda = 0.2, Le = 2.0, Ec = 0.2, \epsilon = 0.1, Pr = 2.0, M = 0.5, \gamma = 0.2, R = 0.1, Nb = 0.5, Nt = 0.2, We = 0.5$

Figure 1 illustrates the increases of the velocity ratio parameter ϵ, and the thickness of the boundary layer increases as the velocity of the stretched sheet surpasses that of the free stream. Figures 2 and 3 illustrate the temperature profile and concentration profile, thereby showing that as ϵ improves, the surface falls at the surface.

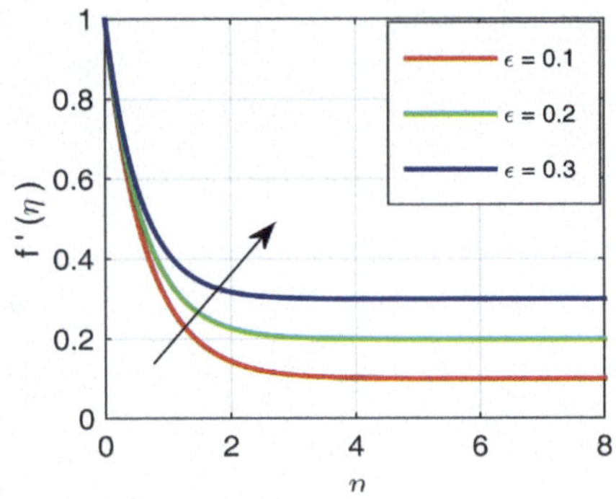

Fig. 1 The profile of velocity $f'(\eta)$ for ϵ

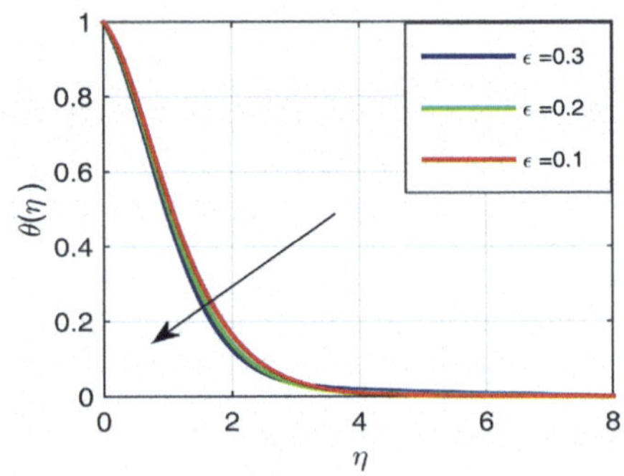

Fig. 2 The profile of temperature $\theta(\eta)$ for ϵ

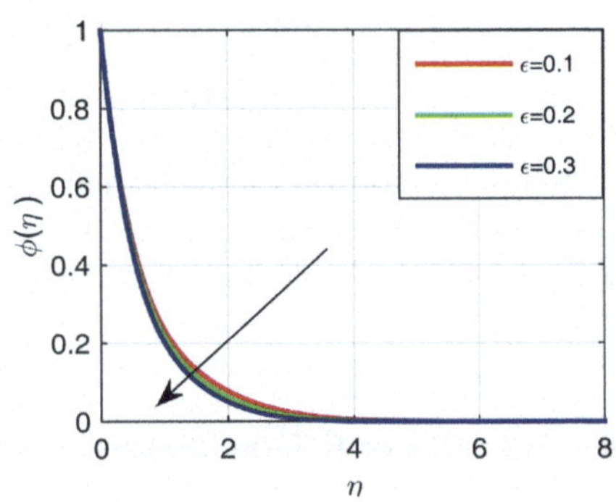

Fig. 3 The profile of concentration $\phi(\eta)$ for ϵ

Figure 4 illustrates, according to the present study, that higher Williamson's parameter values increase the skin friction coefficient rate.

With reference to Fig. 5, which depicts the fluid velocity attitude toward the Weissenberg number, obviously the velocity profile falls as the Weissenberg number rises. This is because a rise in the Weissenberg number lengthens the fluid particles' relaxation times, increasing viscosity and causing resistance to the fluid flow, which in turn limits fluid velocity.

Figure 6 demonstrates how the Weissenberg number increases with the temperature profile. As the Weissenberg number climbs, the thermal boundary layer gets thinner, which causes the temperature profile to rise.

Fig. 4 $f''(0)$ for the λ

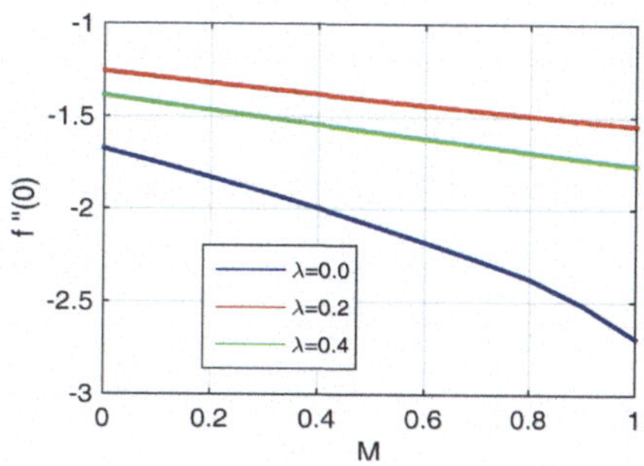

Fig. 5 The profile of velocity $f'(\eta)$ for We

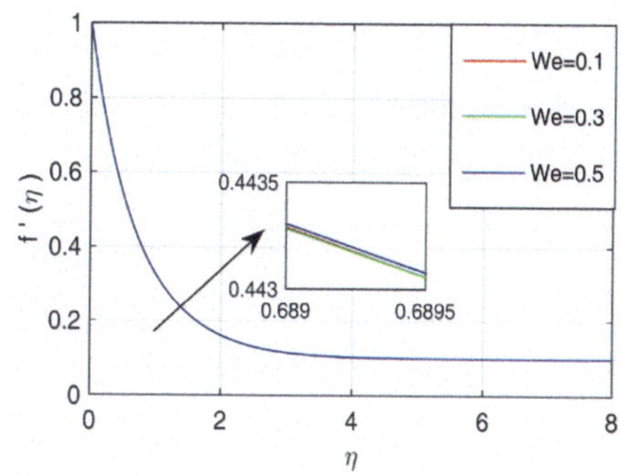

Fig. 6 The profile of temperature $\theta(\eta)$ for We

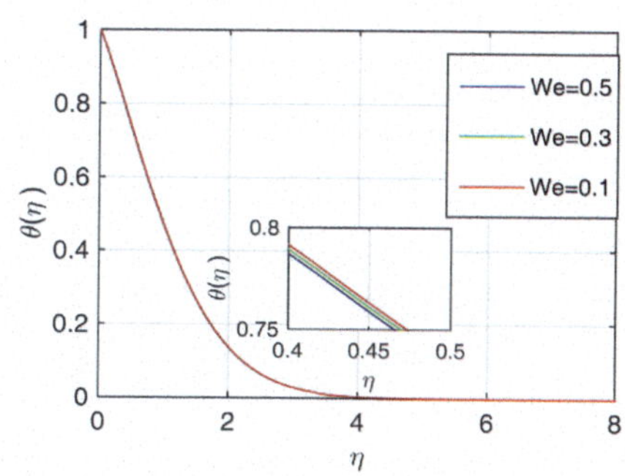

As Fig. 7 illustrates, it is evident that the Nusselt number naturally increases when the Reynolds number and oil's inlet temperature rise. The majority of compact heat exchangers' uses are in industrial heat transfer implications.

Figure 8 explains the Sherwood number on the profile of concentration. It can increase for certain values and then decrease due to the reciprocal property of diffusivity. Due to reciprocal property of diffusivity.

Figure 9 illustrates that as the Lewis number increases, the concentration of nanoparticles decreases. The concentration boundary layer decays when the Lewis number increases because it decreases the molecule diffusivity.

Figure 10 indicates that the fluid's concentration will be suppressed as the chemical reaction parameter increases.

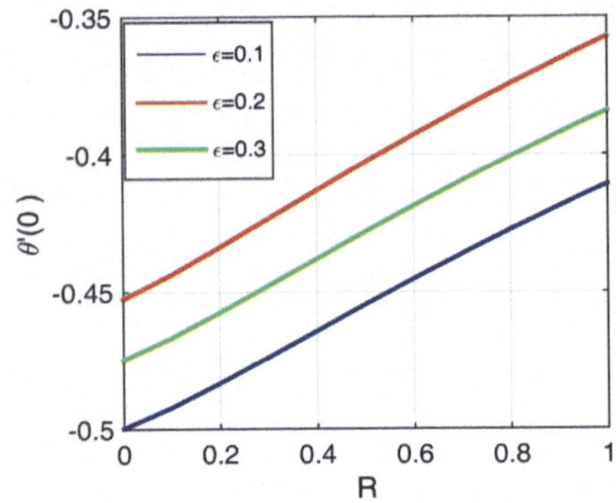

Fig. 7 $\theta(0)$ for ϵ

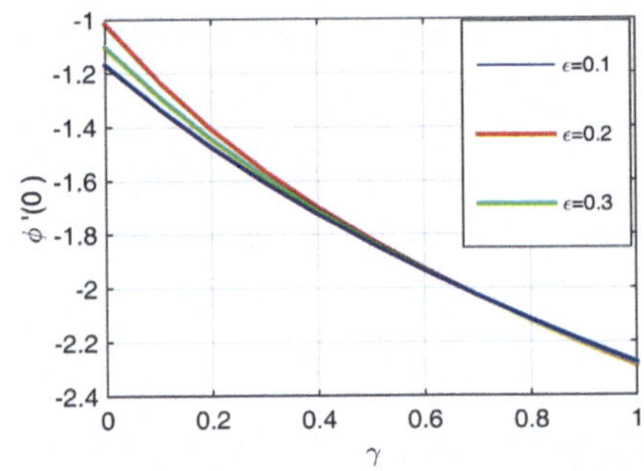

Fig. 8 $\phi'(0)$ for ϵ

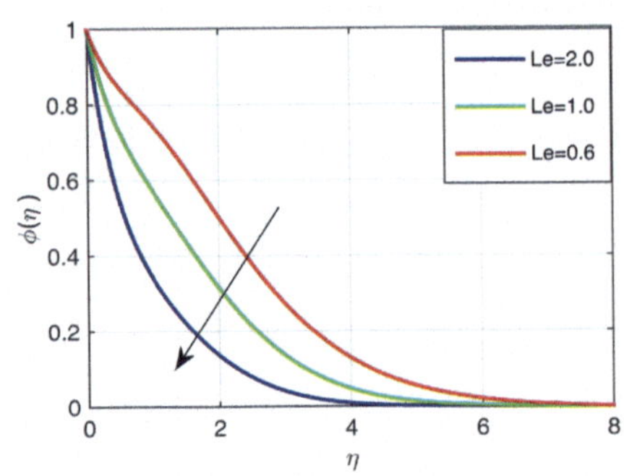

Fig. 9 The profile of concentration $\phi(\eta)$ for Le

5 Conclusion

The following observations are drawn from the current study:

- Shear hood number, Nusselt number, and skin friction coefficient effects accelerate with increasing values of Williamson's parameter and velocity ratio parameters.
- Within the boundary region, the fluid concentration will be suppressed by Lewis numbers and chemical reaction parameters.

Fig. 10 The profile of concentration $\phi(\eta)$ for γ

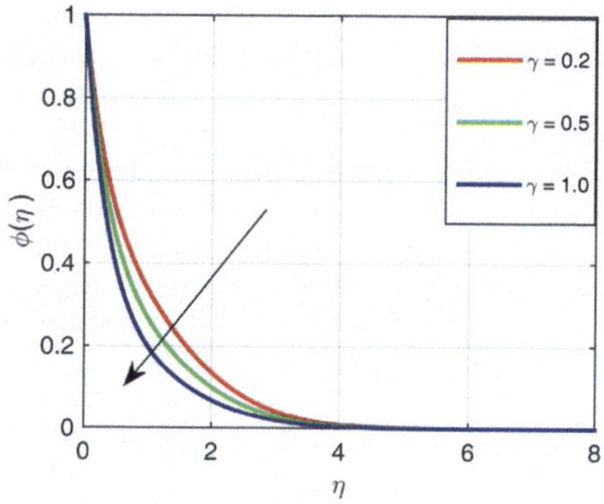

- Both the concentration boundary layer thickness and the temperature profile are improved with an increase in the thermophoresis parameter.
- The temperature profile drops as the Brownian motion parameter increases, whereas the concentration profile shows the opposite effects.
- The velocity profile increases as the Weissenberg number values increase.
- The fluid's temperature decreases as the Weissenberg number increases.

References

1. Sakiadis, B.C.: Boundary-layer behavior on continuous solid surfaces: I. Boundary-layer equations for two-dimensional and axisymmetric flow. Am. Inst. Chem. Eng. **7**(1), 26–28 (1961)
2. Bhatti, M.M., Rashidi: Effects of thermo-diffusion and thermal radiation on Williamson nanofluid over a porous shrinking/stretching sheet. J. Mol. Liq. **221**, 567–573 (2016)
3. Tamilzharasan, B.M., Karthikeyan, S., Kaabar, M.K.A., Yavuz, M., Ozkose, F.: Magneto mixed convection of Williamson nanofluid flow through a double stratified porous medium in attendance of activation energy. Math. Comput. Appl. **27**(3), 46 (2022)
4. Saini, S.K., Agrawal, R., Kaswan, P.: Activation energy and convective heat transfer effects on the radiative Williamson nanofluid flow over a radially stretching surface containing Joule heating and viscous dissipation. Numer. Heat Transfer Part A: Appl. **1**(27), 1040–7782 (2023)
5. Islam, M.R., Reza-E-Rabbi, S., Yousuf Ali, M., Rasel, M.M.H., Ahmmed, S.F.: Numerical simulation of mass and heat transport phenomena of hydromagnetic flow of Casson fluid with sinusoidal boundary conditions. Eng. Rep. **5**(11), e12659 (2023)
6. Hussain, S.M., Shankar Goud, B., Madheshwaran, P., Jamshed, W., Pasha, A.A., Safdar, R., Arshad, M., Ibrahim, R.W., Ahmad, M.K.: Effectiveness of nonuniform heat generation (sink) and thermal characterization of a carreau fluid flowing across a nonlinear elongating cylinder: a numerical study. ACS Omega **7**(29), 25309–25320 (2022)
7. Vishwambhar, Patil, S., Humane, P.P., Patil, A.B.: MHD Williamson nanofluid flow past a permeable stretching sheet with thermal radiation and chemical reaction. Int. J. Model. Simul. **43**(3), 185–199 (2023)
8. Walelign, T., Kebede, T., Haile, E.: Analytical study of heat and mass transfer in unsteady MHD radiant flow of a Williamson nanofluid over stretching sheet with heat generation and chemical reaction. Eng. Phys.-Heat Trans. **49**, 4246–4263 (2020)

9. Sunitha, C., Kumar, P.V., Lorenzini, G., Ibrahim, S.M.: A study of thermally radiant Williamson nanofluid over an exponentially elongating sheet with chemical reaction via homotopy analysis method. CFD Lett. **14**, 68–86 (2022)
10. Bidin, B., Nazar, R.: Numerical solution of the boundary layer flow over an exponentially stretching sheet with thermal radiation. Euro. J. Sci. Res. **33**(4), 710–717 (2009)
11. Ishak, A.: MHD boundary layer flow due to an exponentially stretching sheet with radiation effect. Sains Malaysiana **40**(4), 391–395 (2011)
12. Reddy, Ch.A., Shankar, B.: MHD stagnation point flow in a boundary layer flow of a nanofluid over a stretching sheet in the presence of viscous dissipation and chemical reaction. Mech. Mat. Sci. Eng. **4**, 45–57 (2017)

Impact of Radiation and Dissipation on an Unsteady Nanofluid Flow over an Elongating Sheet

Aparna N. Shendkar and Jagadish V. Tawade

1 Introduction

Magnetohydrodynamics (MHD) is a field of study combined by the fluid mechanics and electromagnetism. The flow of conducting materials is substantially influenced by the electromagnetic force. Fluids easily change shape under external forces, lacking a fixed structure. They are classified as Newtonian and non-Newtonian, with the latter, including honey, blood, greases, oils, and foodstuffs, widely used for diverse practical and industrial applications. Applying external fields alters the flow pattern of conducting fluids, and the magnetization of Newtonian and non-Newtonian fluids has wide-ranging utilization in streams such as astrophysics, plasma physics, nuclear reactor cooling, technology, engineering, and beyond. By introducing nano-solid particles into conventional base fluids such as water, we can overcome the limitations in heat transfer performance typically associated with common fluids. Nanofluids, composed of solid-liquid mixtures with particle sizes ranging from 1 to 100 nanometers, feature diverse nanoparticle types such as metals, carbon, carbides, oxides, nitrides, and more. Recently, nanofluids have emerged as a significant research focus in thermal technologies, demonstrating applications across various fields such as medicine, biomedical industry, power generation in nuclear reactors, engine cooling, refrigeration, detergents, miniaturized electronics, fuel cells, and numerous other physical and engineering domains. Heat dissipation refers to the transfer of internal energy from one substance to another. Studying the impact of heat dissipation in various mediums and boundary-layer flows, particularly across stretching sheets, is crucial due to its widespread applications in diverse

A. N. Shendkar · J. V. Tawade (✉)
Department of Mathematics, Vishwakarma University, Pune, Maharashtra, India
e-mail: jagadish.tawade@vupune.ac.in

I. Rajasingh et al. (eds.), *Applied Computational Mathematics*,
Springer Proceedings in Mathematics & Statistics 480,
https://doi.org/10.1007/978-3-031-77764-6_28

industries. This includes biological and chemical sectors, as well as manufacturing processes such as compression, extrusion, forging, combustion engines, and more.

Sakiadis et al. [1] have studied the behavior of boundary layers on continuous surfaces and developed the fundamental differential and integral momentum equations for the boundary-layer theory applicable to these surfaces. Constant temperature and concentration boundary layers on a continuously moving flat surface have been investigated by L.E. Erickson [2]. Analysis of momentum, heat, and mass transfer on a stretching sheet in the boundary layer which is subjected to suction or blowing was done by P. S. Gupta [3]. Narayana, P. V. [4] has studied the effects of "thermal radiation" on unsteady MHD nanofluid flow on a stretching sheet with uneven heat source/sink. The flow of thin uniform Casson liquid film on an unsteady stretching sheet is studied numerically by Jagadish V Tawade [5]. HAM explores entropy dynamics in an accelerating stretching permeable surface in an unsteady magnetohydrodynamic nanofluid. This study examines water as the base fluid and includes four types of nanoparticles: copper (Cu), copper oxide (CuO), aluminum oxide (Al_2O_3), and titanium dioxide (TiO_2) by Abolbashari Mohammad Hossein [6]. Pal and Dulal [7] have done a study that explores fluid dynamics and heat transfer in an ever-changing, smooth layer near a surface. It focuses on the flow of incompressible, viscous fluid over a continuously expanding and permeable surface, considering uneven heat distribution and the influence of thermal radiation. Alrihieli, Haifaa et al. [8] suggested the possibility of a heated fluid below the lower surface of the extending sheet, contributing to surface warming through convection.

Inspired by the existing research, we aim to investigate the influences of variable uneven heat source, thermal radiation, and transfer of heat in a nanofluid over an elongating sheet. To achieve this, we derive and solve the governing equations of the flow problem using the MATLAB bvp4c software program after converting PDEs to ODEs by using similarity transformation. A comparison of the results was done using graphical representations. This paper is structured in the following sections: formulation of equations and their conversion are done in Sect. 2, graphical representation and detailed discussion are presented in Sect. 3, and concluding the paper with few comments is discussed in Sect. 4.

2 Mathematical Formulation

Two-dimensional unsteady MHD laminar nanofluid flow on a stretching sheet is considered. Nanofluid prepared by using aluminum oxide (Al_2O_3) and titanium dioxide (TiO_2) nanoparticles mixed with base fluid water is used. The plate is horizontal, so the flow is along the x-axis and the y-axis is kept normal to the plate. A uniform transverse magnetic field B is applied in such a way that it will be in the direction to the y-axis. The assumed thermal equilibrium of the base fluid and nanoparticles is suspended, and there is no motion.

In the equation of motion, density depends on temperature buoyancy forces linearly:

Table 1 The thermophysical properties of the base fluid and different nanoparticles

Physical properties	Fluid phase (water)	Cu	CuO	Al_2O_3	TiO_2
C_P(J/kg K)	4179	385	531.8	765	686.2
ρ(kg/m3)	997.1	8933	6320	3970	4250
k(W/m K)	0.613	401	76.5	40	8.9538

- The heat and fluid flows are steady for time t < 0 and at t = 0, they become unsteady.
- The velocity $U_w(x, t)$ of the stretched surface in the direction of the x-axis at t = 0.
- Assuming that the Reynolds number is insignificantly small, allowing the induced magnetic field to be very small compared to the applied magnetic field.
- The mass transfer velocity which is normal to the stretching surface is $v_w(t)$.
- $T_w(x, t)$ which is the surface temperature linear variants with time t and distance x in the form $T_\infty(x, t) = T_\infty + \frac{ax}{(1-ct)^2}$ where T_∞ is the ambient fluid temperature and $T_\infty > T_w$.
- Viscous dissipation is assumed to be negligibly small (Table 1).

The governing equations are as follows:

$$\frac{\partial u}{\partial x} + \frac{\partial v}{\partial y} = 0 \tag{1}$$

$$\frac{\partial u}{\partial t} + u\frac{\partial u}{\partial x} + v\frac{\partial v}{\partial y} = \frac{\mu_{nf}}{\rho_{nf}}\frac{\partial^2 u}{\partial y^2} - \frac{\sigma B^2}{\rho_{nf}}u \tag{2}$$

Here, the nanofluid's kinematic viscosity is defined as μ_{nf}, and ρ_{nf} is the fluid's density.

When using the boundary-layer approximation, the energy equation takes form as follows:

$$\frac{\partial T}{\partial t} + u\frac{\partial T}{\partial x} + v\frac{\partial T}{\partial y} = \frac{k_{nf}}{(\rho C_p)_{nf}}\frac{\partial^2 T}{\partial y^2} - \frac{1}{(\rho C_p)_{nf}}\frac{\partial q_r}{\partial y} + \frac{q'''}{(\rho C_p)_{nf}} + \frac{\nu}{(\rho C_p)_{nf}}\left(\frac{\partial u}{\partial y}\right)^2 \tag{3}$$

Here, k_{nf} stands for the thermal conductivity, and $(C_p)_{nf}$ stands for the specific heat at constant pressure.

For current flow problem, boundary conditions are as follows:

$$u = Uw(x, t) = \frac{ax}{1 - ct}, \quad v = v_w(t) = \frac{v_0}{\sqrt{1 - ct}},$$
$$T = Tw(x, t) = T_\infty + \frac{ax}{(1 - ct)^2} \quad \text{at} \quad y = 0, \tag{4}$$
$$u \to 0 \Longrightarrow T \to T_\infty \quad \text{as} \quad y \to \infty$$

where u and v represent the velocity components along the x- and y-axes

The Rosseland approximation qr of the radioactive heat flux is

$$q_r = -\frac{4\sigma^*}{3K^*}\frac{\partial T^4}{\partial y} \tag{5}$$

where the Stefan-Boltzmann constant is σ^* and K^* is the average absorption coefficient.

Temperature field is denoted by the letter T.

Using Taylor's T series as a starting point, we can expand T^4 as

$$T^4 = T_\infty^4 + 4T_\infty^3(T - T_\infty) + 6T_\infty^2(T - T_\infty)^2 + \ldots. \tag{6}$$

Higher-order words are being overlooked or omitted:

$$T^4 \approxeq -3T_\infty^4 + 4T_\infty^3 T \tag{7}$$

From Eqs. (5) and (7), the radioactive heat flow using the Rosseland approximation is as follows:

$$\frac{\partial q_r}{\partial y} = -\frac{16\sigma^* T_\infty^3}{3K^*}\frac{\partial^2 T}{\partial y^2} \tag{8}$$

Apply similarity transformations that convert coupled PDEs with matching conditions of boundary into ODEs. This process ensures satisfaction of the continuity equation and simplification of both momentum and energy equations into nondimensional ODEs.

The stream function is defined as follows:

$$u = \frac{\partial \Psi}{\partial y}, \qquad v = -\frac{\partial \Psi}{\partial x} \tag{9}$$

Now, use the similarity transformation as follows:

$$\eta = \sqrt{\frac{a}{v_f(1 - ct)}}y, \quad \Psi(x, y) = \sqrt{\frac{v_f a}{(1 - ct)}}xf(\eta), \quad \theta(\eta) = \frac{T - T_\infty}{T_w - T_0} \tag{10}$$

where

Ψ is the stream function, $f(\eta)$ is the dimensionless stream function, and η is the similarity variable.

Hence, the continuity equations become

$$u = \frac{\partial \Psi}{\partial y} = \frac{ax}{(1-ct)} f'(\eta) \quad and \quad v = -\frac{\partial \Psi}{\partial x} = \sqrt{\frac{\nu_f a}{(1-ct)}} f(\eta) \tag{11}$$

By using these similarity transformations above, Eqs. (2) and (3) are converted to ordinary differential equations:

$$a_1 f''' - a_2 \left\{ f'^2 - ff'' + A(f' + \frac{\eta}{2} f'') \right\} - Mf' = 0 \tag{12}$$

$$\frac{a_3}{Pr} \{(1+R)\theta'' + C^* f' + D^* \theta\} - A \left\{ 2\theta + \frac{\eta}{2} \theta' \right\} - f'\theta + f\theta' + \frac{\nu C_p}{k_{nf}} E_c (f'')^2 = 0 \tag{13}$$

Relevant boundary conditions are

$$\begin{aligned} f = f_w, \quad f' = 1, \quad \theta = 1, \quad \text{at} \quad \eta = 0 \\ f' \to 0, \quad \theta \to 0, \quad \text{as} \quad \eta \to \infty \end{aligned} \tag{14}$$

where

$$q''' = \frac{k_{nf} U_w(x,t)}{x \nu_f} \{C^*(T_w - T_\infty) f' + D^*(T_w - T_\infty)\} \tag{15}$$

$$M = \frac{\sigma B^2}{a \rho_f} \text{is the magnetic parameter, } Nu_x = \frac{x q_w}{k_f (T_w - T_\infty)} \text{ is the Nusselt number} \tag{16}$$

$$\begin{aligned} R = \tfrac{16\sigma^* T_\infty^3}{3K k_{nf}} \text{ is the radiation parameter,} \\ C_f = \tfrac{\tau_w 2l k_r}{\rho_f U_w^2}, \text{ is the skin friction coefficient} \end{aligned} \tag{17}$$

$$E_c = \frac{{u_w}^2}{(C_p)_f (T_w - T_\infty)} \text{is the Eckert number, and } Pr = \frac{\nu_f}{\alpha_f} \text{is the Prandtl number.} \tag{18}$$

Equations (12) and (13) including boundary conditions were resolved using the MATLAB bvp4c software program.

3 Results and Discussions

Results are illustrated through graphs for the fluid velocity and temperature profiles of nanofluid prepared by water as the base fluid and Al_2O_3 nanoparticles for various

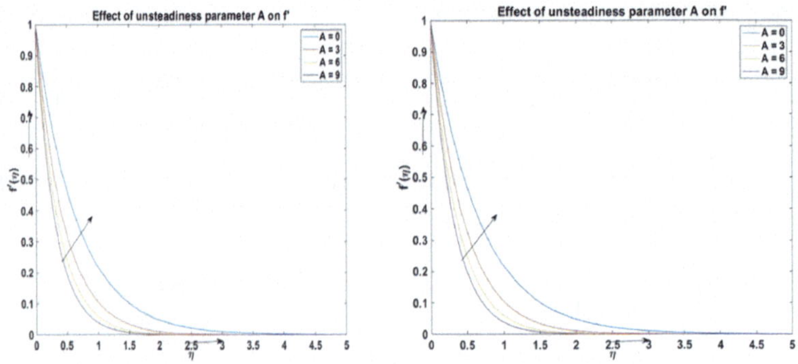

Figure 1-a: Effect of unsteady parameter A on f(η)
for a1 = 1.75, a2 = 1.596, a3 = 2.365, Pr = 7.5, C? = 1, D? = 2.5, M = 2.5, R = 0

Figure 1- b: Effect of unsteady parameter A on Temperature profile θ(η)
for a1 = 1.75, a2 = 1.596, a3 = 2.365, Pr = 5, C? = 1, D? = 2.5, M = 2.5, R = 0

Fig. 1 (**a**) Effect of unsteady parameter A on f′(η) for a1 = 1.75, a2 = 1.596, a3 = 2.365, Pr = 7.5. C? = 1, D? = 2.5, M = 2.5, and R = 0. (**b**) Effect of unsteady parameter A on temperature profile θ(η) for a1 = 1.75, a2 = 1.596, a3 =2,365, Pr = 5, C? = 1, D? = 2.5, M = 2.5, and R = 0

physical parameters such as unsteadiness parameter A, magnetic field parameter M, nanoparticle volume fraction, parameter ϕ, suction parameter f_w, nonuniform heat source/sink parameters (C^* and D^*), Prandtl number Pr, and radiation parameter R.

Figure 1a and b show the influence of unsteady parameter A on the temperature and velocity profile for steady (A = 0) and unsteady (A > 0) cases. As the parameter A rises, the velocity profile decreases, leading to a higher skin friction coefficient. Simultaneously, the temperature decreases with the increasing value of A. This suggests a more rapid cooling rate for higher A values; the cooling process is slow during steady flow.

The above graphs illustrate the effect of magnetic field parameter M on the velocity and temperature profiles. An increase in the magnetic parameter M leads to a decline in velocity, reducing the thickness of the velocity boundary layer. The external magnetic field significantly influences the velocity profile of an electrically conducting fluid, introducing resistance to its motion and a driving force called the Lorentz force, decreasing fluid velocity. With an increase in M, the temperature and thermal boundary-layer thickness increase. This is due to an external magnetic field elevating fluid temperature, inducing swift particle movement, enhancing thermal energy, and thickening the boundary layer, ultimately reducing heat transfer from the sheet (Fig. 2).

The effect of the nanoparticle volume fraction parameter on the velocity and temperature distributions is shown in the above graphs. The velocity profile increases as the increasing values of ϕ, whereas the opposite trend is observed in the temperature profile due to the increase in heat transfer. Physically, as the volume of nanoparticles increases, the thermal conductivity decreases and then results in the increase of the thermal boundary-layer thickness (Fig. 3).

The effect of suction parameter fw on f′ is presented in the above graph for different TiO2 nanoparticles. It is clear that the velocity profiles decrease with

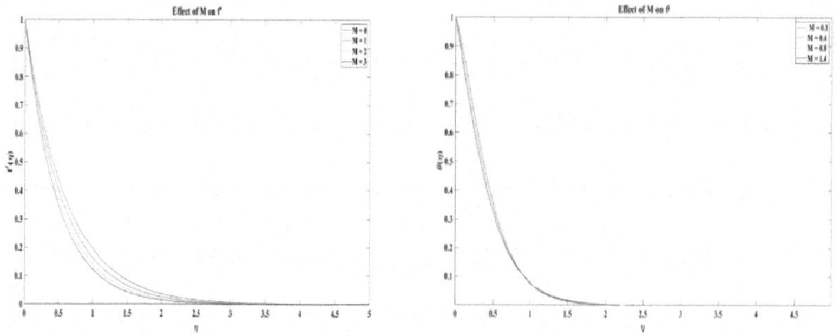

Figure 2-a: The effect of magnetic field parameter M on the velocity f′(η) for A = 2, Pr = 7.5, C* = 1, D* = 2.5, fw = 1.5, R = 0

Figure 2-b: The effect of magnetic field parameter M on Temperature profile θ(η) for A = 1, Pr = 7.5, C* = - 0.05, D* = - 0.05, fw = 1.8, R = 0

Fig. 2 (**a**) The effect of magnetic field parameter M on the velocity f′(η) for A = 2, Pr = 7.5, C* = 1, D* = 2.5, fw = 1.5, and R = 0. (**b**) The effect of magnetic field parameter M on temperature profile θ(η) for A = 1, Pr = 7.5, C* = −0.05, D* = −0.05, fw = 1.8, and R = 0

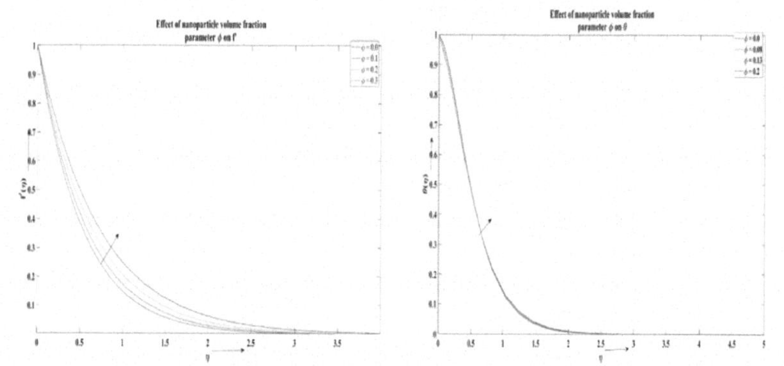

Figure 3-a: The effect of the nanoparticle volume fraction parameter on the Velocity profile f′(η) For A = 0.1, Pr = 6.5, C* = 0.1, D* = 2.5, M = 2.5, fw = 2.6, R = 0

Figure3-b: The effect of the nanoparticle volume fraction parameter on Temperature Profile θ(η) for A = 0.4, Pr = 7.5, C* = 0.1, D* = - 0.05, M = 1, fw = 1.8, R = 0

Fig. 3 (**a**) The effect of the nanoparticle volume fraction parameter on the velocity profile f′(η) for A = 0.1, Pr = 6.5, C* = 0.1, D* = 2.5, M = 2.5, fw = 2.6, and R = 0. (**b**) The effect of the nanoparticle volume fraction parameter on temperature profile θ(η) for A = 0.4, Pr = 7.5, C* = 0.1, D* = −0.05, M = 1, fw = 1.8, and R = 0

increasing values of fw. Due to the stretching surface, the amount of the fluid draws into the surface and as a result, the hydrodynamic boundary layer becomes suppressed by increasing the suction parameter fw (Fig. 4).

4 Conclusion

The current study explores the influence of thermal radiation and nonuniform heat source/sink on the unsteady magnetohydrodynamic nanofluid flow over a stretching

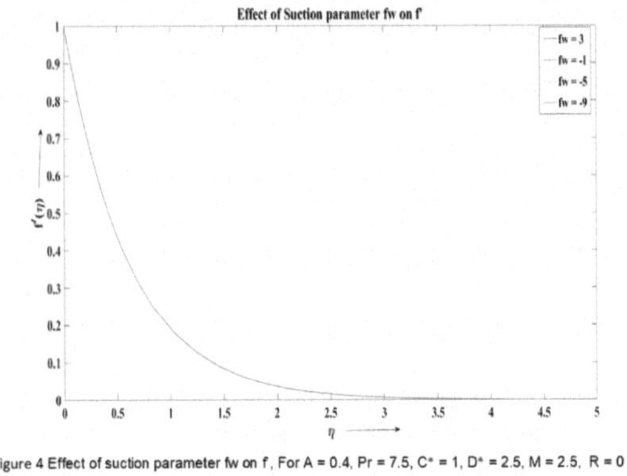

Fig. 4 Effect of suction parameter fw on f′, for A = 0.4, Pr = 7.5, C* = 1, D* = 2.5, M = 2.5, and R = 0

sheet. The bvp4c software is used for solving the associated differential equations. Graphs are employed to explain the effect of different flow parameters on the profiles of fluid velocity and temperature.

The present study provides the following key findings:

- The fluid velocity decreases with increasing values of unsteadiness parameter A, magnetic field parameter M, and suction parameter fw.
- The magnitude of the skin friction coefficient is found to decrease due to increasing values of A, fw, and M.
- The heat transfer rate Nu_x increases due to the increase in A, fw, and Pr whereas opposite effect with M.
- Viscous dissipation enhances fluid temperature and this leads to an expansion of the thermal boundary thickness.

References

1. Sakiadis, B.C.: Boundary-layer behavior on continuous solid surfaces: I. Boundary-layer equations for two-dimensional and axisymmetric flow. AIChE J. **7**(1), 26–28 (1961)
2. Erickson, L.E., Fan, L.T., Fox, V.G.: Heat and mass transfer on moving continuous flat plate with suction or injection. Ind. Eng. Chem. Fund. **5**(1), 19–25 (1966)
3. Gupta, P.S., Gupta, A.S.: Heat and mass transfer on a stretching sheet with suction or blowing. Can. J. Chem. Eng. **55**(6), 744–746 (1977)
4. Narayana, P.V., et al.: Thermal radiation effects on an unsteady MHD nanofluid flow over a stretching sheet with non-uniform heat source/sink. J. Nanofluids **6**(5), 899–907 (2017)
5. Tawade, J.V., et al.: Non-uniform heat source/sink and thermal radiation effects on liquid film flow of Casson fluid thin film over a Stretching Sheet. J. Eng. Sci. **11**, 0377–9254 (2020)
6. Abolbashari, M.H., et al.: Entropy analysis for an unsteady MHD flow past a stretching permeable surface in nano-fluid. Powder Technol. **267**, 256–267 (2014)

7. Pal, D.: Combined effects of non-uniform heat source/sink and thermal radiation on heat transfer over an unsteady stretching permeable surface. Commun. Nonlinear Sci. Numer. Simul. **16**(4), 1890–1904 (2011)
8. Alrihieli, H., Alrehili, M., Megahed, A.M.: Radiative MHD nanofluid flow due to a linearly stretching sheet with convective heating and viscous dissipation. Mathematics **10**(24), 4743 (2022)

Part V
Miscellaneous

Modern Methodology of Medicine Versus Its Traditional One: A Theoretical Graph Approach

Adeline Sheeja and Sharmila Mary Arul

1 Introduction

Different philosophers in different cultural backgrounds developed traditional and modern systems of medicine. Traditional medicine is a system that uses natural remedies to help the body heal itself and the side effects are minimal [1]. It embraces many therapies including herbs, massage, acupuncture, exercise, nutritional counseling, and so on. In rural and semi-urban parts of India, traditional medicine is still the first option of treatment for diseases. The unique feature of traditional medicine is that it avoids chemical drugs and surgeries (only the minor surgeries are done) such as stitching up a small wound, which is being performed by a licensed doctor, and it emphasizes the use of natural agents and physical means such as tissue manipulation, manipulative therapy, and electrotherapy.

In the modern system of medicine, there are several advantages and disadvantages found in it. Modern medicine can prevent incurable diseases. It has really helped in increasing life expectancy of people [2]. The other side of modern medicine is that it contains usage of drugs to treat the diseases, which might lead to side effects in the later period of life [3].

The original version of the chapter has been revised. A correction to this chapter can be found at https://doi.org/10.1007/978-3-031-77764-6_34

A. Sheeja
Champion InfoMetrics Pvt Ltd., Chikkabellanduru, Bengaluru, Karnataka

S. M. Arul (✉)
Department of Mathematics, Saveetha School of Engineering, Saveetha Institute of Medical and Technical Sciences, Chennai, India
e-mail: sharmilamaryarul.sse@saveetha.com

I. Rajasingh et al. (eds.), *Applied Computational Mathematics*, Springer Proceedings in Mathematics & Statistics 480, https://doi.org/10.1007/978-3-031-77764-6_29

2 Overview: Traditional Versus Modern

2.1 In System of Medicine

Traditional system of medicine is otherwise known as alternative, complimentary, indigenous, or folk medicine. The knowledge on this system of medicine was developed over generations within various societies before the era of modern medicine. Among the traditional medicine, it is being found that Ayurveda is the world level oldest medicine found among the Indian medical system in the world [4]. Traditional medicine is cost-effective since it is readily available to the patients. The demerit of traditional medicine is that it is not useful in emergency cases. The scientific research or evidence is limited.

Modern system of medicine is otherwise known as biomedicine, Western medicine, evidence-based medicine, or allopathic medicine. In contrast to traditional medicine, modern medicine has done much in the field of infectious diseases and emergencies to aid cure. In this system of modern medicine, we could find out the disease-causing agent and attack the agent medically. This conventional modern system of medicine is a disease management approach, rather than a preventive approach. Acute illness is being treated well in this system of medicine. Modern medicine is the first to discover that even people with the same genetic inheritance will manifest diseases in different ways.

2.2 A Study on Practices

According to the study conducted in Mizoram in the Lunglei district, documenting the traditional practice and knowledge of medicines based on the age and educational level of traditional practice was conducted starting from the age group of 15–60 and above as shown in Fig. 1a (Dotted line indicates the direction (+ve/−ve) of relationship). The study revealed that there is a positive relationship between age and traditional knowledge and practice, while a negative relationship between educational level and traditional knowledge and practice was observed as shown in

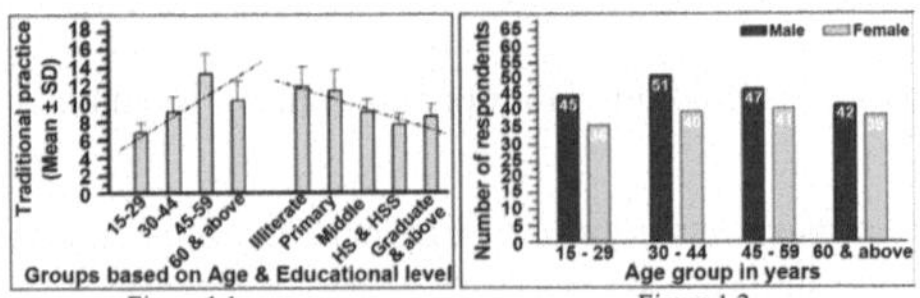

Fig. 1 (**a**) Graph showing the relationship of traditional practice on ethno-medicine with age and educational levels of the respondents. (**b**) The graph showing the male and female respondents according to the age

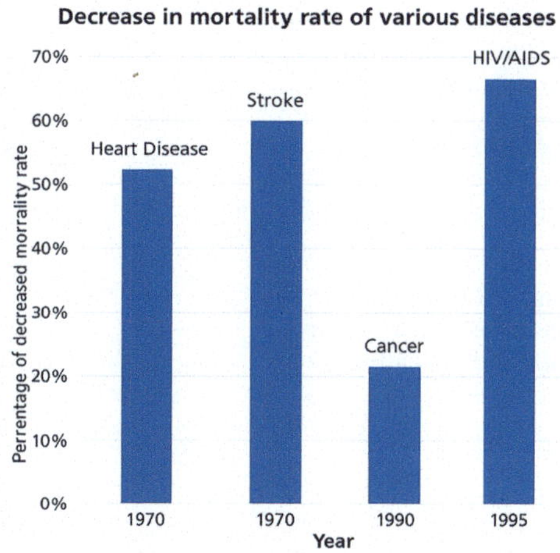

Fig. 2 Decrease in mortality rate of various diseases

Fig. 1b. There are several different systems of traditional medicine practiced in the different parts of the world most widely found in India, China, and Africa.

The modern system of medicine practice has reduced the mortality rate, and the overall treatment cost has lowered also. According to the study conducted on the modern system of treatment, the mortality rate has reduced for heart disease, stroke, cancer, and several other deadly diseases as shown in Fig. 2. Even though there are many advanced technologies found in the modern medical system, we still face several chronic diseases, the side effects of the conventional medicine, and health-based disorders. This system of practice focuses on the diagnosis of, treatment and cure of a particular disease by using pharmaceutical drugs, by surgeries, or by the process of radiation and by other therapeutic modalities.

3 Emerging Therapies: Modern Versus Traditional

Recently, the emerging technique in modern medicine is now being used for giving treatment in the form of therapy in cancer and hepatitis. Some of the therapies such as liposome-mediated gene therapy, photothermal therapy (PTT), monoclonal antibody therapy, stem cell therapy, molecular therapy, chemotherapy, radiotherapy, and electrotherapy are some of the newly emerging therapy-based modern medicine.

Liposome-Mediated Gene Therapy Liposome acts as a vector to deliver a molecular cargo such as DNA for therapeutic benefit. The drug is enclosed in a liposome (a fat bubble) and the outer surface of the material is coated to avoid detection as well as destruction by the immune system. As the coating dissolves slowly, the drug would reach the target site such as cells of oncogene. Figure 3 shows the characterization of optimized liposomal formulation.

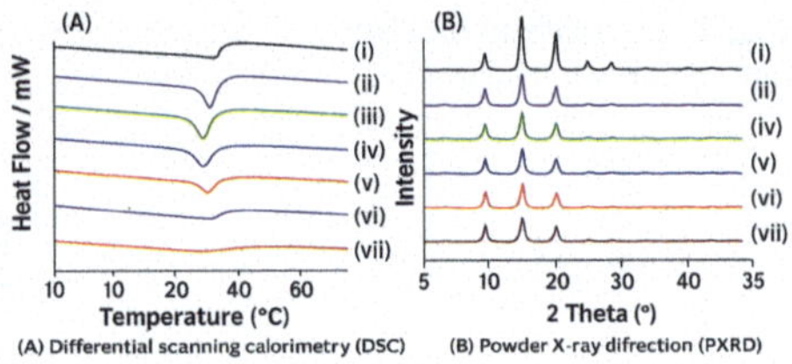

Fig. 3 Characterization of optimized liposomal formulation

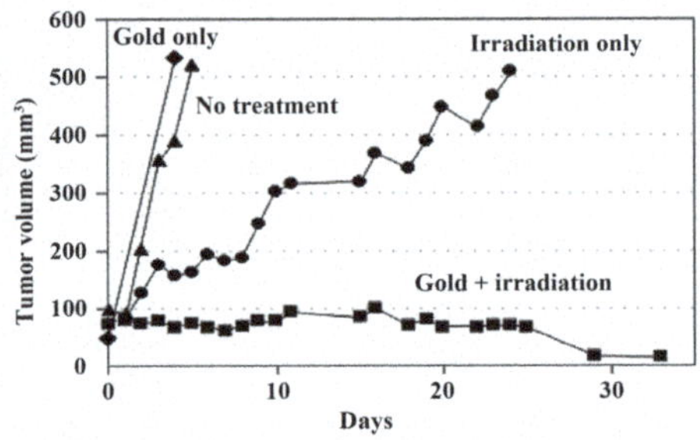

Fig. 4 Comparison of photothermal therapy using gold and irradiation

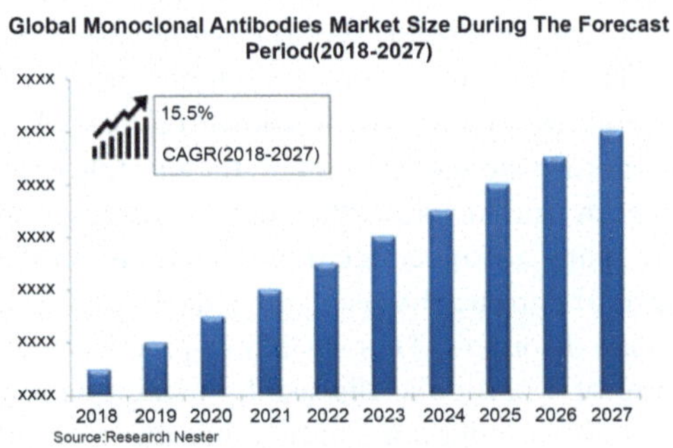

Fig. 5 Growth of monoclonal antibodies in the global market

Photothermal Therapy In this photothermal therapy, gold nanoparticles are used in the treatment against cancer. These gold nanoparticles absorb the incident photons and convert them in the form of heat for the destruction of cancerous cells. Figure 4 shows treatment using only gold, only irradiation, gold along with irradiation, and without any treatment.

Monoclonal Antibody Therapy Monoclonal antibody therapy is another targeted cancer therapy. Even monoclonal antibodies such as rituximab act as an immunotherapy, and they help the immune system fight against cancer. These antibodies are produced in the exogenous condition in a laboratory; they have the ability to mimic, enhance, or restore the immune system attack on the oncogene cells. This therapy is often given along with chemotherapy. The growth prediction of monoclonal antibodies in the global market from the period of 2018 to 2027 is shown in Fig. 5.

Stem Cell Therapy Stem cell therapy is a regenerative medicine since the damaged cells can be replaced with the new stem cells derived from the umbilical cord.

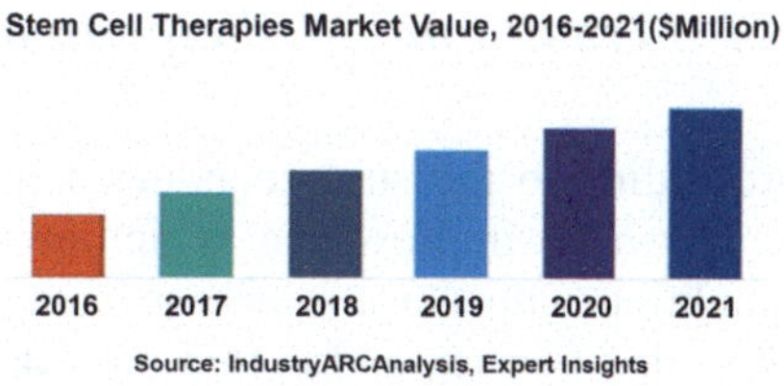

Fig. 6 Growth of stem cell therapy value in the market

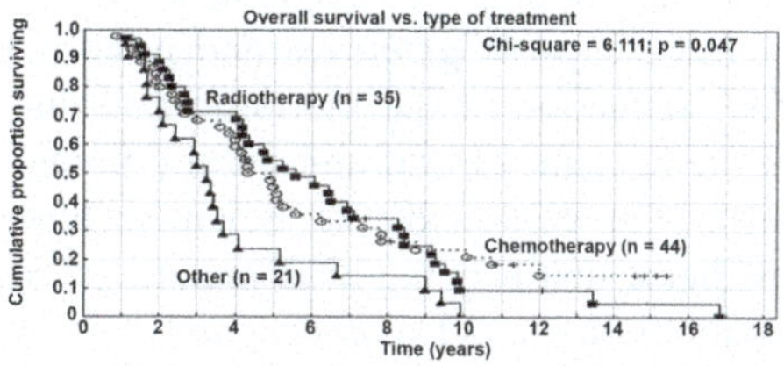

Fig. 7 Overall survival rate for chemotherapy and radiotherapy

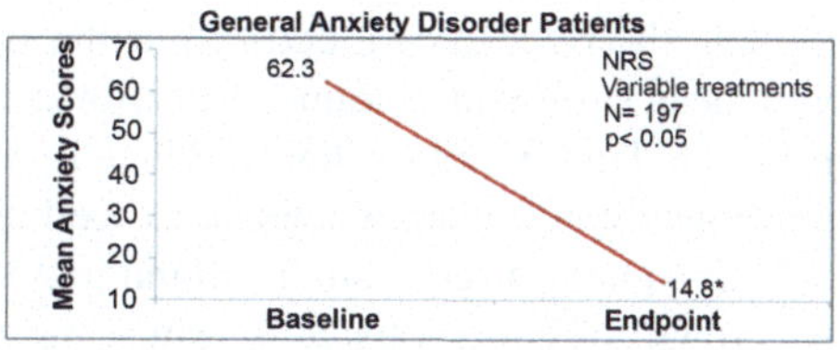

Fig. 8 Cranial electrotherapy stimulation in a patient suffering from acute anxiety disorders

According to the research, the effectiveness of the stem cell therapy is 82.2%. Stem cell therapy was developed around the year 2016; it plays a vital role in the hematopoietic stem cell transplantation. Stem cells can be even stored for future use in the stem cell banks. The graph in Fig. 6 shows the growth in the market value for stem cell therapies.

Chemotherapy and Radiotherapy These therapies are used in the treatment against cancer. Chemotherapy is dealt using the cytotoxic antineoplastic drugs, which kill the cancerous cells. Radiotherapy is otherwise known as radiation therapy; over here the patient undergoes the high doses of radiation through the skin to kill the cancer cells [5]. The survival rate for these treatments is shown in Fig. 7.

Electrotherapy Electrotherapy is also known as e-stim; over here electricity is used for the treatment and it acts as a pain reliever, improves circulation, repairs tissues, improves muscle strengthening, and promotes bone growth for good physical functioning. It can be used for anxiety and depression disorders; in Fig. 8 the graph shows the electrotherapy stimulation for acute anxiety disorder. According to the study, it was observed that electrical muscle stimulation, which is being done for 12 weeks, leads to 30% increase in strength and the mass of the fat is being reduced [6].

3.1 Traditional Medicine in a Modern World

According to the study conducted in the year 2020, it is being found that yoga is known to improve the symptoms of generalized anxiety disorder, which is a condition with chronic nervousness. The study has proven that drinking plenty of both green tea and coffee on a daily basis is found to reduce the risk rate of mortality among the people suffering from type 2 diabetes. It was seen that four or more cups of green tea and more than two cups of coffee were linked to the 63% reduction of the mortality rate due to type 2 diabetes. Traditional medicine is the ancient, oldest, and culture-bound medical practice and it exists before the birth of modern medicine. Scientists have researched and shown that the traditional medicine that is being used in parts of Africa and Asia plays a vital role in treating major diseases such as cancer and diabetes. According to the research done in King's College London, it was observed that India's curry leaf tree has the potential to regulate the release of glucose into the blood stream that helps a diabetic patient who lacks sufficient amount of insulin secretion to cope with excessive sugar level in blood. Traditional healers in Ghana use the extract of the climbing day flower; this flower exhibits antibacterial as well as antifungal properties. The combination of both aquatic weed from Thailand and the Chinese star anise is known to inhibit the growth of cancerous cells. Thus, these are few new emerging therapies in the field of traditional system of medicine where very important benefits are seen [7].

Therapeutic-Based Traditional Treatment Therapeutic traditional treatment plays a vital role in treating many illnesses and ailments. Some of the therapies used in the traditional medicines are herbal therapies, massage, acupuncture, exercise, and manipulation therapy.

Herbal Therapies In this kind of therapy, natural herbal products are used which are extracted from medicinal plants. Herbal medicines are otherwise known as herbalism, botanical medicine, phytomedicine, or phytotherapy. In every part of the traditional medicine, herbal therapies play a vital role. According to the recent estimate of WHO, it was observed that 80% of people worldwide rely on botanical medicines for their primary health.

3.2 Manipulation Therapy

Massage therapy is very beneficial for manipulating soft tissues in the body. In this therapy, the manual technique is used such as rubbing and kneading the body using hands or by using small instruments; it is a very effective therapy for patients suffering from low back pain, anxiety, high blood pressure, migraines, and so on. Massage therapist gives massages by applying strong or gentle pressure in the joints or the muscles in the body to ease pain and tension. This therapy is being widely used by the physiotherapist, physical therapist, and occupational therapist.

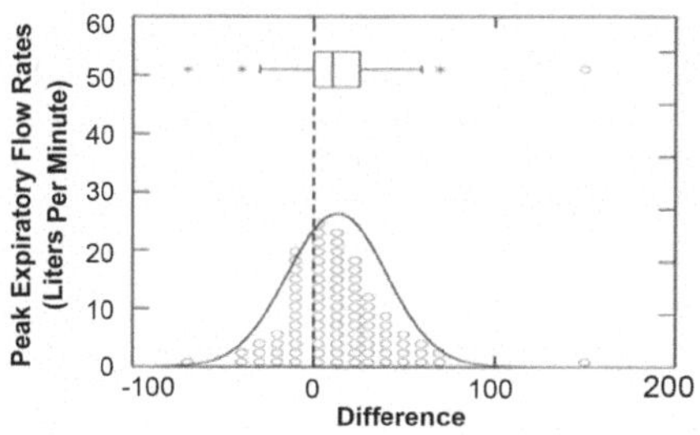

Fig. 9 Graph showing the osteopathic manipulative treatment

In Fig. 9, the osteopathic manipulative treatment graph shows a 95% increase in the probability of expiratory flow rates between 7L per minute and 19L per minute.

Acupuncture Acupuncture therapy treatment is done using thin needles; this therapy is used to cure many ailments and illnesses in cancer patients. The patient prefers this kind of therapy to control pain and to get relief from anxiety, nausea, vomiting, and sleeping problems. According to the research, it was found that acupuncture could reduce inflammation all over the body. Even electro-acupuncture is used in the traditional Chinese medicine.

4 Conclusion

Both modern and traditional medicine play an equal role in the health contribution. The goal of the traditional treatment is to treat as a whole, which means the mind, body, and spirit. Most of the rural community still relies on the traditional remedies; the proper use of this medicine can contribute in achieving the goal of health since the modern healthcare service cannot be fully provided at the primary healthcare level. Modern medicine could prevent the incurable diseases over decades of years it has risen tremendously in curing the deadly diseases.

References

1. Silvanathan, S, Low, B.S.: Current public awareness on the safety of traditional and complementary medicines (T and CM) in Malaysia. Eur. J. Integr. Med. **7**(2), 184–189 (2015)
2. Polonelli, L., Casadevall, A., Han, Y., et al.: The efficacy of acquired humoral and cellular immunity in the prevention and therapy of experimental fungal infec-tions. Med. Mycol. **38**(supplement 1), 281–292 (2000)
3. Marasco, W.A., Sui, J.: The growth and potential of human antiviral monoclonal antibody therapeutics. Nat. Biotechnol. **25**(12), 1421–1434 (2007)
4. Cooper, E.L.: Ayurveda and eCAM: a closer connection. Evidence Based Complement. Altern. Med. **5**(2), 121–122 (2008)

5. Timmerman, R., Paulus, R., Galvin, J., et al.: Stereotactic body radiation therapy for inoperable early stage lung cancer JAMA. **303**, 1070–1076 (2010)
6. Mitchell, F.L.: The Muscle Energy Manual. METpress, East Lansing (1999)
7. WHO Traditional Medicine Strategy: WHO Traditional Medicine Strategy. Altern. Integr. Med. [Internet] **1**(1), 1–78 (2013)

Parikh Matrices of Prouhet Array Morphic Images of Two-Dimensional Arrays

R. Stella Maragatham and V. Nithya Vani

1 Introduction

An interesting and novel connection between words and matrices was developed by Mateescu et al. [8] where a word, expressed in mathematical terms, is a finite sequence whose terms belong to an alphabet that is a finite set of symbols. In fact an upper triangular matrix, called Parikh matrix, was associated with a word over an ordered alphabet with an ordering on its symbols. The entries in the Parikh matrix above the principal diagonal are nonnegative integers that are counts of certain scattered subwords (also simply called subwords) change as that are subsequences of a word over an ordered alphabet. Studies on properties of such matrices have been done intensively (see, e.g., [1–3, 5, 7, 8, 11, 14–17, 19, 20] and references therein). A two-dimensional (2D) $m \times n$ picture array [4] consists of m rows and n columns of symbols taken from an alphabet, with the rows and columns arranged in a rectangular form. For instance, a binary picture array belonging to $\Sigma^{*} = \{a, b\}^{*}$ where Σ^{*} is the set of all picture arrays over Σ is shown in Fig. 1. The picture array can be thought of as describing the English letter L made of the symbol a, where b is interpreted as denoting a blank.

Several morphisms on words have been considered and there are many studies on their properties in the area of combinatorics on words [6]. Prouhet morphism [12, 13] is a natural generalization of the well-known Thue morphism [17]. Prouhet morphism p on a set $\{a, b, c\}$ of three symbols, called ternary alphabet, is defined by

R. Stella Maragatham (✉)
Department of Mathematics, Saveetha School of Engineering, SIMATS, Chennai, India
e-mail: stellamaragathamr.sse@saveetha.com

V. Nithya Vani
Department of Mathematics, Bharath Institute of Science and Technology, Bharath Institute of Higher Education and Research, Bharath University, Chennai, India

I. Rajasingh et al. (eds.), *Applied Computational Mathematics*,
Springer Proceedings in Mathematics & Statistics 480,
https://doi.org/10.1007/978-3-031-77764-6_30

Fig. 1 A picture array in the form L made of two symbols a, b interpreting b's as blanks

$$\begin{matrix} a & b & b & b \\ a & b & b & b \\ a & b & b & b \\ a & b & b & b \\ a & a & a & a \end{matrix}$$

$p(a) = abc$, $p(b) = bca$, $p(c) = cab$. Atanasiu [2] has done a study of subwords and Parikh matrices based on morphisms on words. In [19], Prouhet morphism is considered and derived properties of Prouhet morphic images of binary and ternary words and their corresponding Parikh matrices.

In this paper, we consider the extension of Prouhet morphism on words to rectangular pictures, called Prouhet array morphism, introduced in [9]. Certain properties of images of binary picture arrays (arrays over an ordered binary alphabet) with respect to Prouhet array morphism in the context of Parikh matrices of arrays have been studied in [9]. Here we obtain corresponding results for images of ternary picture arrays (arrays over an ordered ternary alphabet) under this Prouhet array morphism.

2 Preliminaries

We recall certain basic notions needed in subsequent sections. For other notions pertaining to words (not recalled here), we refer to [6, 10]. A finite set Σ, called an alphabet, is a collection of symbols and it is ordered if there is an ordering on its symbols. A finite sequence, called a word, has its terms belonging to an alphabet. The count of the symbols in a word w constitutes the length $|w|$ of w. As an illustration, if $\Sigma = \{a, b\}$, then $w = abaa$ is a binary word with length $|w| = 4$, and if $\Sigma = \{a, b, c\}$, then $w = abacbc$ is a binary word with length $|w| = 6$. The notation λ denotes an empty word that has no symbols. The notation Σ^* denotes the set of all words made of symbols in Σ, including the empty word λ. A subword v of a word w is a subsequence of the word w. If v is a subword of w, the number of times v occurs in w distinctly is denoted by $|w|_v$. As an example, $|aabab|_{ab} = 5$ and $|ababcabc|_{abc} = 9$. In [8], the concept of Parikh matrix defined by a word was developed, and this is done by considering Parikh matrix mapping, which is a mapping associating a $(k+1) \times (k+1)$ upper triangular matrix $\psi_k(w)$, called Parikh matrix, with a word w over an ordered alphabet $\Sigma_k = \{a_1 < a_2 < \ldots < a_k\}$, $(k \geq 2)$. Here our main focus is on binary words over a binary alphabet $\{a < b\}$ and ternary words over a ternary alphabet $\{a < b < c\}$. The Parikh matrices for such words are of the form

$$\psi_2(x) = \begin{pmatrix} 1 & |x|_a & |x|_{ab} \\ 0 & 1 & |x|_b \\ 0 & 0 & 1 \end{pmatrix}, \psi_3(y) = \begin{pmatrix} 1 & |y|_a & |y|_{ab} & |y|_{abc} \\ 0 & 1 & |y|_b & |y|_{bc} \\ 0 & 0 & 1 & |y|_c \\ 0 & 0 & 0 & 1 \end{pmatrix}$$

where x is a word over $\Sigma_2 = \{a < b\}$ and y is a word over $\Sigma_3 = \{a < b < c\}$. We will also denote the Parikh matrix $\psi_k(w)$ of a word w over an ordered alphabet by $M(w)$ (if w is understood, it is also omitted).

The notion of a two-dimensional binary or ternary picture array [15] is now recalled. A binary or ternary array A of size $m \times n$ consists of m rows and n columns with its symbols arranged in the form of a rectangle.

As an illustration, a 3×5 array A over $\{c < d < e\}$ is shown here:

$$A = \begin{matrix} c & d & e & c & d \\ e & d & c & d & c \\ d & c & e & d & c \end{matrix}$$

The notion of row Parikh matrix [15] of a binary or ternary picture array over a corresponding ordered alphabet is now recalled. The corresponding notion of column Parikh matrix of a picture array is similar.

If w_i are the words in the consecutive rows of an $m \times n$ picture array A, then we write $A = rows(w_1, \ldots, w_m)$ with w_1 in the first row and w_m in the last row.

Consider the Parikh matrices $M_1, \ldots, M_m$ of the words $w_1, \ldots, w_m$ in the array $A = rows(w_1, \ldots, w_m)$. Then the row Parikh matrix $M_r(A)$ of the picture array A is given by $M_r(A) = M_1 \oplus M_2 \oplus \cdots \oplus M_m$ where the operation $\oplus$ is defined as follows: If P and Q are Parikh matrices of the same size, then $P \oplus Q$ is a matrix whose diagonal entries are 1, and each of the remaining entries is the sum of the corresponding entries of P and Q. Since we deal with only row Parikh matrix of a picture array, we will refer to $M_r(A)$ simply as the Parikh matrix of the array.

For the notion of row Parikh matrix of a picture array over a general ordered alphabet, the reader is referred to [15].

Example 1 Consider the array $A = \begin{matrix} a & b & b & a \\ b & a & b & b \\ a & a & a & b \end{matrix}$.

The Parikh matrices of the row words $abba$, $baab$, $aaab$ of this array A are given by

$$\begin{pmatrix} 1 & 2 & 2 \\ 0 & 1 & 2 \\ 0 & 0 & 1 \end{pmatrix}, \begin{pmatrix} 1 & 1 & 2 \\ 0 & 1 & 3 \\ 0 & 0 & 1 \end{pmatrix}, \begin{pmatrix} 1 & 3 & 3 \\ 0 & 1 & 1 \\ 0 & 0 & 1 \end{pmatrix}$$

so that the row Parikh matrix of A is

$$M_r(A) = \begin{pmatrix} 1 & 2 & 2 \\ 0 & 1 & 2 \\ 0 & 0 & 1 \end{pmatrix} \oplus \begin{pmatrix} 1 & 1 & 2 \\ 0 & 1 & 3 \\ 0 & 0 & 1 \end{pmatrix} \oplus \begin{pmatrix} 1 & 3 & 3 \\ 0 & 1 & 1 \\ 0 & 0 & 1 \end{pmatrix} = \begin{pmatrix} 1 & 6 & 7 \\ 0 & 1 & 6 \\ 0 & 0 & 1 \end{pmatrix}$$

A property of words, called weak-ratio property, introduced in [14] is as follows: Let x, y be two words over $\Sigma = \{a < b < c\}$. These two words satisfy the weak-ratio property, written $x \sim_{wr} y$, if

$$\frac{|x|_a}{|y|_a} = \frac{|x|_b}{|y|_b} = \frac{|x|_c}{|y|_c} = \alpha, \text{ for some } \alpha \neq 0.$$

Likewise for binary words u, v over $\{a < b\}$, the weak-ratio property is defined.

For arrays, Subramanian et al. [15] extended the weak-ratio property as follows: Let X, Y be two arrays over $\{a < b < c\}$. These two arrays satisfy the weak-ratio property written $X \sim_{wr} Y$ if $\frac{|X|_a}{|Y|_a} = \frac{|X|_b}{|Y|_b} = \frac{|X|_c}{|Y|_c} = \alpha$, for some $\alpha \neq 0$ where $|X|_x$ is the number of symbols x in X.

A condition for the Parikh matrices of two words (respy. two arrays) to commute is given in [7, 18] which is now recalled.

Lemma 1 ([7]) *Let x, y be two words over $\{a < b < c\}$ and satisfy the following:*

(a) The weak-ratio property, namely, $x \sim_{wr} y$ and
(b) $|x|_a|x|_{bc} = |x|_{ab}|x|_c$; $|y|_a|y|_{bc} = |y|_{ab}|y|_c$,

then $M(xy) = M(yx)$.

Theorem 1 ([18]) *Let X and Y be two arrays of sizes $m \times n$ and $k \times l$, respectively, over $\{a < b < c\}$ such that $X = rows(u_1, \ldots, u_m)$ and $Y = rows(v_1, \ldots, v_m)$ where u_i $(1 \leq i \leq m)$ and v_i $(1 \leq i \leq m)$ are words of length m and l over $\{a, b, c\}$, respectively. Let u_i, v_i for $1 \leq i \leq m$ satisfy*

(a) $u_i \sim_{wr} v_i$
(b) $|u_i|_a|u_i|_{bc} = |u_i|_{ab}|u_i|_c$; $|v_i|_a|v_i|_{bc} = |v_i|_{ab}|v_i|_c$

Then $M_r(X \circ Y) = M_r(Y \circ X)$ where $X \circ Y$ is the column catenation [4].

3 Prouhet Morphism and Prouhet Array Morphism

A morphism [6] (also called word morphism or string morphism) f on words is defined as follows: For words $x, y, f(x), f(y)$ are words such that $f(xy) = f(x)f(y)$.

Prouhet morphism [12, 13] is an extension of Thue morphism in a natural way [17]. The definition of this morphism is now recalled but the alphabet is restricted to three letters.

Definition 1 ([12, 13]) The Prouhet morphism $p : \Sigma_3^* \to \Sigma_3^*$ is defined by $p(a) = abc$; $p(b) = bca$; $p(c) = cab$, where $\Sigma_3 = \{a < b < c\}$.

We recall some needed results from [19] relating to Prouhet morphism.

Theorem 2 ([19]) *Let $\Sigma_2 = \{a < b\}$ and $\Sigma_3 = \{a < b < c\}$. For a word x over Σ_2, the following properties hold good:*

(a) $|p(x)|_a = |p(x)|_b = |p(x)|_c = |x|$.
(b) $|p(x)|_{ab} = \frac{1}{2}|x|(|x| + 1) - |x|_b$.
(c) $|p(x)|_{bc} = \frac{1}{2}|x|(|x| + 1)$.
(d) $|p(x)|_{abc} = |x|_a + |x|_a(|x|_a - 1) + \frac{1}{2}|x|_b(|x|_b - 1) + 2|x|_{ab} + |x|_{ba} + \eta$.

where $p : \Sigma_2^* \to \Sigma_3^*$ *is the Prouhet string morphism given in Definition 1 and* $\eta = |x|_{aaa} + |x|_{aab} + |x|_{aba} + |x|_{abb} + |x|_{baa} + |x|_{bba} + |x|_{bbb}$.

Theorem 3 ([19]) *Let* $\Sigma_3 = \{a < b < c\}$. *For a word* γ *over* Σ_3, *we have the following:*

(a) $|p(\gamma)|_a = |p(\gamma)|_b = |p(\gamma)|_c = |\gamma|$.
(b) $|p(\gamma)|_{ab} = \frac{1}{2}|\gamma|_a(|\gamma|_a + 1) + \frac{1}{2}|\gamma|_b(|\gamma|_b - 1) + \frac{1}{2}|\gamma|_c(|\gamma|_c + 1) + |\gamma|_a|\gamma|_b + |\gamma|_b|\gamma|_c + |\gamma|_c|\gamma|_a$.
(c) $|p(\gamma)|_{bc} = \frac{1}{2}|\gamma|_a(|\gamma|_a + 1) + \frac{1}{2}|\gamma|_b(|\gamma|_b + 1) + \frac{1}{2}|\gamma|_c(|\gamma|_c - 1) + |\gamma|_a|\gamma|_b + |\gamma|_b|\gamma|_c + |\gamma|_c|\gamma|_a$.

where $p : \Sigma_3^* \to \Sigma_3^*$ *is the Prouhet string morphism using Definition 1.*

A (string) morphism $f : X^* \to Y^*$ where X, Y are alphabets with $X \neq 0$ and $Y \neq 0$ is uniform if $|f(a)|$ is the same for all a in X.

Uniform morphisms on words have been extended to arrays [6, 19]. Prouhet array morphism was introduced and investigated in [9] for properties of Parikh matrix of image of a binary and a ternary word under Prouhet morphism. We recall now Prouhet array morphism, which is an extension to arrays of Prouhet morphism on words.

Definition 2 Let $\Sigma = \{a, b, c\}$. Let Σ^{++} denote the set of all non-empty arrays over Σ. A Prouhet array morphism $P : \Sigma^{++} \to \Sigma^{++}$ is defined by

$$P(a) = \begin{matrix} a & b & c \\ b & c & a \\ c & a & b \end{matrix} \qquad P(b) = \begin{matrix} b & c & a \\ c & a & b \\ a & b & c \end{matrix} \qquad P(c) = \begin{matrix} c & a & b \\ a & b & c \\ b & c & a \end{matrix}$$

Example 2 Consider the picture array A over $\{a < b\}$ given by $A = \begin{matrix} b & a \\ b & b \end{matrix}$.

The image of A under Prouhet array morphism is

$$P(A) = \begin{matrix} b & c & a & a & b & c \\ c & a & b & b & c & a \\ a & b & c & c & a & b \\ b & c & a & b & c & a \\ c & a & b & c & a & b \\ a & b & c & a & b & c \end{matrix}$$

In order to obtain properties of picture arrays in the context of Prouhet array morphism, two other string morphisms, which are called as p_1, p_2, that are similar to Prouhet morphism were considered in [9]. We recall these.

Definition 3 Let $\Sigma = \{a, b, c\}$. The morphism $p_1 : \Sigma_3^* \to \Sigma_3^*$ is given by $p_1(a) = bca$; $p_1(b) = cab$; $p_1(c) = abc$ and the morphism $p_2 : \Sigma_3^* \to \Sigma_3^*$ by $p_2(a) = cab$; $p_2(b) = abc$; $p_2(c) = bca$

Based on the definition of p_1, p_2, we have the following results (Lemmas 2 and 3).

Lemma 2 ([9]) *Let $\Sigma_2 = \{a < b\}$ and $\Sigma_3 = \{a < b < c\}$. For a word x over Σ_2, we have the following:*

(a) $|p_1(x)|_a = |p_1(x)|_b = |p_1(x)|_c = |x|$.
(b) $|p_1(x)|_{ab} = \frac{1}{2}|x|(|x| + 1) - |x|_a$.
(c) $|p_1(x)|_{bc} = \frac{1}{2}|x|(|x| + 1) - |x|_b$.
(d) $|p_1(x)|_{abc} = \frac{1}{2}|x|_a(|x|_a - 1) + \frac{1}{2}|x|_b(|x|_b - 1) + 2|x|_{ba} + \eta$.

where $p_1 : \Sigma_2^* \to \Sigma_3^*$ is the morphism as in Definition 3 and $\eta = |x|_{aaa} + |x|_{aab} + |x|_{aba} + |x|_{abb} + |x|_{baa} + |x|_{bba} + |x|_{bbb}$.

Lemma 3 ([9]) *Let $\Sigma_2 = \{a < b\}$ and $\Sigma_3 = \{a < b < c\}$. For a word x over Σ_2, we have the following:*

(a) $|p_2(x)|_a = |p_2(x)|_b = |p_2(x)|_c = |x|$.
(b) $|p_2(x)|_{ab} = \frac{1}{2}|x|(|x| + 1)$.
(c) $|p_2(x)|_{bc} = \frac{1}{2}|x|(|x| + 1) - |x|_a$.
(d) $|p_2(x)|_{abc} = |x|_b + \frac{1}{2}|x|_a(|x|_a - 1) + |x|_b(|x|_b - 1) + 2|x|_{ab} + |x|_{ba} + \eta$.

where $p_2 : \Sigma_2^* \to \Sigma_3^*$ is the morphism as in Definition 3 and $\eta = |x|_{aaa} + |x|_{aab} + |x|_{aba} + |x|_{abb} + |x|_{baa} + |x|_{bba} + |x|_{bbb}$.

As a consequence, we have the following result on the number of a, b, c and the number of subwords ab and bc on applying the Prouhet array morphism.

Theorem 4 ([9]) *Let $X = rows(x_1, x_2, \ldots, x_m)$ be an $m \times n$ array over $\{a < b\}$ and then $|P(X)|_a = |P(X)|_b = |P(X)|_c = 3mn$, where P is the Prouhet array morphism.*

Theorem 5 ([9]) *Let $X = rows(x_1, x_2, \ldots, x_m)$ be an $m \times n$ array over $\{a < b\}$ and then $|P(X)|_{ab} = |P(X)|_{bc} = \frac{1}{2}mn(3n + 1)$, where P is the Prouhet array morphism.*

Theorem 6 ([9]) *Let $X = rows(x_1, x_2, \ldots, x_m)$ be an $m \times n$ array over $\{a < b\}$. Then*

$$|P(X)|_{abc} = mn(2n - 1) + \eta_1 + \cdots + \eta_m,$$

where P is the Prouhet array morphism and for all i $(1 \leq i \leq m)$ $\eta_i = |x_i|_{aaa} + |x_i|_{aab} + |x_i|_{aba} + |x_i|_{abb} + |x_i|_{baa} + |x_i|_{bab} + |x_i|_{bba} + |x_i|_{bbb}$.

Corollary 1 ([9]) *Let $X = rows(x_1, x_2, \ldots, x_m)$ be an $m \times n$ array over $\{a < b\}$. Then the Parikh matrix $P(X)$ is given by*

$$\begin{pmatrix} 1 & 3mn & \frac{1}{2}mn(3n+1) & mn(2n-1) + 3\sum_{i=1}^{m}\eta_i \\ 0 & 1 & 3mn & \frac{1}{2}mn(3n+1) \\ 0 & 0 & 1 & 3mn \\ 0 & 0 & 0 & 1 \end{pmatrix}$$

We now state a property related to weak ratio of arrays.

Theorem 7 ([9]) *For non-empty arrays X, Y over $\{a < b\}$, we have $P(X) \sim_{wr} P(Y)$ whenever $X \sim_{wr} Y$, where P is the Prouhet array morphism.*

4 Prouhet Array Morphism and a Ternary Alphabet

In Sect. 3, we have considered words and arrays over a binary alphabet $\{a, b\}$, and properties of the images of picture arrays under Prouhet array morphism, established in [9], have been recalled.

In this section, we obtain analogous properties under Prouhet array morphism of picture arrays over a ternary alphabet $\{a < b < c\}$ and derive formulas for counting subwords ab, bc in the picture array $P(A)$ which A is an array over $\{a, b, c\}$ and P is the Prouhet array morphism. These formulas enable us to obtain conditions for the row Parikh matrix of image of a picture array under Prouhet array morphism, to commute. Based on the definition of p_1 and definition of p_2, we can state the following results for a word over a ternary alphabet (Lemmas 4 and 5)

Lemma 4 *Let $\Sigma_3 = \{a < b < c\}$. For a word x over Σ_3, we have the following:*

1. $|p_1(x)|_a = |p_1(x)|_b = |p_1(x)|_c = |x|$.
2. $|p_1(x)|_{ab} = \frac{1}{2}|x|_a(|x|_a - 1) + \frac{1}{2}|x|_b(|x|_b + 1) + \frac{1}{2}|x|_c(|x|_c + 1) + |x|_a|x|_b + |x|_b|x|_c + |x|_c|x|_a$.
3. $|p_1(x)|_{bc} = \frac{1}{2}|x|_a(|x|_a + 1) + \frac{1}{2}|x|_b(|x|_b - 1) + \frac{1}{2}|x|_c(|x|_c + 1) + |x|_a|x|_b + |x|_b|x|_c + |x|_c|x|_a$.

where $p_1 : \Sigma_3^* \to \Sigma_3^*$ is the string morphism as in Definition 3

Proof Let $w \in \Sigma_3^*$. The proof of statement (i) is analogous to (i) in Theorem 3. The identities $|x|_{ab} + |x|_{ba} = |x|_a \times |x|_b$, $|x|_{bc} + |x|_{cb} = |x|_b \times |x|_c$, and $|x|_{ac} + |x|_{ca} = |x|_a \times |x|_c$ are made use of in obtaining the formulas in the lemma. We have

$$\begin{aligned}|p_1(x)|_{ab} = (|x|_a - 1) + \ldots + 1 + |x|_b + (|x|_b - 1) + \ldots + 1 + |x|_c + (|x|_c - 1) \\ + \ldots + 1 + |x|_{ab} + |x|_{ba} + |x|_{bc} + |x|_{cb} + |x|_{ac} + |x|_{ca}\end{aligned}$$

$$= \frac{1}{2}|x|_a(|x|_a - 1) + \frac{1}{2}|x|_b(|x|_b + 1) + \frac{1}{2}|x|_c(|x|_c + 1)$$
$$+ |x|_a|x|_b + |x|_b|x|_c + |x|_c|x|_a$$

This proves statement (ii).

Statement (iii) can be proved in a similar manner. In fact,

$$|p_1(x)|_{bc} = |x|_a + (|x|_a - 1) + \ldots + 1 + (|x|_b - 1) + \ldots + 1 + |x|_c + (|x|_c - 1)$$
$$+ \ldots + 1 + |x|_{ab} + |x|_{ba} + |x|_{bc} + |x|_{cb} + |x|_{ac} + |x|_{ca}$$
$$= \frac{1}{2}|x|_a(|x|_a + 1) + \frac{1}{2}|x|_b(|x|_b - 1) + \frac{1}{2}|x|_c(|x|_c + 1) + |x|_a|x|_b$$
$$+ |x|_b|x|_c + |x|_c|x|_a.$$

Lemma 5 *Let $\Sigma_3 = \{a < b < c\}$. For a word x over Σ_3, we have the following:*

1. $|p_2(x)|_a = |p_2(x)|_b = |p_2(x)|_c = |x|$.
2. $|p_2(x)|_{ab} = \frac{1}{2}|x|_a(|x|_a + 1) + \frac{1}{2}|x|_b(|x|_b + 1) + \frac{1}{2}|x|_c(|x|_c - 1) + |x|_a|x|_b + |x|_b|x|_c + |x|_c|x|_a$.
3. $|p_2(x)|_{bc} = \frac{1}{2}|x|_a(|x|_a - 1) + \frac{1}{2}|x|_b(|x|_b + 1) + \frac{1}{2}|x|_c(|x|_c + 1) + |x|_a|x|_b + |x|_b|x|_c + |x|_c|x|_a$.

where $p_2 : \Sigma_3^* \rightarrow \Sigma_3^*$ is the string morphism as in Definition 3

Proof Let $w \in \Sigma_3^*$. Statement (i) can be proved analogous to (i) in Theorem 3. The identities $|x|_{ab} + |x|_{ba} = |x|_a \times |x|_b$, $|x|_{bc} + |x|_{cb} = |x|_b \times |x|_c$, and $|x|_{ac} + |x|_{ca} = |x|_a \times |x|_c$ are made use of in obtaining the formulas in the lemma. We have

$$|p_2(x)|_{ab} = |x|_a + (|x|_a - 1) + \ldots + 1 + |x|_b + (|x|_b - 1) + \ldots + 1 + (|x|_c - 1)$$
$$+ \ldots + 1 + |x|_{ab} + |x|_{ba} + |x|_{bc} + |x|_{cb} + |x|_{ac} + |x|_{ca}$$
$$= \frac{1}{2}|x|_a(|x|_a + 1) + \frac{1}{2}|x|_b(|x|_b + 1) + \frac{1}{2}|x|_c(|x|_c - 1)$$
$$+ |x|_a|x|_b + |x|_b|x|_c + |x|_c|x|_a$$

This proves statement (ii).

Statement (iii) can be proved in a similar manner. In fact,

$$|p_2(x)|_{bc} = (|x|_a - 1) + \ldots + 1 + |x|_b + (|x|_b - 1) + \ldots + 1 + |x|_c + (|x|_c - 1)$$
$$+ \ldots + 1 + |x|_{ab} + |x|_{ba} + |x|_{bc} + |x|_{cb} + |x|_{ac} + |x|_{ca}$$
$$= \frac{1}{2}|x|_a(|x|_a - 1) + \frac{1}{2}|x|_b(|x|_b + 1) + \frac{1}{2}|x|_c(|x|_c + 1)$$
$$+ |x|_a|x|_b + |x|_b|x|_c + |x|_c|x|_a$$

Theorem 8 *Let $X = rows(x_1, x_2, \ldots, x_m)$ be an $m \times n$ array over $\{a < b < c\}$ and then $|P(X)|_a = |P(X)|_b = |P(X)|_c = 3mn$, where P is the Prouhet array morphism.*

Proof Let the successive rows of the array X be the words $u_1, u_2, \ldots, u_m$ such that u_1 is in the first row, u_m is in the last row, and $|u_i|_a = p_i$, $|u_i|_b = q_i$, $|u_i|_c = r_i$, $1 \le i \le m$.

We make use of (i) in Theorem 3, and we get $|u_i| = |u_i|_a + |u_i|_b + |u_i|_c$, i.e., $|u_i| = n$.

The statement of Theorem 8 can be proved analogous to Theorem 4.

Theorem 9 *Let $\Sigma_3 = \{a < b < c\}$, for an $m \times n$ array $X = rows(x_1, x_2, \ldots, x_m)$ over Σ_3, and then $|P(X)|_{ab} = |P(X)|_{bc} =$*

$$\frac{3}{2}\sum_{i=1}^{m}|x_i|_a(|x_i|_a+1) + \frac{3}{2}\sum_{i=1}^{m}|x_i|_b(|x_i|_b+1) + \frac{3}{2}\sum_{i=1}^{m}|x_i|_c(|x_i|_c+1)$$

$$+3\sum_{i=1}^{m}|x_i|_a|x_i|_b + 3\sum_{i=1}^{m}|x_i|_b|x_i|_c + 3\sum_{i=1}^{m}|x_i|_c|x_i|_a - \sum_{i=1}^{m}|x_i|$$

where $P : \Sigma_3^* \to \Sigma_3^*$ is the Prouhet array morphism.

Proof On applying Prouhet array morphism P to each row x_i in X, we obtain, corresponding to each row, three consecutive rows of words in $P(X)$, and these are the result of the application of three (string) morphisms p (Prouhet morphism) and two other morphisms p_1, p_2 given by $p_1(a) = bca$, $p_1(b) = cab$, $p_1(c) = abc$ and $p_2(a) = cab$, $p_2(b) = abc$, $p_2(c) = bca$.

We make use of the result (ii) in Theorem 3 in order to derive the remaining equalities.

By (ii) in Theorem 3, we have

$|p(x_i)|_{ab} = \frac{1}{2}|x_i|_a(|x_i|_a+1)+\frac{1}{2}|x_i|_b(|x_i|_b-1)+\frac{1}{2}|x_i|_c(|x_i|_c+1)+|x_i|_a|x_i|_b+|x_i|_b|x_i|_c+|x_i|_c|x_i|_a$

By Lemma 4, we have

$|p_1(x_i)|_{ab} = \frac{1}{2}|x_i|_a(|x_i|_a-1)+\frac{1}{2}|x_i|_b(|x_i|_b+1)+\frac{1}{2}|x_i|_c(|x_i|_c+1)+|x_i|_a|x_i|_b+|x_i|_b|x_i|_c+|x_i|_c|x_i|_a$

By Lemma 5, we have

$|p_2(x_i)|_{ab} = \frac{1}{2}|x_i|_a(|x_i|_a+1)+\frac{1}{2}|x_i|_b(|x_i|_b+1)+\frac{1}{2}|x_i|_c(|x_i|_c-1)+|x_i|_a|x_i|_b+|x_i|_b|x_i|_c+|x_i|_c|x_i|_a$

$$|P(X)|_{ab} = \sum_{i=1}^{m}p(x_i) + \sum_{i=1}^{m}p_1(x_i) + \sum_{i=1}^{m}p_2(x_i)$$

$$= \frac{3}{2}\sum_{i=1}^{m}|x_i|_a(|x_i|_a+1) + \frac{3}{2}\sum_{i=1}^{m}|x_i|_b(|x_i|_b+1) + \frac{3}{2}\sum_{i=1}^{m}|x_i|_c(|x_i|_c+1)$$

$$+3\sum_{i=1}^{m}|x_i|_a|x_i|_b + 3\sum_{i=1}^{m}|x_i|_b|x_i|_c + 3\sum_{i=1}^{m}|x_i|_c|x_i|_a - \sum_{i=1}^{m}|x_i|$$

as $|x_i| = |x_i|_a + |x_i|_b + |x_i|_c$

Likewise, we can derive the formula for $|P(X)|_{bc}$.

We obtain $|P(X)|_{ab} = |P(X)|_{bc}$.

Example 3 Consider the array $X = \begin{matrix} a & c & b & a \\ c & b & a & c \end{matrix}$

$$P(X) = \begin{matrix} a & b & c & c & a & b & b & c & a & a & b & c \\ b & c & a & a & b & c & c & a & b & b & c & a \\ c & a & b & b & c & a & a & b & c & c & a & b \\ c & a & b & b & c & a & a & b & c & c & a & b \\ a & b & c & c & a & b & b & c & a & a & b & c \\ b & c & a & a & b & c & c & a & b & b & c & a \end{matrix}$$

In the array $X = rows(u_1, u_2)$, the words u_1, u_2 in the rows of the array X satisfy $|u_1|_a = 2$, $|u_2|_a = 1$, $|u_1|_b = 1$, $|u_2|_b = 1$ and $|u_1|_c = 1$, $|u_2|_c = 2$

$|u_1| = |u_1|_a + |u_1|_b + |u_1|_c = 2 + 1 + 1 = 4$

$|u_2| = |u_2|_a + |u_2|_b + |u_2|_c = 1 + 1 + 2 = 4$

$$|P(X)|_{ab} = \frac{3}{2}\sum_{i=1}^{m}|x_i|_a(|x_i|_a + 1) + \frac{3}{2}\sum_{i=1}^{m}|x_i|_b(|x_i|_b + 1) + \frac{3}{2}\sum_{i=1}^{m}|x_i|_c(|x_i|_c + 1) +$$

$$3\sum_{i=1}^{m}|x_i|_a|x_i|_b + 3\sum_{i=1}^{m}|x_i|_b|x_i|_c + 3\sum_{i=1}^{m}|x_i|_c|x_i|_a - \sum_{i=1}^{m}|x_i|$$

$= [(\frac{3}{2}\times 2\times 3)+(\frac{3}{2}\times 1\times 2)]+[(\frac{3}{2}\times 1\times 2)+(\frac{3}{2}\times 1\times 2)]+[(\frac{3}{2}\times 1\times 2)+(\frac{3}{2}\times 2\times 3)]+$
$[(3\times 2\times 1)+(3\times 1\times 1)]+[(3\times 1\times 1)+(3\times 1\times 2)]+[(3\times 1\times 2)+(3\times 2\times 1)]-(4+4)$
$= (9 + 3) + (3 + 3) + (3 + 9) + (6 + 3) + (3 + 6) + (6 + 6) - 8$
$= 12 + 6 + 12 + 9 + 9 + 12 - 8 = 52$

Likewise, $|P(X)|_{bc} = 52$

Theorem 10 *Let X and Y be two array sizes $m \times n$ and $m \times l$, respectively, over $\Sigma_3 = \{a < b < c\}$ such that $X = rows(u_1, u_2, \ldots, u_m)$ and $Y = rows(v_1, v_2, \ldots, v_m)$ where u_i $(1 \le i \le m)$ and v_i $(1 \le i \le m)$ are words of length n and l, respectively.*

Let u_i, v_i for $1 \le i \le m$ satisfy

1. $u_i \sim_{wr} v_i$,
2. $|u_i|_b = |u_i|_c$, *and* $|v_i|_b = |v_i|_c$, *and*

then $M_r(P(X) \circ P(Y)) = M_r(P(Y) \circ P(X))$, where $P : \Sigma_3^* \to \Sigma_3^*$ is the Prouhet array morphism.

Proof Since $A = rows(u_1, u_2, \ldots, u_m)$, the application of P on u_i $(1 \le i \le m)$ yields three consecutive rows of words in $P(X)$. Call these rows x_i, y_i and z_i $(1 \le i \le m)$. Then $P(X)$ is an array with $3m$ rows, i.e., $P(X) = rows(x_{11}, y_{11}, z_{11}, \ldots, x_{i1}, y_{i1}, z_{i1}, \ldots, x_{m1}, y_{m1}, z_{m1})$.

Likewise, since $Y = rows(v_1, v_2, \ldots, v_m)$, we have
$P(Y) = rows(x_{12}, y_{12}, z_{12}, \ldots, x_{i2}, y_{i2}, z_{i2}, \ldots, x_{m2}, y_{m2}, z_{m2})$
where x_{i2}, y_{i2} and z_{i2} are the rows of words resulting from the application of P on v_i $(1 \leq i \leq m)$.

In order to prove Theorem 10, it is enough to prove for $1 \leq i \leq m$ that

1. $x_{i1} \sim_{wr} x_{i2}$; $y_{i1} \sim_{wr} y_{i2}$; $z_{i1} \sim_{wr} z_{i2}$
2. $|x_{i1}|_a |x_{i1}|_{bc} = |x_{i1}|_{ab} |x_{i1}|_c$
3. $|x_{i2}|_a |x_{i2}|_{bc} = |x_{i2}|_{ab} |x_{i2}|_c$
4. $|y_{i1}|_a |y_{i1}|_{bc} = |y_{i1}|_{ab} |y_{i1}|_c$
5. $|y_{i2}|_a |y_{i2}|_{bc} = |y_{i2}|_{ab} |y_{i2}|_c$
6. $|z_{i1}|_a |z_{i1}|_{bc} = |z_{i1}|_{ab} |z_{i1}|_c$
7. $|z_{i2}|_a |z_{i2}|_{bc} = |z_{i2}|_{ab} |z_{i2}|_c$

Theorem 10 will then follow from (i) in Lemmas 4 and 5.

Now $|x_{i1}|_a = |x_{i1}|_b = |x_{i1}|_c = |u_{i1}|_a + |u_{i1}|_b + |u_{i1}|_c = |u_i|$.

Likewise, $|x_{i2}|_a = |x_{i2}|_b = |x_{i2}|_c = |v_{i1}|_a + |v_{i1}|_b + |v_{i1}|_c = |v_i|$.

Similarly, $|y_{i1}|_a = |y_{i1}|_b = |y_{i1}|_c = |u_i|$, $|y_{i2}|_a = |y_{i2}|_b = |y_{i2}|_c = |v_i|$ and $|z_{i1}|_a = |z_{i1}|_b = |z_{i1}|_c = |u_i|$, $|z_{i2}|_a = |z_{i2}|_b = |z_{i2}|_c = |v_i|$
so that $x_{i1} \sim_{wr} x_{i2}$ as $u_i \sim_{wr} v_i$ by hypothesis. Similarly, we can show that $y_{i1} \sim_{wr} y_{i2}$ and $z_{i1} \sim_{wr} z_{i2}$. This proves relation (i).

As done in the proof of (ii) and (iii) in Theorem 3, we have $|x_{i1}|_{ab} = \frac{1}{2}|u_i|_a(|u_i|_a+1)+\frac{1}{2}|u_i|_b(|u_i|_b-1)+\frac{1}{2}|u_i|_c(|u_i|_c+1)+|u_i|_a|u_i|_b+|u_i|_c+|u_i|_c|u_i|_a$
$|x_{i1}|_{bc} = \frac{1}{2}|u_i|_a(|u_i|_a + 1) + \frac{1}{2}|u_i|_b(|u_i|_b + 1) + \frac{1}{2}|u_i|_c(|u_i|_c - 1) + |u_i|_a|u_i|_b + |u_i|_b|u_i|_c + |u_i|_c|u_i|_a$.

As done in the proof of (ii) and (iii) in Lemma 4, we have

$$\begin{aligned}|y_{i1}|_{ab} &= \frac{1}{2}|u_i|_a(|u_i|_a - 1) + \frac{1}{2}|u_i|_b(|u_i|_b + 1) + \frac{1}{2}|u_i|_c(|u_i|_c + 1) + |u_i|_a|u_i|_b\\ &+ |u_i|_b|u_i|_c + |u_i|_c|u_i|_a|y_{i1}|_{bc}\\ &= \frac{1}{2}|u_i|_a(|u_i|_a + 1) + \frac{1}{2}|u_i|_b(|u_i|_b - 1) + \frac{1}{2}|u_i|_c(|u_i|_c + 1)\\ &+ |u_i|_a|u_i|_b + |u_i|_b|u_i|_c + |u_i|_c|u_i|_a.\end{aligned}$$

As done in the proof of (ii) and (iii) in Lemma 5, we have

$$\begin{aligned}|z_{i1}|_{ab} &= \frac{1}{2}u_i|_a(|u_i|_a + 1) + \frac{1}{2}|u_i|_b(|u_i|_b + 1) + \frac{1}{2}|u_i|_c(|u_i|_c - 1) + |u_i|_a|u_i|_b\\ &+ |u_i|_b|u_i|_c + |u_i|_c|u_i|_a\end{aligned}$$

and

$$\begin{aligned}|z_{i1}|_{bc} &= \frac{1}{2}|u_i|_a(|u_i|_a - 1) + \frac{1}{2}|u_i|_b(|u_i|_b + 1) + \frac{1}{2}|u_i|_c(|u_i|_c - 1) + |u_i|_a|u_i|_b\\ &+ |u_i|_b|u_i|_c + |u_i|_c|u_i|_a.\end{aligned}$$

Hence, $|x_{i1}|_a|x_{i1}|_{bc} = |x_{i1}|_a[\frac{1}{2}|u_i|_a(|u_i|_a+1)+\frac{1}{2}|u_i|_b(|u_i|_b+1)+\frac{1}{2}|u_i|_c(|u_i|_c-1)+|u_i|_a|u_i|_b+|u_i|_b|u_i|_c+|u_i|_c|u_i|_a]$.

Using $|u_i|_b = |u_i|_c$ from (ii) in statement of Theorem 10 and $|x_{i1}|_a = |x_{i1}|_c$,

$$\begin{aligned}|x_{i1}|_a|x_{i1}|_{bc} &= |x_{i1}|_c[\frac{1}{2}|u_i|_a(|u_i|_a+1)+\frac{1}{2}|u_i|_b(|u_i|_b-1)+\frac{1}{2}|u_i|_c(|u_i|_c+1)\\ &\quad +|u_i|_a|u_i|_b+|u_i|_b|u_i|_c+|u_i|_c|u_i|_a]\\ &= |x_{i1}|_c|x_{i1}|_{ab}\\ &= |x_{i1}|_{ab}|x_{i1}|_c\end{aligned}$$

This proves relation (ii).

Likewise, the remaining relations (iii), (iv), (v), (vi), and (vii) can be obtained. Using [18, Theorem 1], it follows that $M_r(P(X)\circ P(Y)) = M_r(P(Y)\circ P(X))$.

As a consequence of Theorem 10, we obtain the following corollary.

Corollary 2 *Let X and Y be two arrays satisfying the hypothesis of Theorem 10, and then the arrays $P(X)\circ P(Y)$ and $P(Y)\circ P(X)$ are row M-ambiguous [18].*

Theorem 11 *For non-empty arrays X, Y over $\{a < b < c\}$, we have $P(X) \sim_{wr} P(Y)$ whenever $X \sim_{wr} Y$, where P is the Prouhet array morphism.*

Proof Let the arrays X and YS have sizes $m \times n$ and $k \times l$, respectively.

By hypothesis $X \sim_{wr} Y$ so that $\frac{|X|_a}{|Y|_a} = \frac{|X|_b}{|Y|_b} = \frac{|X|_c}{|Y|_c} = \alpha$ for some $\alpha \neq 0$. Let $u_1, u_2, \ldots, u_m$ be the words in the consecutive rows of the array X (starting from the first row) such that $|u_i|_a = p_i$, $|u_i|_b = q_i$, $|u_i|_c = f_i$, $|u_i| = n$, $1 \le i \le m$.

Hence, $|X|_a = \sum_{i=1}^{m}|u_i|_a = \sum_{i=1}^{m} p_i$, $|X|_b = \sum_{i=1}^{m}|u_i|_b = \sum_{i=1}^{m} q_i$ and $|X|_c = \sum_{i=1}^{m}|u_i|_c = \sum_{i=1}^{m} f_i$.

Likewise, let $v_1, v_2, \ldots, v_k$ be the words in the consecutive rows of the arrays Y such that $|v_i|_a = r_i$, $|v_i|_b = s_i$, $|v_i|_c = t_i$, $|v_i| = l$, $1 \le i \le k$ so that $|Y|_a = \sum_{i=1}^{k} r_i$, $|Y|_b = \sum_{i=1}^{k} s_i$, and $|Y|_c = \sum_{i=1}^{k} t_i$.

Hence, $\alpha = \frac{|X|_a}{|Y|_a} = \frac{|X|_b}{|Y|_b} = \frac{|X|_c}{|Y|_c} = \frac{|X|_a+|X|_b+|X|_c}{|Y|_a+|Y|_b+|Y|_c} = \frac{\sum_{i=1}^{m}(p_i+q_i+f_i)}{\sum_{i=1}^{k}(r_i+s_i+t_i)} = \frac{\sum_{i=1}^{m}|u_i|}{\sum_{i=1}^{k}|v_i|} = \frac{\sum_{i=1}^{m} n}{\sum_{i=1}^{k} l} = \frac{mn}{kl}$.

Hence, $\frac{|P(X)|_a}{|P(Y)|_a} = \frac{|P(X)|_b}{|P(Y)|_b} = \frac{|P(X)|_c}{|P(Y)|_c}$.

Using Theorem 8, $|P(X)|_a = |P(X)|_b = |P(X)|_c = 3mn$.

Similarly, $|P(Y)|_a = |P(Y)|_b = |P(Y)|_c = 3kl$

$\frac{|P(X)|_a}{|P(Y)|_a} = \frac{|P(X)|_b}{|P(Y)|_b} = \frac{|P(X)|_c}{|P(Y)|_c} = \frac{3mn}{3kl} = \alpha$.

Since $|X|_a = \alpha|Y|_a$, $|X|_b = \alpha|Y|_b$ and $|X|_c = \alpha|Y|_c$, as $X \sim_{wr} Y$.

Thus, $P(X) \sim_{wr} P(Y)$.

5 Conclusion

Properties of Parikh matrix of Prouhet morphic image of a word are obtained in the ternary case. It will be interesting to obtain properties of Parikh matrix corresponding to other kinds of array morphisms.

References

1. Atanasiu, A.: Binary amiable words. Int. J. Found. Comp. Sci. **18**, 387–400 (2007)
2. Atanasiu, A.: Morphisms on Amiable Words. Annals of Bucharest University. LIX, pp. 99–111 (2010)
3. Atanasiu, A.: Parikh matrices, amiability and istrial morphism. Int. J. Found. Comput. Sci. **21**, 1021–1033 (2010)
4. Giammarresi, D., Restivo, A.: Two-dimensional languages. In: Rozenberg, G., Salomaa, A. (eds.) Handbook of Formal Languages, vol. 3, pp. 215–267. Springer, Berlin, Heidelberg (1997)
5. Isawasan, P., Venkat, I., Subramanian, K.G., Sarmin, N.H.: Morphic images of binary words and Parikh matrix. In: AIP Conference Proceedings, vol. 1605, pp. 708–713 (2014)
6. Lothaire, M.: Combinatorics on Words. Addison-Wesley, Reading (1983)
7. Mahalingam, K., Subramanian, K.G.: Product of Parikh matrices and commutativity. Int. J. Found. Comput. Sci. **23**, 207–223 (2012)
8. Mateescu, A., Salomaa, A., Salomaa, K., Yu, S.: A sharpening of the Parikh mapping. Theor. Inf. Appl. **35**, 551–564 (2001)
9. Nithya Vani, V., Stella Maragatham, R., Subramanian, K.G.: Prouhet array morphism and Parikh matrices of arrays. Int. J. Pure Appl. Math. **120**, 29–37 (2018)
10. Rozenberg, G., Salomaa, A. (eds.): Handbook of Formal Languages, vol. 1–3. Springer, Berlin/Heidelberg/New York (1997)
11. Salomaa, A.: Criteria for the matrix equivalence of words. Theor. Comp. Sci. **411**, 1818–1827 (2010)
12. Seebold, P.: On some generalizations of the Thue-Morse morphism. Theor. Comput. Sci. **292**, 283–298 (2003)
13. Seebold, P., Lyndon factorization of Prouhet words. Theor. Comput. Sci. **307**, 179–197 (2003)
14. Subramanian, K.G., Huey, A.M., Nagar, A.K.: On Parikh matrices. Int. J. Found. Comput. Sci. **20**, 211–219 (2009)
15. Subramanian, K.G., Mahalingam, K., Abdullah, R., Nagar, A.K.: Two-dimensional digitized picture arrays and Parikh matrices. Int. J. Found. Comput. Sci. **24**, 393–408 (2013)
16. Subramanian, K.G., Isawasan, P., Venkat, I.: Parikh matrices and Istrail morphism. Malay. J. Fund. Appl. Sci. **9**(1), 5–9 (2013)
17. Subramanian, K.G., Isawasan, P., Venkat, I., Khader, A.T.: Parikh matrices of words under a generalized Thue morphism. In: AIP Proceedings 11th International Conference on Numerical Analysis and Applied Mathematics, pp. 2516–2519 (2013)
18. Subramanian, K.G., Isawasan, P., Venkat, I., Khader, A.T.: Properties of Parikh matrices of picture array. In: Proceedings of the 3rd International Conference on Computer Engineering and Mathematical Sciences (ICCEMS 2014), pp. 651–657 (2014)
19. Subramanian, K.G., Venkat, I., Isawasan, P., Nagar, A.K.: On Prouhet morphisms and Parikh matrices. Paper presented at the 4th International Conference on Mathematical Sciences. Universiti Kebangsaan Malaysia (2016)
20. Teh, W.C., Kwa, K.H.: Core words and Parikh matrices. Theor. Comput. Sci. **582**, 60–69 (2015)

Graph Analysis on Physiological Aging: A Theoretical Investigation

Adeline Sheeja and Sharmila Mary Arul

1 Introduction

Aging is divided into two types, namely, primary aging, which is inevitable and caused by the cellular structure and its function, and secondary aging, which is caused by stress, strain, diseases, and consuming junk food which leads to being obese, thereby bringing secondary aging at a faster rate, and might be caused by several environmental factors such as consuming alcohol, smoking, and excessive pollution.

Aging is usually a natural process which is inevitable but this phenomena of aging [4] could be prolonged if a person takes care of his/himself biologically and physiologically by consuming hygienic food and avoiding exposure to the UV rays. The functional status of a person is being determined by the biological age. When there is a drop in the biological age, the physiological age tends to appear much older.

The original version of the chapter has been revised. A correction to this chapter can be found at https://doi.org/10.1007/978-3-031-77764-6_34

A. Sheeja
Champion InfoMetrics Pvt Ltd., Chikkabellanduru, Bengaluru, Karnataka

S. M. Arul (✉)
Department of Mathematics, Saveetha School of Engineering, Saveetha Institute of Medical and Technical Sciences, Chennai, India
e-mail: sharmilamaryarul.sse@saveetha.com

I. Rajasingh et al. (eds.), *Applied Computational Mathematics*,
Springer Proceedings in Mathematics & Statistics 480,
https://doi.org/10.1007/978-3-031-77764-6_31

2 Methodology

2.1 Physiological Aging of Cardiovascular System

In this section [3], we deal with the normal physiological changes that take place in the vasculature and the heart, which in turn affect the response of the cardiac system as shown in Fig. 1.

The aging of the heart is due to deposits of the "aging pigment" lipofuscin. As we get older, cardio responsiveness also declines and this is due to the physiological aging of the cardiac system. Higher heart rate is called tachycardia and lower heart rate is called bradycardia. During vigorous exercise, the stroke volume may double because of venous return; when we do cardiac activities, there will be an increase in heart rate supplying oxygen to all our parts of the body, and this can keep our circulatory system in a healthier manner. Aging of the heart may occur rapidly due to the following reasons:

1. **Hypertension**: It is one of the most common circulatory diseases. A rise in the level of systolic blood pressure or a decline in the diastolic blood pressure takes place. The normal blood pressure of a person is 120/80 mmHg. When there is a blood pressure rise above 150 mm Hg or 90 mmHg, it is said to be the condition called hypertension. Uncontrolled hypertension leads to the damage of several organs in our body system.
2. **Atherosclerosis**: Growth of plaques within the arterial lumen contains cholesterol, fibers, dead muscles, and platelets. It is termed as atherosclerosis. This plaques could form blood clot within the arteries and it affects the regular flow of blood, and when the normal regular flow of blood is affected, the heart has to pump the blood with a force, causing the cardiac muscle to overwork and weaken and ultimately causing aging in an accelerated level.

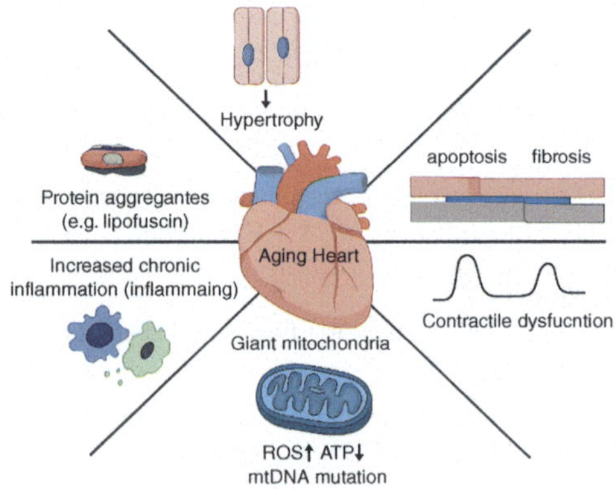

Fig. 1 Aging of the heart

2.2 *Physiological Aging of Respiratory System*

The major role of the respiratory system is gaseous exchange. It regulates blood pH, controls BP, and provides nonspecific immune defense mechanism. The thorax becomes stiffer with age owing to calcification of the coastal cartilage and causes reduced rib excursion.

1. **Lung volume**: Physiological aging of the lung is associated with dilation of alveoli and enlargement of air spaces. The lung contains the intercostal muscles, which are of two types. They are the external intercostal muscles and the internal intercostal muscles. These muscles can be found in between the ribs and diaphragm and play a major role in producing pressure gradient. As we age, our respiratory muscle strength also tends to decline with age. This can even result in cough; therefore, the defense mechanism of the lung also declined. Further, it causes a decline in the flow of air and gas exchange; thus, there is a reduction in the lung function as shown in Fig. 2.
2. **Gas exchange**: The gas exchange takes place in the alveoli; the alveoli are found inside the lungs. Thin alveoli contain the diffusion membrane that is made up of three layers. They are the thin squamous epithelial cells, the endothelium of the alveolar capillaries, and the basement substance between them. The major role of the alveoli is the exchange of gases, and it takes place in between the blood and the cells. Therefore, the cells take in oxygen and release carbon dioxide. As we age, the diffusion capacity in the lung is decreased. This is because there is a rise in the alveolar-arterial oxygen gradient. The lung contains the pleural fluid, and it is responsible for reducing friction.

2.3 *Physiological Aging of Genitourinary System*

The kidneys are part of the urinary system, which includes the uterus, bladder, and urethra. Muscle changes and changes in the reproductive system can affect bladder

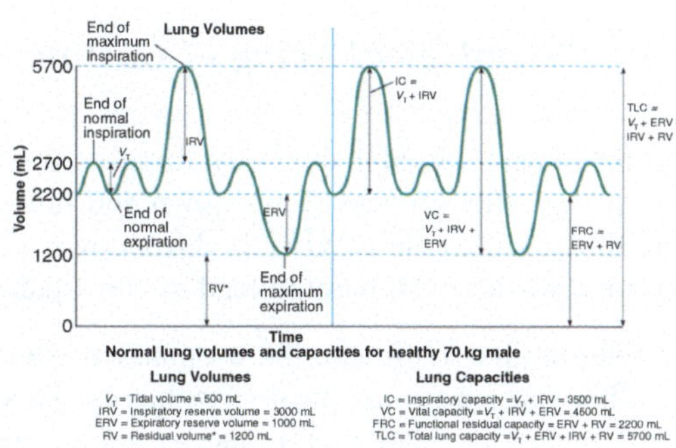

Fig. 2 Graph showing lung volume and capacity

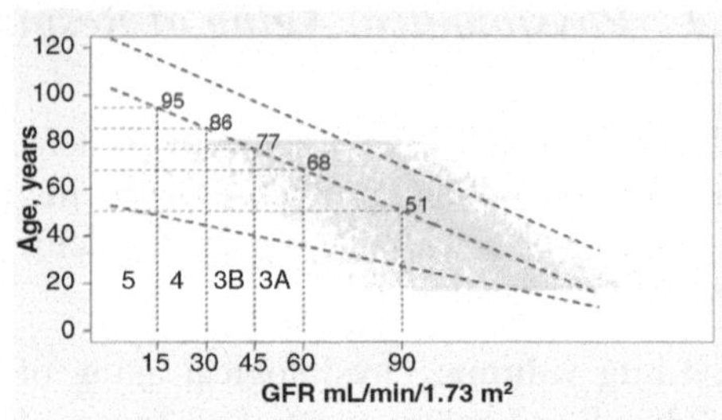

Fig. 3 Graphical representation of GFR w.r.t. age

control. As we age, the kidneys and the bladder change, and this causes the kidneys to filter blood more slowly.

1. **Changes in the kidneys that occur with age**: As we get older, the amount of tissue in a kidney decreases. The kidney contains millions of nephrons that play a major role in filtering waste material from the blood. When age declines, these numbers of filtering, units (nephron), gradually decrease. This causes the decrease in the blood supply in the vessels and the vessels could become hardened. This can lead to the slow filtration of blood as shown in Fig. 3.
2. **Changes in the bladder that occur with age**: Due to the process of aging, there is a change taking place in the wall of the bladder. The bladder contains the elastic tissue. As we age, it becomes tough, and when it becomes tough, the bladder becomes less stretchy. Therefore, the muscles of the bladder get weaken and unable to hold urine. This is because of the weakened muscles. The urinary bladder should be emptied at the proper interval of time. If it is prolonged for a wide interval of time, the kidney reabsorbs the urine that is toxic to the human body. This can lead to several problems such as urinary tract infections (UTIs) or chronic kidney diseases.
3. **Changes in the prostate that occur with age**: This occurs in the males. The urethra gets obstructed by an enlarged prostate gland. In case it is left untreated, it may lead to several problems such as urinary retention and kidney damage.

2.4 Physiological Aging of Gastrointestinal System

For the intake of food in our body, there are several muscle contractions taking place. As we age, this process slows down and this causes the food to move at a slower rate through the colon. When it slows down, plenty of water will be absorbed from the food waste; thus, it could lead to constipation.

1. **Esophagus**: With age, the esophageal contraction strength gets decreased. It is otherwise known as presbyesophagus. This is the term given for a dysfunction of the esophageal motor due to aging. However, diseases that interfere with

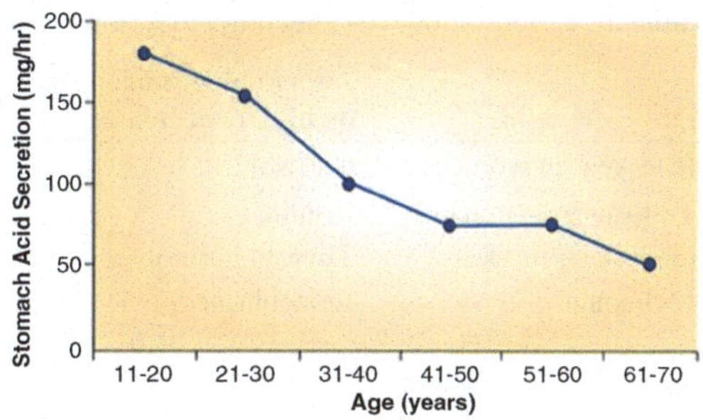

Fig. 4 Graphical analysis of stomach acid secretion w.r.t. age

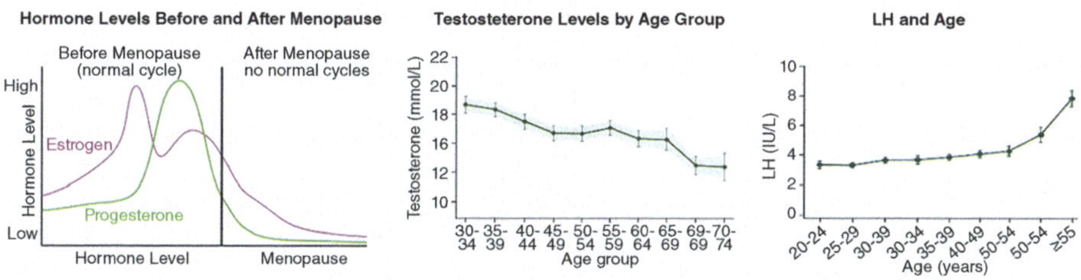

Fig. 5 Graphical representation of various hormonal changes due to aging

esophageal contractions can affect some older adults. Aspiration and problems with dysphagia are the complications faced.

2. **Stomach**: The stomach lining's capacity gets depleted as the age decreases. As per the study, it is noted that the effect of aging on serum gastrin in humans remains controversial, noting a decrease in fasting gastrin levels and others showing no pepsinogen secretion decreases slightly with aging but is not as great as that of acid secretion. Recently, Kim et al. have noted an age-related decrease in bicarbonate secretion in response to luminal acid. The stomach secretes juices such as HCL acid which are responsible for the breakdown of food particles and killing the pathogens such as bacteria and pepsin; thus, due to the process of aging, these working factors tend to slow down as shown in Fig. 4. Atrophic gastritis is the common condition that results in decreased acid secretion.
3. **Intestine**: Aging does not have much effect on the small intestine. However, the level of the lactase gets declined; therefore, dairy products could not be digested by the older adults; thus, they will be broken down by the pathogens such as bacteria in the colon and in turn produce gases. Reduction in the mucosal surface area is thought to be responsible for the impaired calcium absorption.
4. **Pancreas, liver, and gallbladder**: In the digestive system, the pancreas is known to be the second largest gland. Due to aging, there will be a decline in the overall weight of the pancreas and certain tissues might be replaced by fibrosis. The structural and microscopic changes are also visible in the liver as well as in the gallbladder. Physiological aging of endocrine system:

With aging most of the hormone level decreases as shown in Fig. 5. Hormones play a major role in our body. They act as a messengers that control and coordinate

Table 1 Various hormones affecting physiological aging

Decreases aging	Aging remains unchanged or slightly decreases	Increases aging
Estrogen (in women)	Cortisol	Follicle-stimulating hormone (FSH)
Testosterone (in men)	Insulin	Luteinizing hormone (LH)
Growth hormone	Thyroid hormones	Norepinephrine
Melatonin	Epinephrine	Parathyroid hormone

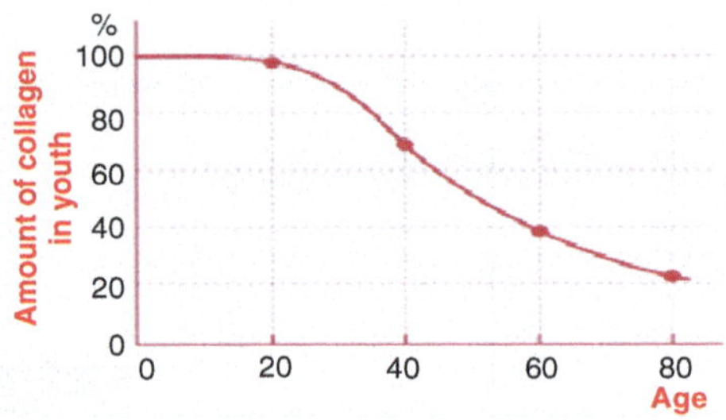

Fig. 6 A graph showing decrease in collagen with age

the activities of our body. Hormones are the chemical substances that are synthesized by our body system. Table 1 shows how various hormones affect physiological aging [1, 2].

Effects of aging on the skin Figure 6 shows the decline in the elastic tissue collagen in a person with respect to their age. Sun, harsh weather, and bad habits such as smoking can cause aging at a faster rate. Gravity plays an important role. When the skin becomes less elastic, gravity makes the eyebrows and eyelids droop. If excess of ultraviolet (UV) rays enters into the skin, it might damage certain fibers such as elastin. Smoking can lead your skin to get wrinkles at a faster rate; thus, the person of the same age who does not smoke tends to appear younger than the person who smokes. This is because oxygen molecules remain unstable or it might be due to over-reactiveness. Over-using of soaps or bathing in hot water can cause your skin to dry and cause faster aging. Skin aging depends on several factor such as lifestyle, diet, heredity, and other personal habits. **Epidermis:** It acts as a protective outer barrier of the skin. The epidermis contains several layers such as skin cells, pigments, and proteins. Due to the aging process, the number of pigment melanocytes decreases. Aging skin looks translucent. **Dermis:** The dermis lies below the epidermis and above the subcutaneous layer. The dermis consists of blood vessels, nerve fibers, hair follicles, and oil glands. The dermis helps the epidermis by enriching nutrients to it. When collagen production in our body gets lower, it could lead to fine lines and wrinkles in the skin. In order to overcome this, there is a therapy that can be used to stimulate and synthesize collagen. This therapy is known as red light therapy. **Physiological aging of nervous system:** Aging is also associated with neurological disorder. Alzheimer's and Parkinson's diseases are the most common neurodegenerative diseases that are known to cause aging. Oxidative stress, apoptosis, is the major cause of aging in the nervous system.

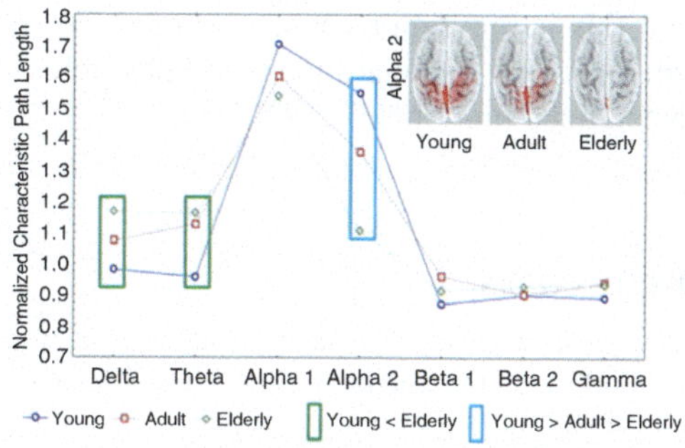

Fig. 7 Analysis of nerve signals for different age groups

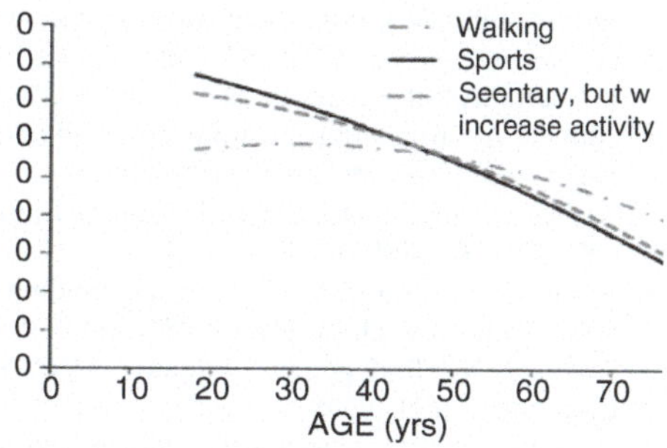

Fig. 8 Decrease in exercise with aging

The aging process modulates the brain connectivity network. Based on the study conducted (Fig. 7) in accordance with respective age from (15–45 years) 36 young were evaluated, from (50–70 years) 46 adult were evaluated, and (>70 years) 31 adult were evaluated. Likewise the EEG evaluation is being done on 113 healthy volunteers and the final report is presented in the pattern young>adult>elder. EEG helps to study the physiological features of the nervous system. The basic goal of neuroscience is to understand the connection between the structural and functional connectivity of the brain. The young brain has much greater capacity to adopt change and learn far better than an adult. **Role of exercise in physiological aging:** By experience, we know that physical activity can boost up your body with a positivity. When no exercise is being given to the body, the aging process can take place in an accelerated level as shown in Fig. 8. A scientist performed an experiment with people who do physical activity 3 hours per week and others who do it less than 60 minutes a week. He collected the sample skin from both people who do physical activity 3 hours per week and 60 minutes per week. In his research, he came with the conclusion that those who exercise 3 hours per week had healthier skin.

3 Conclusion

According to the study conducted in 2010, 29 million female between age 20 and 24 tend to age faster. For 0.8%–1% of aging people happens after the age of 30 due to age or misuse of the body in an unhygienic way. This results in tissue damage and this is the major cause observed for old age. The theoretical graph approach interacts with the anatomical and physiological features of aging pertaining to their specialized function.

References

1. Holloszy, J.O.: The biology of aging. Mayo. Clin. Proc. **75**(Suppl), S3–8 (2000); discussion S8–9
2. Manor, B., Lipsitz, L.A.: Physiologic complexity and aging: implications for physical function and rehabilitation. Prog. Neuropsychopharmacol. Biol. Psychiatr. **45**, 287e93 (2013)
3. Ferrari, A.U., Radaelli, A., Centola, M.: Invited review: aging and the cardiovascular system. J. Appl. Physiol. Bethesda. Md. 1985 **95**, 2591e7 (2003)
4. Ahmed, A., Tollefsbol, T.: Telomeres and telomerase: basic science implications for aging. J. Am. Geriatr. Soc. **49**, 1105–1109 (2001)
5. Alexeyev, M.F., LeDoux, S.P., Wilson, G.L.: Mitochondrial DNA and aging. Clin. Sci. (Lond.) **107**, 355–364 (2004)
6. Allen, J.A., Coombs, M.M.: Covalent binding of polycyclic aromatic compounds to mitochondrial and nuclear DNA. Nature **287**, 244–245 (1980)
7. Ames, B.N.: Endogenous oxidative DNA damage, aging, and cancer. Free Radic. Res. Commun. **7**, 121 (1989)
8. Anisimov, V.N.: Life span extension and cancer risk: myths and reality. Exgerontol. **36**, 1101–1136 (2001)
9. Arking, R., Burde, V., Graves, K., et al.: (2000) Forward and reverse selection for longevity in Drosophila is characterized by alteration of antioxidant gene expression and oxidative damage patterns. Exp. Gerontol. **35**, 167–185
10. Aronheim, J.C.: Handbook of Prescribing Medications for Geriatric Patients, p ix. Little, Brown & Co, New York (1992)
11. Miller, R.A.: Genes against aging. J. Gerontol. A Biol. Sci. Med. Sci. **67A**, 495–502 (2012)
12. West, G.B., Bergman, A.: Toward a systems biology framework for understanding aging and healthspan. J. Gerontol. A Biol. Sci. Med. Sci. **64**, 2005–2008 (2009). GenAge

Multiset Quadripartioned Neutrosophic Cubic Topological Space

J. Sharmila and F. Nirmala Irudayam

1 Introduction

Fuzzy sets, which were introduced by Zadeh, deal with possibilistic uncertainty, connected with imprecision of states, perceptions, and preferences [1]. K. Atanassor [2, 3] in 1983 divided the idea of intuitionistic fuzzy set on a universe X as a generation of fuzzy set. Here besides the degree of membership, a degree of nonmembership for each element is also defined. This paper's goal is to build multiset quadripartitioned neutrosophic cubic topological space, a novel topological space generalization, and investigate its potential applications. Neutrosophic set and topological space have been studied by Salama and Albow. The idea of neutrosophic cubic topological space with quadripartitions in multiple sets has been used to explore various aspects of spatial objects. Multiset quadripartitioned neutrosophic cubic topological qualities, such as closeness, connectedness, compactness, continuous, and separation axioms, have been appropriately applied in this paper, and its applications are being studied.

2 Preliminaries

Definition 1 ([5]) Let X be a non-empty fixed set. A neutrosophic set (NS) A is an object having the form $B = \{< y, T_B(x), I_B(x), F_B(x) >: y \in Y\}$, where $T_B(x)$, $I_B(x)$, and $F_B(x)$ represent the degree of membership function (namely, $\mu_B(x)$), the degree of indeterminacy function (namely, $\sigma_B(x)$), and the degree

J. Sharmila · F. N. Irudayam (✉)
Department of Mathematics, Nirmala College for Women, Coimbatore, India

I. Rajasingh et al. (eds.), *Applied Computational Mathematics*,
Springer Proceedings in Mathematics & Statistics 480,
https://doi.org/10.1007/978-3-031-77764-6_32

of nonmembership function (namely, $\gamma_B(x)$), respectively, of each element x $\in X$ to the set A and $0 \leq \sup T_B(x) + \sup_B(x) + \sup F_B(x) \leq 3$.

Definition 2 ([4]) Let Y be a space of points with a generic element in Y denoted by y. An interval neutrosophic set (INS) B in Y is characterized by truth membership function T_B, indeterminacy membership function I_B, and false membership function F_B. For each point y in Y, T_B(y), I_B(y), F_B(y) $\subseteq [0, 1]$. When Y is continuous, an INS B can be written as

$$\text{B} = \int_y < T(y), I(y), F(y) > y, y \in Y$$

When Y is discrete, an INS B can be written as

$$\text{B} = \sum_{i=1}^{n} < T(y_i), I(y_i), F(y_i) > /y_i, y \in Y$$

Definition 3 ([5]) Let Y be a non-empty set. A neutrosophic cubic set (NCS) in Y is denoted by a pair of $\mathfrak{B} = (B, \Lambda)$, where

$$\text{B} = \{\langle x : B_T(x), B_I(x), B_F(x)\rangle\, x \in X\}$$

is an interval valued neutrosophic set in X and

$$\Lambda = \{\langle x : \lambda_T(x), \lambda_I(x), \lambda_F(x) > x \in X\}$$

is a neutrosophic set in X.

Definition 4 ([3]) A quadripartitioned is defined as dividing or distributing something into four equal pieces or by dividing a quantity or number into its fourth part.

Definition 5 ([3]) Let Y be a non-empty set. A quadripartitioned neutrosophic cubic set (QNCS) in Y is a pair $\mathfrak{B} = (B, \Lambda)$ where

$$\text{B} = \{\langle y : B_T(y), B_C(y), B_I(y), B_F(y) > /y \in Y\}$$

is a quadripartitioned interval neutrosophic set in Y, where B_T, B_C, B_I, B_F are the degrees of truth, contradiction, ignorance, and falsity membership functions, respectively, and $\Lambda = \{< y : \Lambda_T(y), \Lambda_C(y), \Lambda_I(y), \Lambda_F(y) > /y \in Y\}$ is a quadripartitioned neutrosophic set in Y, $\Lambda_T(y), \Lambda_C(y), \Lambda_I(y), \Lambda_F(y)$ are the degrees of truth, contradiction, ignorance, and falsity membership functions, respectively. $0 \leq \text{Sup}\, B_T(y) + \text{Sup}\, B_C(y) + \text{Sup}\, B_I(y) + \text{Sup}\, B_F(y) \leq 4$.

3 Multiset Quadripartitioned Neutrosophic Cubic Set

Definition 6 A multiset quadripartitioned neutrosophic cubic set (MQNCS) represents a set in which one or more members are repeated either with the same or different quadripartitioned neutrosophic cubic components.

Definition 7 Let X be a MQNCS, and a non-empty family τ subsets of Z_X is said to be a multiset quadripartitioned neutrosophic cubic topological space if the following axioms hold:

(i) $N_0, N_1 \in Z_X$.
(ii) $A \cap B \in \tau$, for $A, B \in \tau$.
(iii) $\bigcup_{i \in \theta} A_i \in \tau, \forall \{A_i : i \in \theta\} \in \tau$.

In this case, the pair (Z_X, τ) is called a multiset quadripartitioned neutrosophic cubic topological space (MQNCTS in short), and any multiset quadripartitioned neutrosophic cubic set in τ is known as open multiset quadripartitioned neutrosophic cubic set in Z_X. The elements of τ^C are called closed multiset quadripartitioned neutrosophic cubic set in Z_X.

Definition 8 Let (Z_X, τ) be a multiset quadripartitioned neutrosophic cubic topological space. The space Z_X is said to be connected if Z_X cannot express the union of two disjoint multiset quadripartitioned neutrosophic cubic sets.

Definition 9 Let (Z_X, τ) be a multiset quadripartitioned neutrosophic cubic topological space. The space Z_X is said to be compact if every open cover of Z_X has a finite subcover.

Definition 10 Let (Z_X, τ_1) and (Z_X, τ_2) be two multiset quadripartitioned neutrosophic cubic topological spaces. The MQNCS function $f : (Z_X, \tau_1) \to (Z_X, \tau_2)$ is said to be continuous if for each open multiset quadripartitioned neutrosophic cubic set W of τ_2, the multiset quadripartitioned neutrosophic cubic set $f^{-1}(W)$ is an open subset of τ_1.

Proposition 1

1. *Every finite multiset quadripartitioned neutrosophic cubic set is compact.*
2. *Let* f *be a continuous function from a multiset quadripartitioned neutrosophic cubic topological space* (Z_X, τ_1) *to another multiset quadripartitioned neutrosophic cubic topological space* (Z_X, τ_2)*; the function* f *is said to be homomorphism if* $f(A \cup B) = f(A) \cup f(B)$, *where* A, B $\in \tau_1$ *and* f(A), f(B) $\in \tau_2$.

4 Distance Measure on Multiset Quadripartitioned Neutrosophic Cubic Set

Definition 11 Let X be a universal set. Let A $= \{(x_i, \alpha_A(x_i), \beta_A(x_i), \gamma_A(x_i), \delta_A(x_i)\}$, B $= \{x_i, \alpha_B(x_i), \beta_B(x_i), \gamma_B(x_i), \delta_B(x_i)\}$ be two multiset quadripartitioned neutrosophic cubic sets on X. Define a mapping d:MQNCS $\times$ MQNCS $\rightarrow$ [0, 1] as

$$\mathrm{d}(A, B) = \frac{5}{3n}\sum_{i=1}^{n} \frac{\sin\{\frac{\pi}{6}|\alpha_A(x_i)-\alpha_B(x_i)|\}+\sin\{\frac{\pi}{6}|\beta_A(x_i)-\beta_B(x_i)|\}+\mathrm{Sin}\{\frac{\pi}{6}|\gamma_A(x_i)-\gamma_B(x_i)|\}+\mathrm{Sin}\,\frac{\pi}{6}|\delta_A(x_i)-\delta_B(x_i)|\}}{1+in\{\frac{\pi}{6}|\alpha_B(x_i)-\alpha_B(x_i)|\}+\mathrm{Sin}\{\frac{\pi}{6}|\beta_A(x_i)-\beta_B(x_i)|\}+\mathrm{Sin}\{\frac{\pi}{6}|\gamma_A(x_i)-\gamma_B(x_i)|\}+\mathrm{Sin}\,\frac{\pi}{6}|\delta_A(x_i)-\delta_B(x_i)|\}}$$

Theorem 1 *The following properties are true for the multiset quadripartitioned neutrosophic cubic sets C,D,E:*

$$\mathrm{d(C, D)} \geq 0, \forall C, D \in MQNCS(X)$$

Proof If C, D $\in$ MQNCS(X), then $0 \leq \alpha_C(x_I) \leq 1, 0 \leq \beta_C(x_I) \leq 1, 0 \leq \gamma_C(x_i) \leq 1, 0 \leq \beta_C(x_I) \leq 1$
$\Rightarrow 0 \leq |\alpha_C(x_I) - \alpha_D(x_I)| \leq 1, 0 \leq |\beta_C(x_I) - \beta_D(x_I)| \leq 1, 0 \leq |\gamma_C(x_I) - \gamma_D(x_I)| \leq 1, 0 \leq |\delta_C(x_I) - \delta_D(x_I)| \leq 1.$
$\Rightarrow 0 \leq \mathrm{Sin}\left\{\frac{\pi}{6}|\alpha_C(x_I) - \alpha_D(x_I)|\right\} \leq \frac{1}{2}, \quad 0 \leq \mathrm{Sin}\left\{\frac{\pi}{6}|\beta_C(x_I) - \beta_D(x_I)|\right\} \leq \frac{1}{2}, \quad 0 \leq \mathrm{Sin}\left\{\frac{\pi}{6}|\gamma_C(x_I) - \gamma_D(x_I)|\right\} \leq \frac{1}{2}, \quad 0 \leq \mathrm{Sin}\left\{\frac{\pi}{6}|\delta_C(x_I) - \delta_D(x_I)|\right\} \leq \frac{1}{2}.$

Then, $0 \leq \mathrm{Sin}\left\{\frac{\pi}{6}|\alpha_C(x_I) - \alpha_D(x_I)|\right\} + \mathrm{Sin}\left\{\frac{\pi}{6}|\beta_C(x_I) - \beta_D(x_I)|\right\} + \mathrm{Sin}\left\{\frac{\pi}{6}|\gamma_C(x_I) - \gamma_D(x_I)|\right\}$

$$+\,\mathrm{Sin}\left\{\frac{\pi}{6}|\delta_C(x_I) - \delta_D(x_I)|\right\} \leq \frac{3}{2} \quad \text{and}$$

$0 \leq 1 + \mathrm{Sin}\left\{\frac{\pi}{6}|\alpha_C(x_I) - \alpha_D(x_I)|\right\} + \mathrm{Sin}\left\{\frac{\pi}{6}|\beta_C(x_I) - \beta_D(x_I)|\right\} + \mathrm{Sin}\left\{\frac{\pi}{6}|\gamma_C(x_I) - \gamma_D(x_I)|\right\}$

$$+\,\mathrm{Sin}\left\{\frac{\pi}{6}|\delta_C(x_I) - \delta_D(x_I)|\right\} \leq \frac{5}{2}$$

From the previous two expressions, we have

$$0 \leq \frac{5}{3}\frac{\mathrm{Sin}\{\frac{\pi}{6}|\alpha_C(x_I)-\alpha_D(x_I)|\}+\mathrm{Sin}\{\frac{\pi}{6}|\beta_C(x_I)-\beta_D(x_I)|\}+\mathrm{Sin}\{\frac{\pi}{6}|\gamma_C(x_I)-\gamma_D(x_I)|\}+\mathrm{Sin}\{\frac{\pi}{6}|\delta_C(x_I)-\delta_D(x_I)|\}}{1+\mathrm{Sin}\{\frac{\pi}{6}|\alpha_C(x_I)-\alpha_D(x_I)|\}+\mathrm{Sin}\{\frac{\pi}{6}|\beta_C(x_I)-\beta_D(x_I)|\}+\mathrm{Sin}\{\frac{\pi}{6}|\gamma_C(x_I)-\gamma_D(x_I)|\}+\mathrm{Sin}\{\frac{\pi}{6}|\delta_C(x_I)-\delta_D(x_I)|\}}$$

This implies

$$0 \leq \frac{5}{3n}\sum_{i=1}^{n} \frac{\mathrm{Sin}\{\frac{\pi}{6}|\alpha_C(x_I)-\alpha_D(x_I)|\}+\mathrm{Sin}\{\frac{\pi}{6}|\beta_C(x_I)-\beta_D(x_I)|\}+\mathrm{Sin}\{\frac{\pi}{6}|\gamma_C(x_I)-\gamma_D(x_I)|\}+\mathrm{Sin}\{\frac{\pi}{6}|\delta_C(x_I)-\delta_D(x_I)|\}}{1+\alpha_C(x_I)-\alpha_D(x_I)|\}+\mathrm{Sin}\{\frac{\pi}{6}|\beta_C(x_I)-\beta_D(x_I)|\}+\mathrm{Sin}\{\frac{\pi}{6}|\gamma_C(x_I)-\gamma_D(x_I)|\}+\mathrm{Sin}\{\frac{\pi}{6}|\delta_C(x_I)-\delta_D(x_I)|\}} \leq 1$$

Thus, $0 \leq \mathrm{d(A, B)} \leq 1$.
Hence, the theorem is proved.

5 Methodology for Metric Distance Measure Multiset Quadripartitioned Neutrosophic Cubic Set

This section presents a methodological solution employing sine metric for quadripartitioned neutrosophic cubic multi-attribute decision-making problems. The procedures listed below are required for the suggested methodological approach in order to choose the appropriate attributes and alternatives.

Step 1: Problem field selection

Consider the multi-attribute decision-making problem with one conditioned attributes $K_1, K_2, K_3, \ldots\ldots\ldots\ldots\ldots K_l$, n decision attributes $\{P_1, P_2, P_3 \ldots\ldots\ldots P_n\}$; m alternatives are $\{F_1, F_2, F_3 \ldots\ldots\ldots F_m\}$ (Tables 1 and 2).

m and a $= 1, 2$. Here all the attributes k_{xy} and p_{ay}(x $= 1, 2$; y $= 1, 2$) are all quadripartitioned neutrosophic cubic sets.

Step 2: Distance measure for alternatives P_i and decision attributes F_j by using the following distance measure:

$$\mathrm{d}\left(P_i, F_j\right) = \frac{5}{3n}\sum_{l=1}^{k} \mathrm{Sin}\left\{\frac{\pi}{6}\left|\alpha_{P_i}(k_l) - \alpha_{F_j}(k_l)\right|\right\} + \mathrm{Sin}\left\{\frac{\pi}{6}\left|\beta_{P_i}(k_l) - \beta_{F_j}(k_l)\right|\right\} + \sum_{l=1}^{k} \mathrm{Sin}\left\{\frac{\pi}{6}\left|\gamma_{P_i}(k_l) - \gamma_{F_j}(k_l)\right|\right\} + \sum_{l=1}^{k} \mathrm{Sin}\left\{\frac{\pi}{6}\left|\delta_{P_i}(k_l) - \delta_{F_j}(k_l)\right|\right\}$$

where i $= 1, 2, 3 \ldots\ldots$ m, j $= 1, 2, 3 \ldots\ldots$ n, and K $= \{k_i\}_{i=1}^{s}$.

Step 3: Enter the estimated distance measures for the alternatives and attributes in the following table (Table 3):

Table 1 Multi-attribute decision-making problem with one conditioned attributes with m alternatives

	F_1	F_2			F_m
K_1	(k_{11})	(k_{12})	.	.	(k_{1m})
K_2	(k_{21})	(k_{22})			(k_{2m})
.	.	.			.
.	.	.			.
K_l	(k_{l1})	(k_{l2})			(k_{lm})

Table 2 Multi-attribute decision-making problem with one conditioned attributes and n decision attributes

	K_1	K_2	.	.	K_l
P_1	(p_{11})	(p_{21})	.	.	(p_{1l})
P_2	(p_{21})	(p_{22})	.	.	(p_{2l})
.	.	.	.	.	.
.	.	.	.	.	.
P_n	(p_{n1})	(p_{11})	.	.	(p_{nl})

Table 3 Multi-attribute decision-making problem with one n decision attributes; m alternatives

F_j	F_1	F_2	.	.	.	F_n
P_i	d (F_1, P_1)	d (F_1, P_2)	.	.	.	d (F_n, P_1)
P_1	d (F_2, P_2)	d (F_2, P_2)	.	.	.	d (F_n, P_2)
P_2	.	.	.	.	.	
.	.	.	.	.	.	d (F_n, P_m)

Step 4: Problem decision
From the distance measure table, choose the attributes F_j, j = 1, 2 . . . . . . n for the alternatives P_i, i = 1, 2 . . . . .m by which the lowest distance measure value between the alternatives P_i and the attributes F_j and then conclude the attribute F_j is the best attribute for alternative P_i.

6 Application of Sin Metric Distance Measure Quadripartitioned Neutrosophic Cubic Set

Nowadays, a significant multi-criteria decision-making challenge for individuals, particularly on the career and technical fronts, revolves around the selection of laptops. Laptops play a vital role in professional environments, making their selection a critical decision. This section explores a real-life application involving the selection of an appropriate laptop based on its features and price, using the sine metric distance measure within the framework of quadripartitioned neutrosophic cubic sets.

Step 1: Selection of the problem field
Let K = $\{k_1, k_2, k_3, k_4\}$ be the kind of laptops, F = $\{F_1, F_2, F_3, F_4\}$ be the features of the laptops, and P = $\{P_1, P_2, P_3, P_4\}$ be the price of laptops. Tables 4 and 5 presents information about the kinds of laptops and features. Here the example is denoted by degree of membership, degree of ignorance membership, contradiction of the membership, anddegree of nonmembership.

Table 4 Ranges of feature values for different Laptop

Features of Kinds of laptops	F_1	F_2	F_3	F_4
K_1	([0.1, 0.2], [0.1, 0.3] [0.2, 0.4], [0.5, 0.6])	([0.2, 0.3], [0.2, 0.4] [0.3, 0.4], [0.3, 0.6])	([0.0, 0.1], [0.0, 0.2] [0.0, 0.3], [0.0, 0.4])	([0.3, 0.5], [0.4, 0.5] [0.2, 0.5], [0.1, 0.5])
K_2	([0.4, 0.6], [0.4, 0.5] [0.4, 0.7], [0.4, 0.8])	([0.5, 0.6], [0.4, 0.6] [0.3, 0.6], [0.2, 0.6])	([0.1, 0.3], [0.1, 0.2] [0.2, 0.3], [0.3, 0.7])	([0.0, 0.3], [0.1, 0.3] [0.2, 0.3], [0.3, 0.4])
K_3	([0.1, 0.4], [0.1, 0.2] [0.1, 0.3], [0.1, 0.5])	([0.2, 0.5], [0.2, 0.6] [0.2, 0.4], [0.2, 0.3])	([0.3, 0.4], [0.3, 0.6] [0.3, 0.5], [0.3, 0.7])	([0.1, 0.5], [0.1, 0.6] [0.1, 0.8], [0.1, 0.7])
K_4	([0.5, 0.6], [0.5, 0.7] [0.5, 0.8], [0.5, 0.9])	([0.0, 0.1], [0.2, 0.3] [0.1, 0.2], [0.3, 0.4])	([0.4, 0.5], [0.6, 0.7] [0.7, 0.8], [0.2, 0.3])	([0.1, 0.4], [0.2, 0.4], [0.3, 0.4], [0.0, 0.4])

Table 5 Specific feature values for different Laptops

Features of kinds of laptops	F_1	F_2	F_3	F_4
K_1	(0.1, 0.2, 0.4, 0.6)	(0.5, 0.9, 0.1, 0.5)	(0.2, 0.2, 0.3, 0.4)	(0.7, 0.7, 0.3, 0.3)
K_2	(0.1, 0.2, 0.2, 0)	(0.6, 0, 0, 0)	(0, 0, 0, 0.1)	(0.3, 0.2, 0.1, 0)
K_3	(0.3, 0.3, 0.3, 0.3)	(0.5, 0.5, 0.5, 0)	(0.2, 0.1, 0.1, 0)	(0.1, 0, 0, 0.4)
K_4	(0.3, 0.2, 0.9, 0.1)	(0.1, 0, 0, 0.3)	(0.6, 0.5, 0.2, 0.1)	(0.3, 0.6, 0.7, 0.2)

Table 6 Price ranges for different Laptops

Kinds of price of laptops	K_1	K_2	K_3	K_4
P_1	([0.1, 0.3], [0.0, 0.2] [0.1, 0.4], [0.2, 0.3])	([0.1, 0.2], [0.0, 0.3] [0.2, 0.4], [0.2, 0.3])	([0.3, 0.4], [0.6, 0.7] [0.7, 0.8], [0.8, 0.9])	([0.1, 0.5], [0.1, 0.6] [0.2, 0.3], [0.3, 0.4])
P_2	([0.3, 0.5], [0.5, 0.6] [0.2, 0.3], [0.1, 0.4])	([0.0, 0.4], [0.1, 0.3] [0.3, 0.4], [0.5, 0.6])	([0.2, 0.4], [0.4, 0.6] [0.6, 0.8], [0.0, 0.2])	([0.3, 0.5], [0.1, 0.3] [0.5, 0.7], [0.7, 0.9])
P_3	([0.1, 0.2], [0.4, 0.5] [0.6, 0.7], [0.0, 0.1])	([0.2, 0.3], [0.1, 0.3] [0.0, 0.3], [0.4, 0.6])	([0.2, 0.3], [0.3, 0.4] [0.4, 0.5], [0.1, 0.4])	([0.1, 0.3], [0.2, 0.6], [0.3, 0.7], [0.1, 0.6])
P_4	([0.8, 0.9], [0.1, 0.8] [0.0, 0.1], [0.1, 0.2])	([0.0, 0.4], [0.0, 0.3] [0.4, 0.5], [0.4, 0.6])	([0.0, 0.2], [0.0, 0.3] [0.0, 0.6], [0.0, 0.7])	([0.4, 0.5], [0.4, 0.6], [0.3, 0.5], [0.3, 0.6])

Table 7 Specific price values for different Laptops

Kinds of price of laptops	K_1	K_2	K_3	K_4
P_1	(0.1, 0.3, 0.2, 0)	(0.4, 0.5, 0.6, 0.6)	(0.4, 0.5, 0.1, 0.2)	(0.1, 0.3, 0.4, 0.5)
P_2	(0, 0.1, 0.2, 0.7)	(0, 0.2, 0.2, 0.4)	(0.6, 0.7, 0.2, 0.1)	(0.3, 0.4, 0.5, 0)
P_3	(0.8, 0.7, 0, 0)	(0, 0, 0.2, 0.3)	(0.1, 0.2, 0.4, 0.6)	(0.1, 0.6, 0, 0.7)
P_4	(0.1, 0.3, 0.2, 0.2)	(0.4, 0.4, 0.1, 0.3)	(0.8, 0.6, 0.4, 0.2)	(0, 0, 0, 0.1)

Step 2: Distance measure for the alternatives and attributes

The distance measure between each price and features with reference to the laptops (Tables 6 and 7):

$$\mathrm{d}\left(P_i, F_j\right) = \frac{5}{3n}\sum_{l=1}^{k} \mathrm{Sin}\left\{\frac{\pi}{6}\left|\alpha_{P_i}(k_l) - \alpha_{F_j}(k_l)\right|\right\} + \mathrm{Sin}\left\{\frac{\pi}{6}\left|\beta_{P_i}(k_l) - \beta_{F_j}(k_l)\right|\right\}$$

$$+\sum_{l=1}^{k} \mathrm{Sin}\left\{\frac{\pi}{6}\left|\gamma_{P_i}(k_l) - \gamma_{F_j}(k_l)\right|\right\} + \sum_{l=1}^{k} \mathrm{Sin}\left\{\frac{\pi}{6}\left|\ \delta_{P_i}(k_l)\right.\right.$$

$$\left.\left. -\delta_{F_j}(k_l)\ \right|\right\}$$

where i = 1, 2, 3 . . . ..m, j = 1, 2, 3 . . . ..n, and K = $\{k_i\}_{i=1}^{S}$.

The distance measure between each price and features with reference to the laptops in interval neutrosophic values:

$d(F_1, P_1) = 0.5739305165$, $d(F_1, P_2) = 0.3112891872$, $d(F_1, P_3) = 0.0.3666926311$

$d(F_1, P_4) = 0.3546245421$, $d(F_2, P_1) = 0.6705700108$, $d(F_2, P_2) = 0.2896884343$

$d(F_2, P_3) = 0.2744346591$, $d(F_2, P_4) = 0.3394397722$, $d(F_3, P_1) = 0.351580711$

$d(F_3, P_2) = 0.22736127048$, $d(F_3, P_3) = 0.2706533498$, $d(F_3, P_4) = 0.34041796064$

$d(F_4, P_1) = 0.296661370$, $d(F_4, P_2) = 0.0288600289$, $d(F_4, P_3) = 0.2877122877$

$d(F_4, P_4) = 0.2624528250$

The distance measure between each price and features with reference of the laptops in neutrosophic values:
$d(F_1, P_1) = 0.331088664$, $d(F_1, P_2) = 0.232487922$, $d(F_1, P_3) = 0.046536796$,
$d(F_1, P_4) = 0.91071428$, $d(F_2, P_1) = 0.391806722$, $d(F_2, P_2) = 0.3354978355$,
$d(F_2, P_3) = 0.2799422799$, $d(F_2, P_4) = 0.3900850056$, $d(F_3, P_1) = 0.4015700483$,
$d(F_3, P_2) = 0.3219947491$, $d(F_3, P_3) = 0.3060780216$, $d(F_3, P_4) = 0.6758405309$,
$d(F_4, P_1) = 0.5630866708$, $d(F_4, P_2) = 0.3940615836$, $d(F_4, P_3) = 0.2916666667$,
$d(F_4, P_4) = 0.6224167162$

Step 4: Problem decision

Based on the above calculations, the smallest measure values for each parameter are as follows:

For P1, the smallest measure value is 0.296661370, corresponding to Feature F4.

For P2, the smallest measure value is 0.0288600289, corresponding to Feature F4.

For P3, the smallest measure value is 0.2706533498, corresponding to Feature F3.

For P4, the smallest measure value is 0.2624528250, corresponding to Feature F4.

Thus, selecting a suitable laptop based on price and features includes P1, P2, P4, and Feature F4.

In another round of calculations:

For P1, the smallest measure value is 0.331088664, corresponding to Feature F1.

For P2, the smallest measure value is 0.232487922, corresponding to Feature F1.

For P3, the smallest measure value is 0.046536796, corresponding to Feature F1.

For P4, the smallest measure value is 0.3900850056, corresponding to Feature F2.

Therefore, the final selection of a suitable laptop depends on P1, P2, P3, and Feature F1, as they have the smallest measure values. Additionally, P4 is associated with Feature F2, which can also be considered in specific scenarios.

7 Conclusion

Selection of the best laptop depends on the price and the features are P_1, P_2, P_3, F_1 and P_1, P_2, P_3, F_1.

References

1. Arokia Dasan, M., Little Flower, V.F., Tubax, X.: Multi-Attriute Decision-Making Problems in career determination using single-valued neutrosophic distance measure, Research Square
2. Das, R., Tripathy, B.: Neutrosophic multiset topological space. Neutrosophic Sets Syst. **35**, 142–152 (2020)
3. Sharmila, J., Nirmala Irudayam, F.: Quadripartitioned neutrosophic cubic set. Int. J. Creative Res. Thoughts (IJCRT) 11, 2320–2882 (2023)
4. Jun, Y.B., Kim, C.S., Yang, K.O.: Cubic sets. Ann. Fuzzy Math. Inform. **4**(1), 8398 (2012)
5. Jun, Y.B., Smarandache, F., Kim, C.S.: Neutrosophic cubic set. New Math. Nat. Comput. **13**(1), 41–54 (2017)

A Theoretical Graph Analysis on EEG-Controlled Bionic Arm with Importance to Senses of Human

A. Kaviarasu and Sharmila Mary Arul

1 Introduction

The most important part of the human body is our brain. Next to our brain, the human hand plays a major role in our body. The human hand is one of the complex and fascinating parts of our body. It is significantly used for the interaction between the external world and our body. It also helps humans to work and is the most unique part of our body and its functionality too plays a unique role in humans compared to the other living organisms in this planet because of its flexible nature to work effectively. The most versatile organ is the human hand because the thumb finger alone is opposable to the other four fingers in our hand. This versatility makes us unique from other primates who possess also five fingers without opposition in positioning. This makes our hand more versatile than other primates. This helps us to do more functionality works. The thumb tends to take more responsibility in our hand. Bionic arm plays a major role in replacing the amputee hand limb. Research studies show that the lack of sense of human touches is a deficiency in the field of bionic limb. Without the sense of touch, there is no use in the bionic hand. EEG receives the data from the nerve impulses and then the bionic limb analyzes the data by using machine learning. This leads to learn the user's movement, which is utilized by the bionic hand to flexibly move. This project consumes lesser cost of building than other bionic limbs, including the sensory feedback of touches. It will reduce the price of bionic hand.

A. Kaviarasu
Master of Science in Robotics, Northeastern University, Boston, MA, USA

S. M. Arul (✉)
Department of Mathematics, Saveetha School of Engineering, Saveetha Institute of Medical and Technical Sciences, Chennai, India
e-mail: sharmilamaryarul.sse@saveetha.com

I. Rajasingh et al. (eds.), *Applied Computational Mathematics*,
Springer Proceedings in Mathematics & Statistics 480,
https://doi.org/10.1007/978-3-031-77764-6_33

2 Overview

In later investigations, the World Health Organization (WHO) revealed that nearly 15% are suffering from many types of handicap [1]. The number of patients who lose limbs are increasing because of many different reasons [2]. In spite of availability of the bionic arm, people do not feel comfort as much as a human arm [3]. The primary objective for these kinds of problems is to reconstruct a bionic arm suitable to users. Thus, we have theoretically developed a bionic arm that can be controlled by the mind and sense the feeling of touch to the users [4]. This bionic arm has the capability to lift nearly 45 pounds which is constructed with 26 joints. This can be more helpful for amputees. Although the human hand possesses unique features, many accidents occur, leading to the loss of its versatile functionality for numerous individuals. In addition to that, most of the people lose their hand due to disabilities such as diseases and infections Most people lose their hand due to the following reasons:

- Diseases such as peripheral vascular disease (PVD), diabetes, blood clots, or osteomyelitis.
- Injuries, particularly arm limb.
- Trauma is the most common cause of upper extremity [7] amputation, which is up to 75%.
- Surgeries conducted to remove tumors from bones and muscles.
- Thickening of nerve tissues, called a neuroma.

They not only lose their hand but also lose their whole life. This is brutal for amputees. They cannot even go to work to earn a living and support their families. More than 30% of amputees is suffering from depression. The following effects of amputation are based on physical:

Physical Effects
- Infection may take place
- Fatigue
- Deep vein thrombosis
- Muscle contractures
- Stump and phantom limb pain
- Mobility and dexterity issues

Emotional Effects
- Traumatic effects
- Adapting to amputation
- Body image concerns due to abnormal appearances
- Inability to be involved in social activities
- Difficulty moving forward
- Decreased self-esteem
- Increased dependency on others
- Significant levels of social isolation

Table 1 Estimates of amputee population (11/09/2008)

Name of the category	Counts or percentages
Incidence of surgical removal of limb	15 per 10,000
Counts in amputees globally	10 million
Rate of hand amputees	Nearly 0.3
Arm amputees globally	3 million
Arm amputees in economically developing nation	2.4 million

Table 2 Incidence of surgical removal of limb

Name of the part of the body	Percentage
Hand/wrist	5%
Shoulder	8%
Above the elbow and elbow disarticulation	28%
Below the elbow	59%

Table 3 Number of arm amputations in developing countries by level

Name of the part of the body	Number of cases (in million)
Hand/wrist	0.1
Shoulder	0.2
Above the elbow and elbow disarticulation	0.7
Below the elbow	1.5

Nowadays, our world has seen a number of cases of amputation. Many of them have been suffering from hand amputation. Such amputation challenges their day-to-day life. This often causes anxiety, which leads to depression. Anxiety is the most brutal factor of amputation. They are often seeking for a better life to create (Tables 1 and 2):
According to the statistics, which has been surveyed by the Amputee Coalition™, taken in the United States in 1996, nearly 1.285 million people lost their limb in the United States. The birth commonness of innate limb deficiency in 1996 was 25.64 per 100,000 live births. The commonness rate is most elevated among individuals of age 65 years and older, 19.4 per 1000.

When an arm or limb is amputated severely or lost, the bionic hand plays a major role in replacing the amputated or lost hand. The main objective of the bionic limb is to replace the defected natural arm of amputees. This provides them a comfortable life and ease their sorrows due to the affected arm. This bionic arm is made for reproducing sensory motor abilities to the natural arm (Table 3).

The bionic arm is a device that resembles the function of the natural hand. It is designed to support and supplement the function of the existing limb or body part. This bionic arm is the exchange or improvement of the natural hand by electromechanical methods and works very closely with the natural hand of humans.

We are going to exchange this defected arm with a bionic arm with the sense of touch. This helps them control all their work through the bionic arm by using

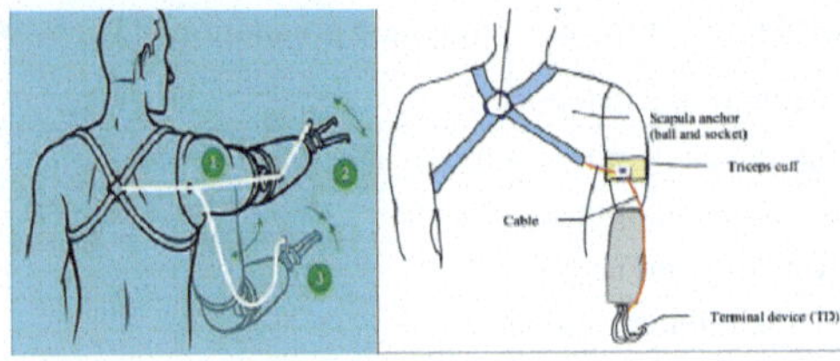

Fig. 1 Body-powered bionic hand

electroencephalography (EEG), and also it has the ability to sense the surroundings, shape, temperature, etc. by using electromyography (EMG) (Fig. 1).

3 Classification of Bionic Hand

Bionic hand can be classified as follows: (1) Body-powered bionic hand (2) Externally powered bionic hand

3.1 Body-Powered Bionic Hand

A body-powered bionic hand depends on the system of cable along with manual controls. This type is also used to regulate the limb itself. It can be used to control the bionic arm through other parts of the body such as the shoulders, elbows, and chest. Advantages of body-powered bionic arms:

- They have a tendency to be cheaper than myoelectric [8] types .
- They do not rely on an external power source to work.
- Body-powered bionic arms also make a superb secondary device.
- They are great for use in more rugged environments.
- Body-powered devices produce very little noise.
- Other advantages include a low starting expense, low maintenance costs, and simpler fixes.

Disadvantages of body-powered bionic arms:

- This feature typically comes with a sacrifice in fine motor skills and precision, so it is not the best option for everyone.
- Enhancing armpit squeeze.
- Problem of compression neuropathy.
- High restriction in framework mechanism.
- High restriction in grabbing spaces.
- Higher need for movement of the whole body.

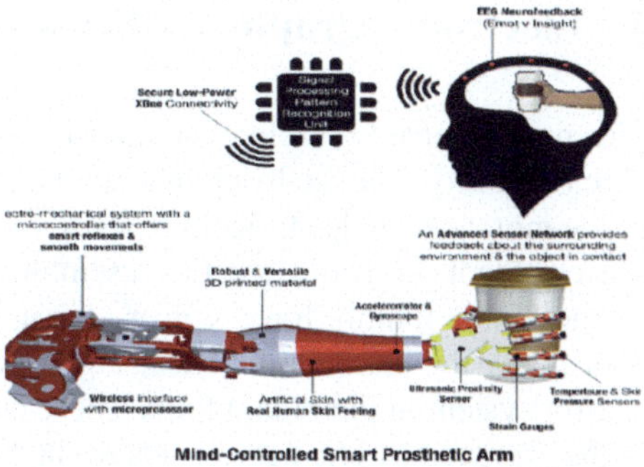

Fig. 2 Externally powered bionic arm by EEG control

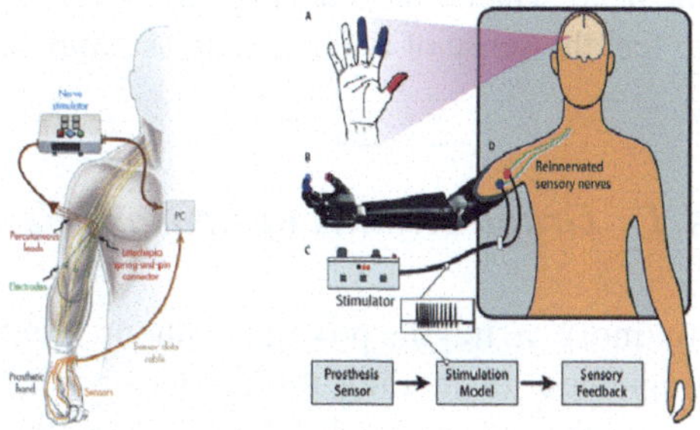

Fig. 3 Process of EMG nerve stimulator

- Increased prospects for health issues.
- Higher need for careful handling by patients.

Thus, we are not going to consider this type of bionic hand in this project model due to the disadvantages of body-powered bionic hands (Fig. 2).

3.2 *Externally Powered Bionic Hand*

In this project, we are using this method to develop bionic hands. The electrical process controls externally powered bionic hands. These are controlled by the electric power and are not controlled by the strength of the muscles of the patients. The bionic hand utilizes the signals, which are received from the brain waves from the brain scalp through electroencephalography (EEG). Electrodes are placed on the head to receive the signals. This stimulates a motor that controls a bionic hand such as the fingers, wrist, and elbows. This motor modulates the momentum of the bionic hand, such as contraction or relaxation of the elbow, or extending and holding of the fingers of the bionic hand. This method is more efficient than the body-powered method. Bionic hands gave improved flexibility and higher grabbing power because of the battery-driven motor that is used to operate these bionic hands. This method is very useful for fast responses due to the availability of the microprocessor. This ability helps them to control their bionic arm voluntarily (Fig. 3).

4 Electromyography for Sense of Feelings

This method enables biological muscles and bionic hands to interface neurally with high reliability. The feedback of artificial [5, 6] sensory enhances the exact function of the motor and helps to feel the sense of touch of an object via a bionic arm. This sensory feedback is designed to resemble the natural neural [9] movement system. At the bionic thumb, force sensors are used to increase the sense of touch, which is made by the bionic arm. These signals are converted into electrical signals by a control system in the bionic limb. These electrical signals are transferred to a nerve in the arm stump. The nerve leads to the brain to collect the data from the nerve in the hand. Then it feels the pressure levels against the hand. This enables the patients to feel the sense of touches such as temperature, shape, and grasping the object.

4.1 *Graph Results of Sensory Feedbacks by EMG*

To move a fragile product without breakage, EMG is used to understand the fragileness of holding the product by feeling the fragility. The following graph shows the sensory feedback to improve the holding nature of object: The following graph shows the performance to analyze various objects (Figs. 4 and 5):

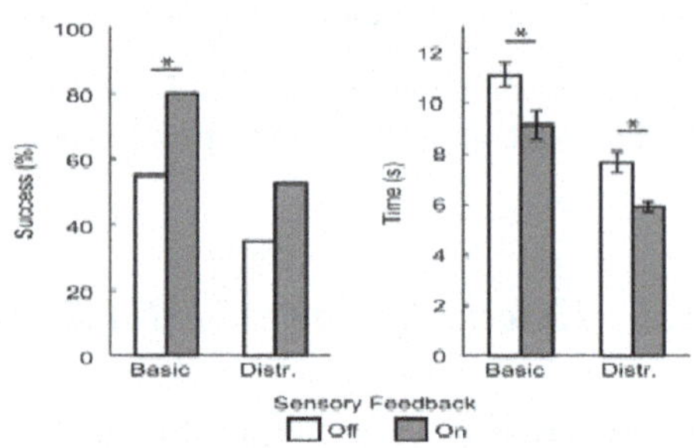

Fig. 4 Sensory feedback to improve the holding nature of object

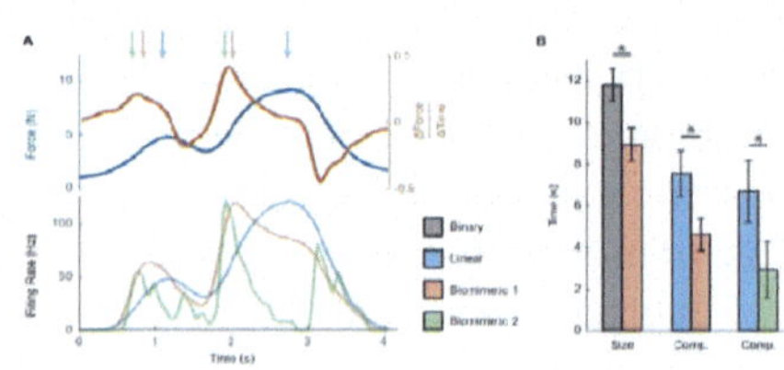

Fig. 5 The performance to analyze various object

5 System Techniques

5.1 Sketch of Bionic Arm

The EEG deployed the bionic hand with the sense of touch and EMG-controlled technology (Fig. 6). This system consists of EEG headset, EMG technology, embedded sensors, Arduino mega board, Raspberry Pi board, XBee module, 3d printed bionic arm, and servomotors. The EEG headset, which communicates and controls the bionic hand, receives the signals of brain waves from the brain through the electrodes that are placed on the head. These signals are transferred to the signal-processing unit through the Bluetooth communication channel, which consists of Wi-Fi communication and XBee module. The signal-processing unit sent signals to the controlling components, which are embedded, in the bionic hand. The controlling components consist of the Arduino board and Raspberry Pi 4 board where both interacted with each other. A total of nine servomotors is installed in the bionic hand. These servomotors are controlled by the Arduino mega board. This part performs a communication signal from the user to the embedded sensor in the bionic hand. After receiving the data from the controlling components, it is transferred to the bionic to make action as per the user's guide. The 3D printed bionic hand is constructed by using lightweight, high-strength materials. It can handle fragile products, soft products, hard materials, etc. The bionic arm can receive the sensation of touches through the EMG device technology, which is also inbuilt in the bionic arm. The EMG device receives the signals through the sensors, which are placed

Fig. 6 Bionic arm materials

Fig. 7 Torque versus speed

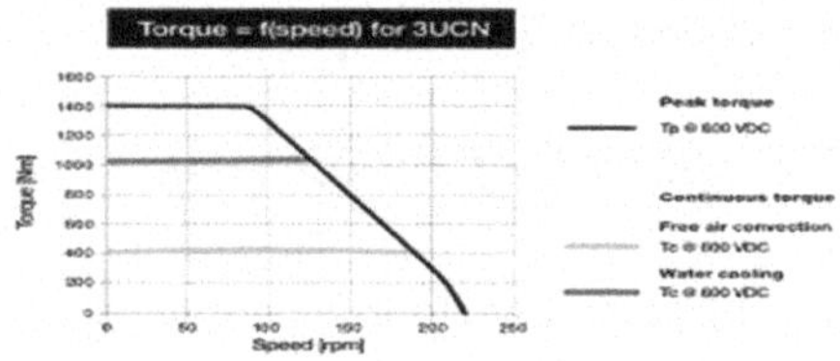

in the bionic arm, and transfers it into the nerves of the human hand. The nervous system will recognize the signal and decode the digital signal to their understanding language exactly similar to the natural nervous system. Here machine learning plays a major role for training the patients to particular arm activities in agreement with the set of mind states. It is done a number of times to learn the machine's activities to get the ideal results as much as possible (Fig. 7).

6 Mechanism of Bionic Hand

In this project, the bionic arm needs some mechanism controls based on the mechanics of the servomotors. To carry the loads by the bionic arm, we have to calculate its momentum and torque, which shows its capability to carry the load. The following formula (1) is used here for a static body:

$$+M = (L1 \times \cos(\theta 1) + L2 \times \cos(\theta 2)) \times 9.8 \tag{1}$$

Here,

- M is the momentum;
- $L1$ is the length between the force and the first angle $\theta 1$;
- $L2$ is the length between the force and the second angle $\theta 2$.

The following formula (2) is used here for a dynamic body, neglecting frictional force:

$$M = F \times l + J \times \alpha \tag{2}$$

Here,

- F is the force applied on the bionic hand;
- l is the length of the bionic arm;
- J is the polar moment of inertia;
- α is the angular acceleration or polar acceleration.

The equation (2) used here for dynamic motion changes from one region to another region such as the wrist, finger, elbow, or forearm.

7 Conclusion

The bionic hand developed here is greatly related to the purposes of the field of medicine as well as other areas. These bionic hands will change the lives of amputees with great potential, especially army soldiers who lost their hands in wars. These bionic hands are not only useful for amputee patients but also for many industrial and commercial applications. The bionic hand, which advances with the sense of touch, plays a major role as a great helper or caregiver for amputees. The bionic hand we developed possesses the feeling of touch, giving a second life to amputees with greater potential. These bionic arms can be used for many purposes in household areas such as cooking, bathing, and dressing, as part of their daily life activities. In this project, we have used a combination of advanced technology of EEG with EMG. This combination makes a significant advancement in our project, allowing it to enter the modern technical world.

References

1. Hawking, S.W.: World Report on Disability. World Health Organization, Geneva (2011)
2. NBC News: Limb loss a grim, growing global crisis, 20 Mar 2010. http://haitiamputees.nbcnews.com/news/2010/03/19/4040341-limb-loss-agrim-growing-globalcrisis
3. LeBlanc, M.: Give Hope-Give a Hand, 14 Jan 2011. https://web.stanford.edu/class/engr110/Newsletter/lecture03a2011.html
4. Moreton, C.: London 2012 Olympics: Oscar Pistorius finally runs in Games after five year battle, 4 Aug 2012. http://www.telegraph.co.uk/sport/olympics/athletics/9452280/London2012-Olympics-OscarPistorius-finally-runs-in-Games-after-five-yearbattle.html
5. Bennett Wilson, A. Jr.: Retrieved Oct 17, 2015 (n.d.). http://www.oandplibrary.org/alp/chap01-01.asp
6. Kuiken, T.A., Li, G., Lock, B.A., Lipschutz, R.D., Miller, L.A., Stubblefield, K.A., Englehart, K.: Targeted muscle reinnervation for real-time myoelectric control of multifunction artificial arms. JAMA J. Am. Med. Assoc. **301**, 619–628 (2009)
7. Kung, T.A., Bueno, R.A., Alkhalifah, G.K., Langhals, N.B., Urbanchek, M.G., Cederna, P.S.: Innovations in prosthetic interfaces for the upper extremity. Plast. Reconstr. Surg. **132**, 1515–1523 (2013)
8. Schiefer, M., Tan, D., Sidek, S.M., Tyler, D.J.: Sensory feedback by peripheral nerve stimulation improves task performance in individuals with upper limb loss using a myoelectric prosthesis. J. Neural Eng. **13**, 016001 (2016)
9. Tan, D.W., Schiefer, M.A., Keith, M.W., Anderson, J.R., Tyler, D.J.: A neural interface provides long-term stable natural touch perception. Sci. Transl. Med. **6**, 257ra138 (2014)

Correction to: Applied Computational Mathematics

Indra Rajasingh, Sharmila Mary Arul, and Diana Grace Thomas

Correction to:
I. Rajasingh et al. (eds.), *Applied Computational Mathematics*, Springer Proceedings in Mathematics & Statistics 480, https://doi.org/10.1007/978-3-031-77764-6

The book was inadvertently published with low-resolution figures on pages 248–251, 253, and 270–275. These have now been replaced with high-resolution versions.

Fig. 1

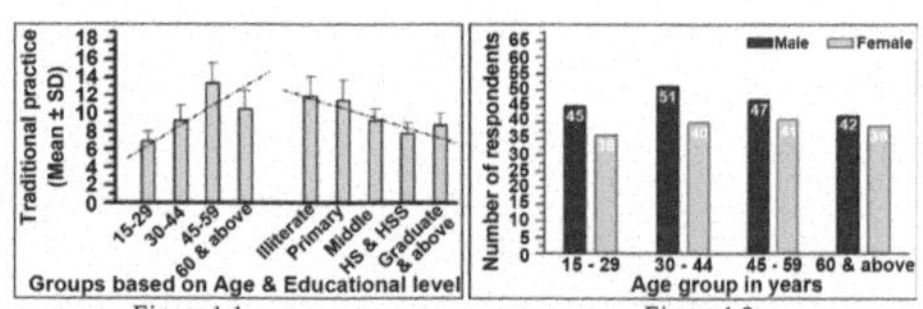

Figure 1.1 Figure 1.2

The updated versions of these chapters can be found at
https://doi.org/10.1007/978-3-031-77764-6_29
https://doi.org/10.1007/978-3-031-77764-6_31

I. Rajasingh et al. (eds.), *Applied Computational Mathematics*,
Springer Proceedings in Mathematics & Statistics 480,
https://doi.org/10.1007/978-3-031-77764-6_34

Fig. 2

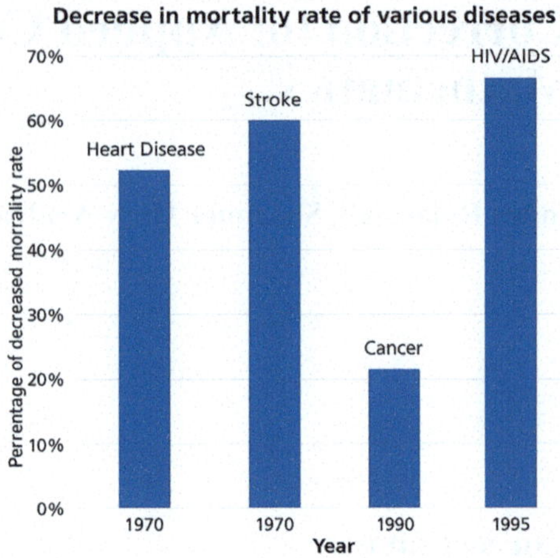

Fig. 3

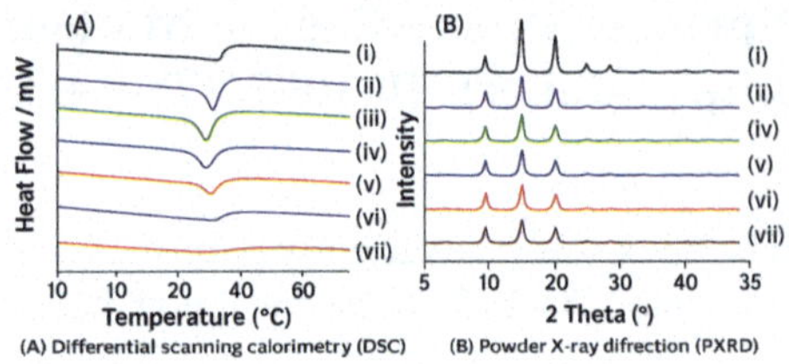

Fig. 4

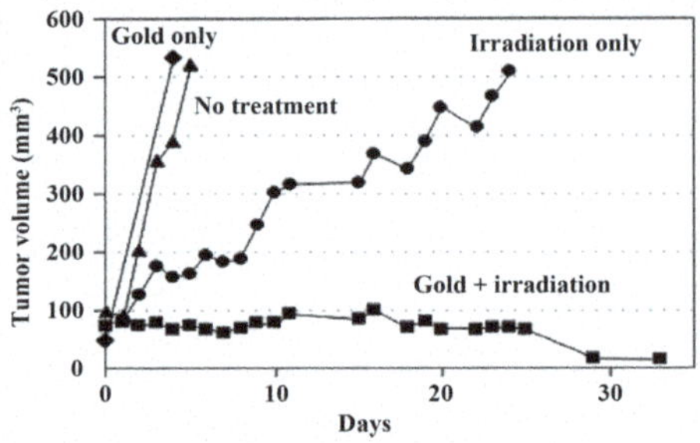

Fig. 5

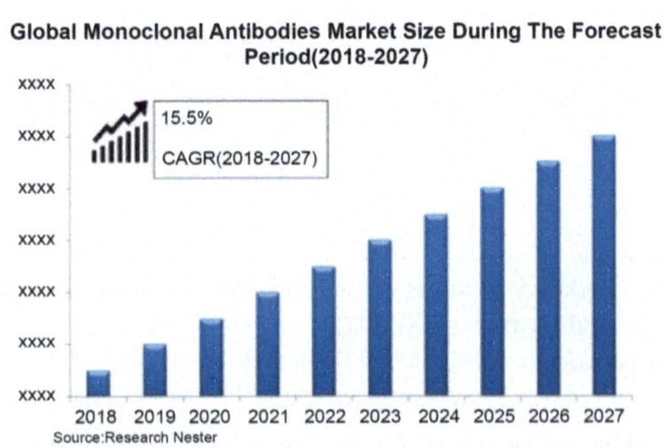

Fig. 6

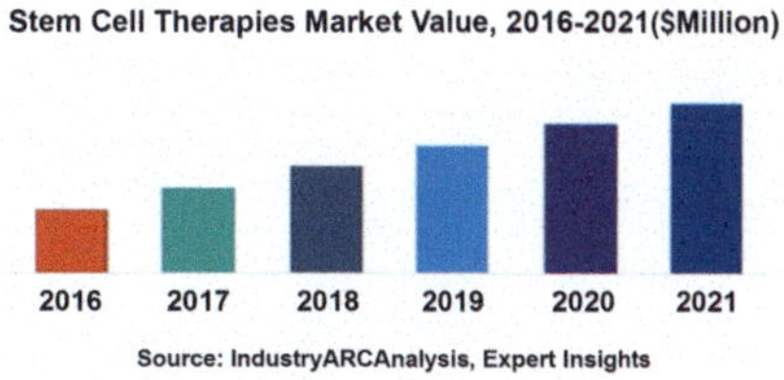

Fig. 7

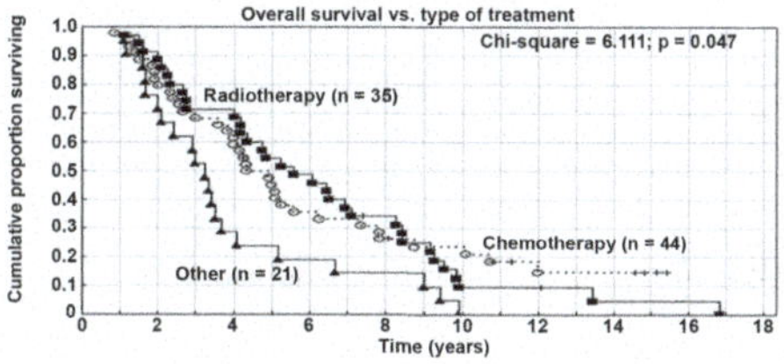

Fig. 8

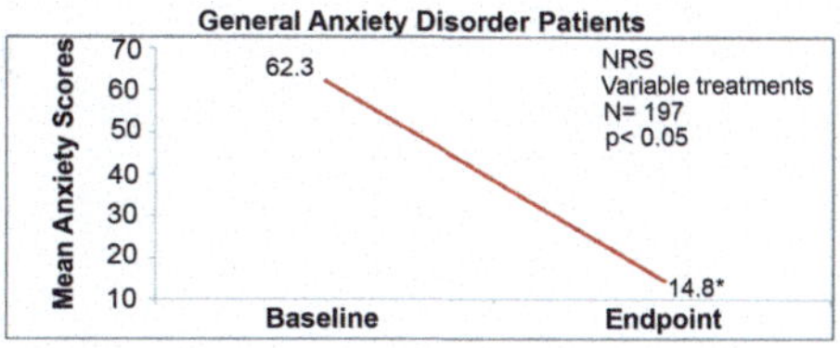

Fig. 9

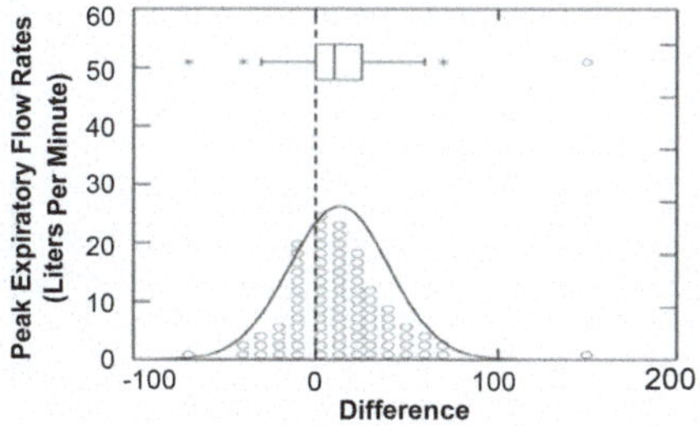

Fig. 10

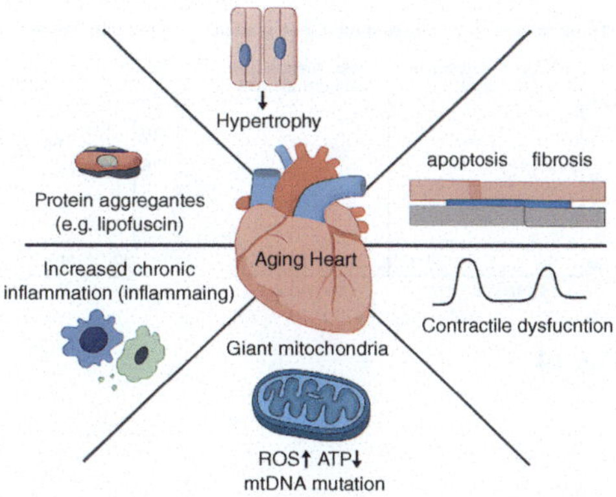

Fig. 11

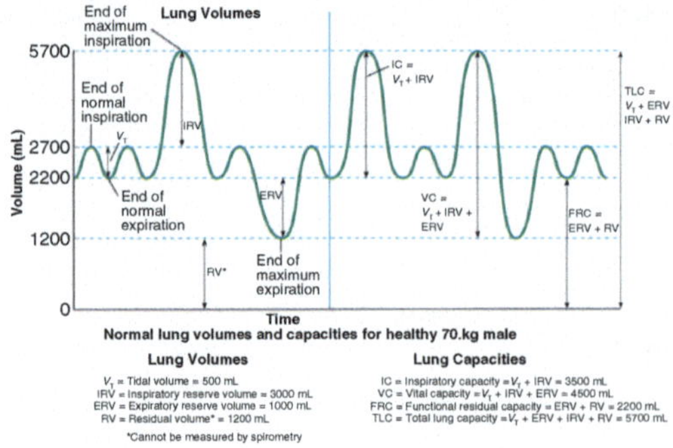

Fig. 12

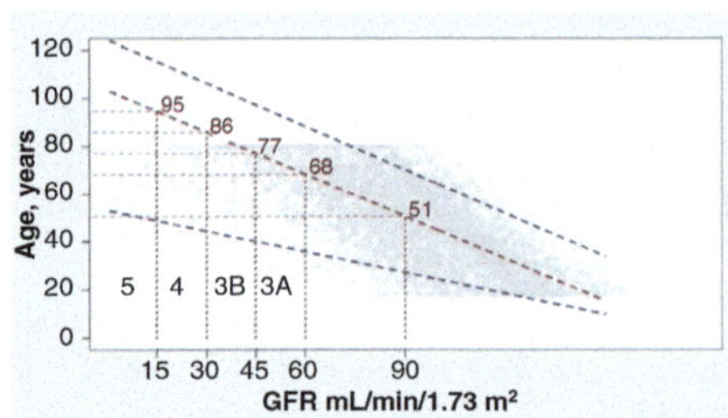

Fig. 13

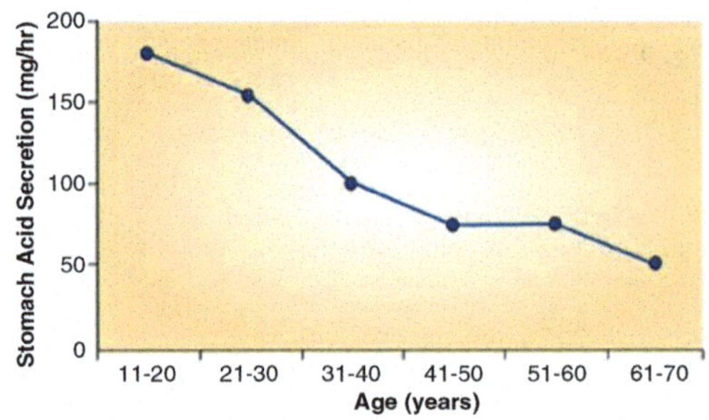

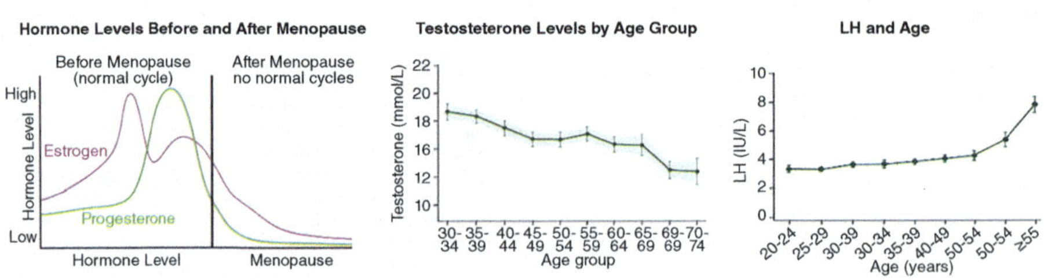

Fig. 14

Fig. 15

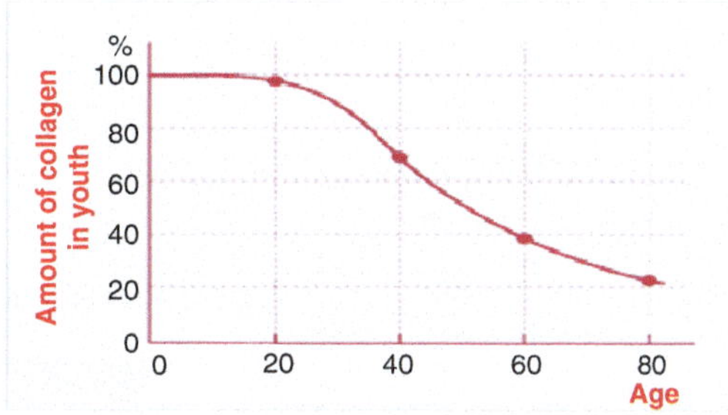

Fig. 16

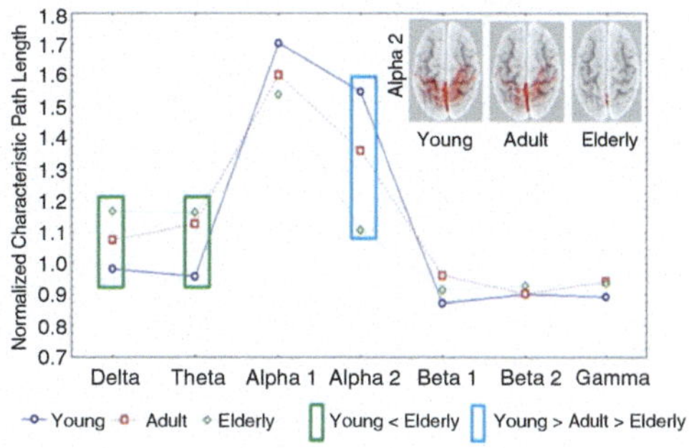

Fig. 17

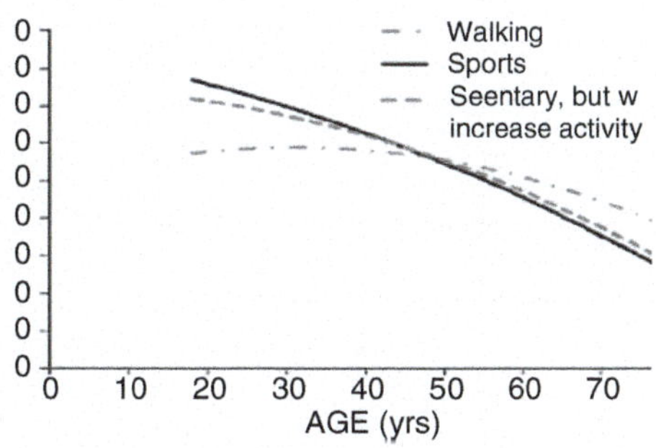

Index

I. Rajasingh et al. (eds.), *Applied Computational Mathematics*, Springer Proceedings in Mathematics & Statistics 480,
https://doi.org/10.1007/978-3-031-77764-6

The manufacturer's authorised representative in the EU is Springer Nature Customer Service Centre GmbH, Europaplatz 3, 69115 Heidelberg, Germany. If you have any concerns regarding our products, please contact ProductSafety@springernature.com

Printed and bound by CPI Group (UK) Ltd, Croydon, CR0 4YY
30/09/2026
02225568-0008